Energy, Environment, and Sustainability

Series Editor

Avinash Kumar Agarwal, Department of Mechanical Engineering, Indian Institute of Technology Kanpur, Kanpur, Uttar Pradesh, India

AIMS AND SCOPE

This books series publishes cutting edge monographs and professional books focused on all aspects of energy and environmental sustainability, especially as it relates to energy concerns. The Series is published in partnership with the International Society for Energy, Environment, and Sustainability. The books in these series are edited or authored by top researchers and professional across the globe. The series aims at publishing state-of-the-art research and development in areas including, but not limited to:

- Renewable Energy
- Alternative Fuels
- Engines and Locomotives
- Combustion and Propulsion
- Fossil Fuels
- Carbon Capture
- Control and Automation for Energy
- Environmental Pollution
- Waste Management
- Transportation Sustainability

Review Process

The proposal for each volume is reviewed by the main editor and/or the advisory board. The chapters in each volume are individually reviewed single blind by expert reviewers (at least four reviews per chapter) and the main editor.

Ethics Statement for this series can be found in the Springer standard guidelines here https://www.springer.com/us/authors-editors/journal-author/journal-author-helpdesk/before-you-start/before-you-start/1330#c14214

More information about this series at http://www.springer.com/series/15901

Avinash Kumar Agarwal · Hardikk Valera ·
Martin Pexa · Jakub Čedík
Editors

Methanol

A Sustainable Transport Fuel for SI Engines

Springer

Editors
Avinash Kumar Agarwal
Department of Mechanical Engineering
Indian Institute of Technology Kanpur
Kanpur, Uttar Pradesh, India

Hardikk Valera
Department of Mechanical Engineering
Indian Institute of Technology Kanpur
Kanpur, Uttar Pradesh, India

Martin Pexa
Quality and Dependability of Machines
Czech University of Life Sciences Prague
Prague, Czech Republic

Jakub Čedík
Quality and Dependability of Machines
Czech University of Life Sciences Prague
Prague, Czech Republic

ISSN 2522-8366 ISSN 2522-8374 (electronic)
Energy, Environment, and Sustainability
ISBN 978-981-16-1226-8 ISBN 978-981-16-1224-4 (eBook)
https://doi.org/10.1007/978-981-16-1224-4

This Springer imprint is published by the registered company Springer Nature Singapore Pte Ltd.
The registered company address is: 152 Beach Road, #21-01/04 Gateway East, Singapore 189721, Singapore

Preface

Due to the increasing human population on Earth, the energy demands of the transportation sector are raising. The increased intensity of transportation causes higher levels of environmental stress. This opens a challenge for researchers in order to decrease the environmental effects and ensure sufficient renewable energy sources. The quality of human life depends to a large extent on the availability of transportation of both people and goods. Also, the increasing amount of transportation poses a severe risk to human health mainly due to pollutants formation during the combustion of fossil fuels.

The International Society for Energy, Environment, and Sustainability (ISEES) was founded at the Indian Institute of Technology Kanpur (IIT Kanpur), India, in January 2014 with an aim to spread knowledge/awareness and catalyze research activities in the fields of Energy, Environment, Sustainability, and Combustion. The society's goal is to contribute to the development of clean, affordable, and secure energy resources and a sustainable environment for the society, spread knowledge in the above-mentioned areas, and create awareness about the environmental challenges, which the world is facing today. The unique way adopted by the society was to break the conventional silos of specializations (Engineering, science, environment, agriculture, biotechnology, materials, fuels, etc.) to tackle the problems related to energy, environment, and sustainability in a holistic manner. This is quite evident by the participation of experts from all fields to resolve these issues. The ISEES is involved in various activities such as conducting workshops, seminars, conferences, etc. in the domains of its interests. The society also recognizes the outstanding works done by young scientists and engineers for their contributions in these fields by conferring them awards under various categories.

The Fourth International Conference on 'Sustainable Energy and Environmental Challenges' (IV-SEEC) was organized under the auspices of ISEES from November 27–29, 2019, at NEERI, Nagpur. This conference provided a platform for discussions between eminent scientists and engineers from various countries including India, USA, China, Italy, Mexico, South Korea, Japan, Sweden, Greece, Czech Republic, Germany, Netherland, and Canada. In this conference, eminent speakers from all over the world presented their views related to different aspects of energy, combustion, emissions, and alternative energy resource for sustainable development and a cleaner environment. The conference presented one high voltage plenary talk by Mrs. Rashmi Urdhwareshe, Director, Automotive Research Association of India (ARAI), Pune.

The conference included 28 technical sessions on topics related to energy and environmental sustainability including 1 plenary talk, 25 keynote talks, and 54 invited talks from prominent scientists, in addition to 70+ contributed talks, and 80+ poster presentations by students and researchers. The technical sessions in the conference included Fuels; Engine Technology and Emissions; Coal and Biomass Combustion/Gasification; Atomization and Sprays; Combustion and Modelling; Alternative Energy Resources; Water and Wastewater Treatment; Automobile and Other Environmental Applications; Environmental Challenges and Sustainability; Nuclear Energy and Other Environmental Challenges; Clean Fuels and Other Environmental Challenges; Water Pollution and Control; Biomass and Biotechnology; Waste to Wealth; Microbiology; Biotechnological and Other Environmental Applications; Waste and Wastewater Management; Cleaner Technology and Environment; Sustainable Materials and Processes; Energy, Environment, and Sustainability; Technologies and Approaches for Clean; and Sensors and Materials for Environmental and Biological Processes and Environmental Sustainability. One of the highlights of the conference was the Rapid Fire Poster Sessions in (i) Engine/Fuels/Emissions, (ii) Environment, and (III) Biotechnology, where 50+ students participated with great enthusiasm and won many prizes in a fiercely competitive environment. 300+ participants and speakers attended this three-day conference, where 12 ISEES books published by Springer, Singapore, under a special dedicated series "Energy, Environment, and Sustainability" were released. This was the third time in a row that such a significant and high-quality outcome has been achieved by any society in India. The conference concluded with a panel discussion on "Balancing Energy Security, Environmental Impacts and Economic Considerations: Indian Perspective", where the panellists were Dr. Anjan Ray, CSIR-IIP Dehradun; Dr. R. R. Sonde, Thermax Ltd.; Prof. Avinash Kumar Agarwal, IIT Kanpur; Dr. R. Srikanth, National Institute of Advanced Studies, Bengaluru; and Dr. Rakesh Kumar, NEERI Nagpur. The panel discussion was moderated by Prof. Ashok Pandey, Chairman, ISEES. This conference laid out the roadmap for technology development, opportunities, and challenges in Energy, Environment, and Sustainability domain. All these topics are very relevant for the country and the world in the present context. We acknowledge the support received from various funding agencies and organizations for the successful conduct of the Fourth ISEES conference IV-SEEC, where these books germinated. We would therefore like to acknowledge SERB, Government of India (Special thanks to Dr. Sandeep Verma, Secretary); NEERI Nagpur (Special thanks to Dr. Rakesh Kumar, Director), CSIR; and our publishing partner Springer (Special thanks to Swati Mehershi).

The editors would like to express their sincere gratitude to a large number of authors from all over the world for submitting their high-quality work in a timely manner and revising it appropriately at a short notice. We would like to express our special thanks to Dr. Josef Bradáč, Assoc. prof. Vladimír Hönig, Prof. František Bauer, Dr. Jiří Bradna, Dr. Rastislav Bernát, Assoc. prof. Jozef Krilek, Assoc. prof. Marián Kučera, Prof. Ocktaeck Lim, Prof. Andreas Bartl, Dr. Prabhas Mishra, Prof.

Burak Zincir, Dr. R. S. Bharj, Dr. Aditya Saurabh, Dr. Parampreet Singh, Dr. Szabó György, Dr. Prashant, who reviewed various chapters of this monograph and provided their valuable suggestions to improve the manuscripts.

This book is focussed on methanol and its utilization in the transportation sector, namely spark-ignition (SI) engines. The book contains chapters regarding methanol production, presenting a variety of production technologies from different feedstocks. A few chapters of the book present methanol as a sustainable fuel for SI engines in transportation. The potential of methanol utilization in transportation in SI engines is discussed, and the challenges, limitations, aspects related to its utilization and current global use of methanol are also presented. The book also contains chapters related to pollutant formation and exhaust emissions from methanol-fuelled SI engines, and one chapter is focussed specifically on formaldehyde emissions, which poses one of the greatest challenges of methanol use in IC engines. The simulations of methanol combustion are also presented. Chapters include recent results and focus more on current trends of methanol utilization in spark-ignition engines. In this book, readers will obtain the idea about the production aspects of methanol, its potential as a sustainable fuel, its utilization in SI engine, and the effect of methanol and its utilization techniques on engine performance, combustion, exhaust emissions, efficiency, and other important parameters. Some chapters of this book are reviews of state of the art and some present original research. We hope that the book would be of great interest to the professionals and post-graduate students involved in alternative fuels, spark-ignition engines, and environmental research.

Kanpur, India — Avinash Kumar Agarwal
Kanpur, India — Hardikk Valera
Prague, Czech Republic — Jakub Čedík
Prague, Czech Republic — Martin Pexa

Contents

Utilization Aspects

Editors and Contributors

About the Editors

Prof. Avinash Kumar Agarwal joined the Indian Institute of Technology (IIT) Kanpur, India in 2001 after working as a post-doctoral fellow at the Engine Research Center, University of Wisconsin at Madison, USA. His interests are IC engines, combustion, alternate and conventional fuels, lubricating oil tribology, optical diagnostics, laser ignition, HCCI, emissions and particulate control, and large bore engines. Prof. Agarwal has published 290+ peer reviewed international journal and conference papers, 42 edited books, 78 books chapters and has 10000+ Scopus and 15300+ Google scholar citations. He is a Fellow of SAE (2012), Fellow of ASME (2013), Fellow of ISEES (2015), Fellow of INAE (2015), Fellow of NASI (2018), Fellow of Royal Society of Chemistry (2018), and a Fellow of American Association of Advancement in Science (2020). He is recipient of several prestigious awards such as Clarivate Analystics India Citation Award-2017 in Engineering and Technology, NASI-Reliance Industries Platinum Jubilee Award-2012; INAE Silver Jubilee Young Engineer Award-2012; Dr. C. V. Raman Young Teachers Award: 2011; SAE Ralph R. Teetor Educational Award -2008; INSA Young Scientist Award-2007; UICT Young Scientist Award-2007; INAE Young Engineer Award-2005. Prof. Agarwal received Prestigious Shanti Swarup Bhatnagar Award-2016 in Engineering Sciences. For his outstanding contributions, Prof. Agarwal is conferred upon Sir J. C. Bose National Fellowship (2019) by SERB.

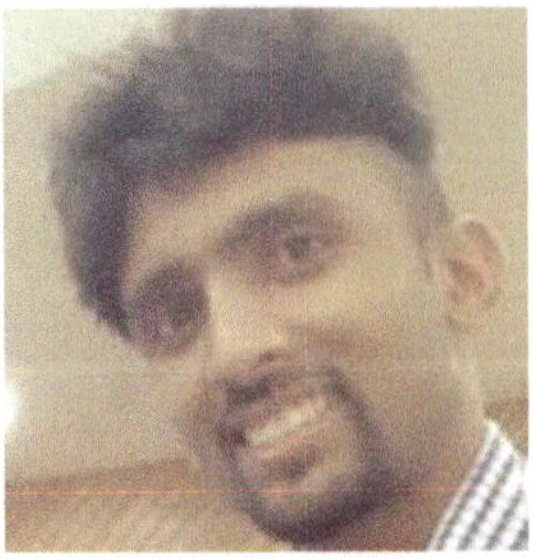

Hardikk Valera is pursuing Ph.D. from Engine Research Laboratory (ERL), Department of Mechanical Engineering, Indian Institute of Technology (IIT) Kanpur. He has completed his M.Tech. and B.Tech. from National Institute of Technology (NIT) Jalandhar, India, and Ganpat University, respectively. His research interests include methanol fuelled SI engines, methanol fuelled CI engines, optical diagnostics, fuel spray characterization and emission control from engines.

Prof. Martin Pexa is associated professor in the field of energetics at the Czech University of Life Sciences Prague, Department for Quality and Dependability of Machines. His research work is focused on the issue of diagnosis and maintenance of machines, especially in the field energy and diagnosis of motor vehicles. As part of his activities, he regularly reviews the contributions into databases Web of Science and Scopus. He also publishes contributions in international scientific journals and conferences. In the mentioned problematics, he teaches students of doctoral, master and bachelor degree programs.

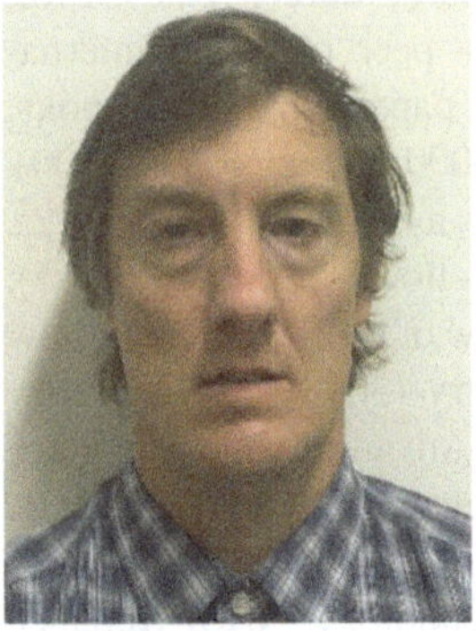

Dr. Jakub Čedík is assistant professor at the Czech University of Life Sciences Prague, Department for Quality and Dependability of Machines. As a part time job he works as a research worker at Research Institute of Agriculture Engineering, p.r.i. in Prague. In the research field he focuses on the issues of biofuels, exhaust emissions, energy demands and operation of internal combustion engines of vehicles and agriculture machinery, issues of agricultural machinery and motor vehicles. He is author or co-author of more than 40 peer-reviewed scientific publications. He teaches courses on Operability of Machines, Quality, Dependability and Logistics of Machines, Technology of Maintenance and Repairs of Machines. He also participate in the review process of peer-reviewed publications, diploma and bachelor theses.

Contributors

Avinash Kumar Agarwal Engine Research Laboratory, Department of Mechanical Engineering, Indian Institute of Technology Kanpur, Kanpur, India

Puneet Bansal Department of Mechanical Engineering, Ganpat University, Mehsana, Gujarat, India

Sayan Biswas University of Minnesota Twin Cities, Minneapolis, MN, USA

Govinda Rao Budda Department of Mechanical Engineering, Gayatri Vidya Parishad College of Engineering (A), Visakhapatnam, India

Jakub Čedík Department for Quality and Dependability of Machines, Czech University of Life Sciences, Prague, Czech Republic

Ilya Chatskiy Generation Project Division, Electric Corporation and Dor Chemicals (Joint Venture), Haifa, Israel

Chris Chatterton Singapore Office, Methanol Institute, Singapore, Singapore

Boris Chudnovsky Generation Project Division, Electric Corporation and Dor Chemicals (Joint Venture), Haifa, Israel

Gregory Dolan Washington DC Office, Methanol Institute, Washington, DC, USA

Vyaas Gururajan Argonne National Laboratory, Lemont, IL, USA

Kim-Pui Koo Fuel Injection Technology, Dongguan, China

Alex Lazebnikov Generation Project Division, Electric Corporation and Dor Chemicals (Joint Venture), Haifa, Israel

Ramnarayan Meena Department of Mechanical Engineering, Ganpat University, Mehsana, Gujarat, India

Muhammad Nihal Naseer National University of Sciences & Technology (NUST), Islamabad, Pakistan

Martin Pexa Department for Quality and Dependability of Machines, Czech University of Life Sciences, Prague, Czech Republic

Tahir Abdul Hussain Ratlamwala National University of Sciences & Technology (NUST), Islamabad, Pakistan

Mordechai Reshef Generation Project Division, Electric Corporation and Dor Chemicals (Joint Venture), Haifa, Israel

Harold Sun GAC R&D Center USA, Farmington Hills, MI, USA

Hardikk Valera Engine Research Laboratory, Department of Mechanical Engineering, Indian Institute of Technology Kanpur, Kanpur, India

Wesley Wang Fuel Injection Technology, Dongguan, China

Yanju Wei Xi'an Jiaotong University, Xi'an, China

Datta Bharadwaz Yellapragada Department of Mechanical Engineering, Gayatri Vidya Parishad College of Engineering (A), Visakhapatnam, India

Asad A. Zaidi National University of Sciences & Technology (NUST), Islamabad, Pakistan

Kai Zhao Beijing Office, Methanol Institute, Beijing, China

General

Introduction of Methanol: A Sustainable Transport Fuel for SI Engines

Avinash Kumar Agarwal, Hardikk Valera, Martin Pexa, and Jakub Čedík

1 Introduction

The increased demand for energy sources in all fields of human activity, and especially in the transportation sector, leads to a search for a suitable alternative fuel, which could substitute currently most common fossil fuels. Besides gaining energy sources, the environmental effects of biofuels are also considered. The increasing number of vehicles encourages the active search for alternative sources and new pathways for fuel production (e.g. catalytic fast pyrolysis or hydrotreating [1]). Current steps lead to an increased amount of alternative fuels in transportation. The electrification in road transport or agriculture sectors has a number of disadvantages (low action radius, recharging time, distribution network, etc.) and until these disadvantages are eliminated, the potential of electrification would remain rather limited [2–4]. From the viewpoint of alternative fuels, these fuels that will bridge the period till the electrification issues are resolved. For SI engines, the alcohol fuels such as methanol, ethanol, or butanol seems to be suitable alternative fuels. Each of these fuels can be produced by a different technology from different feedstocks [5]. This predetermines the regions in the world for the utilization of a specific alternative fuel. Methanol is one of the fuels, which is found to be suitable for all regions on the world. This book presents an important aspects of the production of methanol and its utilization as a fuel in SI engines in transportation. The first section of this monograph presents an Introduction chapter, which introduces all sections of the book and presents the important aspects covered.

A. K. Agarwal (✉) · H. Valera
Engine Research Laboratory, Department of Mechanical Engineering, Indian Institute of Technology Kanpur, Kanpur 208016, India
e-mail: akag@iitk.ac.in

M. Pexa · J. Čedík
Department for Quality and Dependability of Machines, Czech University of Life Sciences, Prague, Czech Republic

A. K. Agarwal et al. (eds.), *Methanol*, Energy, Environment, and Sustainability,
https://doi.org/10.1007/978-981-16-1224-4_1

The second section of the book, focussed on production aspects, describes in its first chapter that methanol has lots of supporting characteristics to be used in the engine such as lower production costs, high octane number, impressive cold working temperature, good emission characteristics, high latent heat of vapourization, enhancing thermal efficiency, high flame speed, and low combustion temperature. This fuel can be used directly and in the form of blends with conventional fuels for powering the engines. Methanol can be produced from various feedstock such as coal, natural gas, biomass, carbon dioxide (CO_2) etc. through various processes like reforming the gas with steam, production with methanotrophic bacteria, gasification process, carbon dioxide hydrogenation, synthesis through syngas, direct oxidation of methane, etc. [6]. Although methanol has many adaptive properties to be used as a fuel individual and with levels of blending; however, there are many challenges too with methanol used as a fuel such as its toxicity, fire safety concern, low cetane number, lower heating value, high kinematic viscosity, low lubricity, and high corrosiveness. The second chapter of this section focuses on methanol as one of the attractive alternative fuels for combustion in utility boilers, gas turbines, and engines. Traditional methods of reducing oxides of nitrogen (NOx) emissions such as modification of the firing system and/or post-combustion treatment of the flue gas to remove NOx are very expensive. Hence, before the implementation of these expensive measures for emission reduction, it is desirable to evaluate all other low-cost alternatives. Existing experience has shown that with minor system modifications, methanol can easily be used as an alternative fuel that significantly decreases NOx and other harmful emissions. On top of that, methanol has the additional advantage of being a universal chemical and a fuel feedstock material, thus representing the ideal candidate for being both as a chemical energy storage material and a substituting fossil raw material. The use of methanol has become an important solution for emissions compliance due to its unique constituents and combustion characteristics. The clean burning characteristics of methanol are expected to lead to clean boiler pressure parts or turbine blades thus lowering maintenance compared to any other fuel oil burning. The following chapter describes the methanol synthesis. The last chapter of this section also strives to highlight the methanol synthesis process by performing thermodynamic analysis. A deliberate energy analysis of methanol synthesis by reverse water gas shift reactor is performed. The thermochemical processing is utilized to convert methane and carbon dioxide to methanol. Based on thermodynamic modelling, concentrations of different constituents over the cycle are determined. It is verified that the optimum state of methanol synthesis is accompanied by the lowest level of carbon monoxide (CO). Moreover, the concentration of carbon dioxide does not affect the methanol synthesis significantly during transesterification reactions.

The third section is focussed on methanol as a fuel. The availability of methanol from a wide variety of feedstock and its characteristics make it an undeniable alternative for SI engines [7]. Methanol is very rich in oxygen content approximately 50 wt%, which helps in cleaner combustion. In the first chapter of this section, various properties of methanol, which makes it a viable SI engine fuel are discussed in detail. The limitations of methanol and its blends in the SI engines are analyzed. The technical solutions available to overcome these limitations are presented in this chapter. The effects of methanol usage and its blends on SI engine performance, combustion, and emissions are reviewed in detail. The requirement of catalytic converters in methanol fuelled engines and design modification required for catalytic converters for methanol operations are discussed in the present chapter. The following chapter of this section is focussed on methanol fuelled vehicles in China. From the 2010s, Chinese methanol fuelled vehicles transferred to a new stage, marked by a Chinese national methanol vehicle pilot program, which involves four provinces and municipality (Shanghai) and over 1,000 vehicles from passenger cars to buses and heavy-duty trucks, aiming to further figure out the key concerns from different parties before any industry promotional policy was made. In 2019, a promotional policy for methanol fuelled vehicles was issued by eight Chinese ministries and administrations, which paves a milestone for the future development of methanol fuelled vehicles in China. The last chapter of the last section explains the formation of both regulated and unregulated pollutants and their effects on human health. Also, it explains the factors in detail that lead to the preference for methanol as a fuel for internal combustion engines (ICEs). It also summarizes the various preferable techniques for methanol utilization in the transport sector.

The last section of this book is focussed on utilization aspects. The following chapter focusses on low-temperature combustion, a desirable mode of fuel utilization that ensures low particulate matter (PM) and NOx emissions. Low-temperature combustion kinetics of five different methanol blended gasoline containing 10%, 20%, 30%, 40%, and 50% of methanol by volume, respectively, were compared with neat gasoline. Methanol addition in gasoline enhanced low-temperature combustion behaviour compared to pure gasoline due to the improved evaporation and mixing characteristics. Unlike the combustion efficiency and power, which showed marginal improvement, carbon monoxide (CO) and NOx emissions dropped significantly with the increase in methanol/gasoline ratio. The effect of methanol/gasoline ratio on different low-temperature chemical pathways was investigated to find an optimum value in terms of combustion performance and emission. The next chapter in the section is focussed on methanol flames. In this chapter, the chemical kinetics effects on fundamental laminar combustion processes are investigated for 20% methanol blended gasoline and 20% hydrogen blended methanol for a range of equivalence ratios, $0.5 < \Phi<2$ using a detailed reaction mechanism comprising of 2027 species and 8619 reactions. Fundamental laminar combustion properties: flame speed, flame thickness, minimum ignition energy, ignition, and extinction strain rates of the two fuel blends were compared with base fuels, pure methanol, gasoline, and hydrogen. Mean extinction strain rates were nearly twice the ignition strain rates indicated a hysteresis behaviour of laminar flames. The knowledge obtained from this study can be leveraged in creating tailored fuel blends with superior laminar combustion competencies enabling higher efficiency and lesser pollutant emission for automotive applications. The following chapter of the last section is focussed on emissions of formaldehyde. As an important intermediate product of methanol combustion, the formaldehyde emission of methanol engines imposes great threats on the public health, and more importantly, the targeted method has not been specially developed to measure the engine-out formaldehyde emissions. This chapter introduces four parts relating to methanol engine-out formaldehyde: (i) the ordinary detection method of airborne formaldehyde and the latest measuring methods of the ones from the engine, (ii) the emission mechanism from hydrocarbon and methanol combustion, (iii) the emission characteristics from methanol/gasoline blends fuelled engine, and (iv) the conversion efficiencies of the three-way catalyst. The last chapter of this section is also focussed on aldehydes. Aldehyde is a highly toxic pollutant, and aldehyde emission out of alcohol fuel combustion could be considerably higher than spark-ignited gasoline engines. The corrosion and durability of methanol fuel components were also concerning for the mass production of methanol fuelled vehicles. Over the past decade, an automotive manufacturer in China made tremendous investigations for the brake thermal efficiency and emission improvement potentials of a methanol fuelled SI engine for passenger car application and to demonstrate the performance and China V emission compliance over its useful life of 160,000 km. The study also shows that a methanol fuelled engine has a brake thermal efficiency advantage over a gasoline engine, and conventional three-way oxidation catalysts for gasoline engine can be used to meet the China V emission standard, despite deteriorations over its useful lifetime. Most studies show that the methanol fuelled engines have excessive aldehyde emissions. However, there is a study indicating that the tailpipe aldehyde emission can be mitigated with a conventional gasoline three-way oxidation catalyst, using gasoline cold start assistance at the beginning of the transient emission cycle.

This monograph is focussing on the utilization of methanol as an alternative fuel for SI engines in the areas: (I) General, (II) Production aspects, (III) Methanol as a Fuel, and (IV) Utilization Aspects. Specific topics covered in the manuscript include the following:

- Introduction of Methanol: A Sustainable Transport Fuel for SI Engines
- Methanol as an alternative fuel in Internal Combustion Engine: Scope, Production, and Limitations
- Methanol as A Low-Cost Alternative Fuel for the Reduction of Emissions
- Energy Analysis of Methanol Synthesis via Reverse Water Gas Shift Reactor
- Methanol—A sustainable fuel for SI engine
- A Brief Review on Methanol-Fuelled Vehicles (MFV) in China and Implementation Policy
- Regulated and Unregulated Emissions from Methanol Fuelled Engines
- Low-Temperature Combustion Kinetics of Methanol Blended Gasoline and Methanol Synthesized Dimethyl Ether
- Ignition and Extinction of Hydrogen and Gasoline Blended Methanol Flames
- Measurement, Mechanism, and Characteristics of Emission from Methanol/Gasoline Blends Fuelled Engine
- Application of methanol as a clean and efficient alternative fuel for passenger cars.

References

1. Kim Y, Thomas AE, Robichaud DJ, Iisa K, John PCS, Etz BD, Fioroni GM, Dutta A, McCormick RL, Mukarakate C, Kim S (2020) A perspective on biomass-derived biofuels: from catalyst design principles to fuel properties. J Hazard Mater 400:
2. Weiss M, Cloos KC, Helmers E (2020) Energy efficiency trade-offs in small to large electric vehicles. Environ Sci Europe 32(1):1–17
3. Kostopoulos ED, Spyropoulos GC, Kaldellis JK (2020) Real-world study for the optimal charging of electric vehicles. Energy Rep 6:418–426
4. Valera H, Agarwal AK (2020) Future automotive powertrains for India: methanol versus electric vehicles. In: Singh A, Sharma Y, Mustafi N, Agarwal A. (eds) Alternative fuels and their utilization strategies in internal combustion engines. Energy, environment, and sustainability. Springer, Singapore
5. Worldwatch Institute (2012) Biofuels for transport: global potential and implications for energy and agriculture. Taylor and Francis, New York
6. Valera H, Agarwal AK (2019) Methanol as an alternative fuel for diesel engines. In: Agarwal A, Gautam A, Sharma N, Singh A. (eds) Methanol and the alternate fuel economy. Energy, environment, and sustainability. Springer, Singapore
7. Valera H, Singh AP, Agarwal AK (2020) Prospects of methanol-fuelled carburetted two wheelers in developing countries. In: Advanced combustion techniques and engine technologies for the automotive sector. Energy, environment, and sustainability. Springer, Singapore

Production Aspects

Methanol as an Alternative Fuel in Internal Combustion Engine: Scope, Production, and Limitations

Puneet Bansal and Ramnarayan Meena

Abbreviations

IC	Internal Combustion
ON	Octane Number
CN	Cetane Number
CO	Carbon Monoxide
HC	Hydrocarbons
STP	Standard temperature and pressure
GEM	Gasoline–ethanol–methanol

1 Introduction

The growth of any country is strongly affected by energy demand and its supply since energy is the driven force for every productive work. One of the greatest challenges of the world today is energy supply. Fossil fuels, which are currently the main sources of energy, besides causing a negative impact on the environment, have increased in price over the last few years. Studies suggest that nearly 11 billion tonnes of fossil fuels are consumed globally every year. Among the fossil fuels, crude oil is depleting at the rate of 4 billion tonnes per year, so it may completely vanish from earth in the next 53 years with the current consumption rate and without any further new finding of such reserve. Natural gas may service about 51 years and coal for the next 150 years [1]. In India, fossil fuels including coal, crude oil, and natural gases provide 28%, 40%, and 20% of energy, respectively, out of which 75% of this total energy

P. Bansal (✉) · R. Meena
Department of Mechanical Engineering, Ganpat University, Mehsana, Gujarat, India
e-mail: punitbansal80@gmail.com

A. K. Agarwal et al. (eds.), *Methanol*, Energy, Environment, and Sustainability,
https://doi.org/10.1007/978-981-16-1224-4_2

is utilized for cooking and lighting of India's urban areas. A concern is that fossil fuels are being used up at an increasing rate, and will soon run out. Limited reserve and continuously increasing the demand for such fuels are the main factors to have a high cost per use. This leads to a severe economic imbalance in developing countries like India. Transport is one the most important sectors that build the economy and up till now it mostly relies on fossil fuel but as the reserve of such fuels are limited but energy demand is increasing exponentially, so there is an urgent need to have some option which can solve this problem of using fossil fuel with least possible greenhouse gas, air pollution, and supportive characteristics to be used in Internal combustion engines, lower production cost, and produced by biomass or non-fossil elements. As far as transportation is concerned, alcohols (Methanol and ethanol) can play an important role to fulfil the demand as fuels since they reflect some of the favourable properties, engine performance, and exhaust emissions.

2 Scope of Methanol to Be Used in IC Engine

There are many alternative options available to deal with the dependency on fossil fuels like bio-diesel, vegetable oils, and alcohols. Among them, alcohols stand out the best suitable replacement of oils used in IC engine as fuels. Methanol may widely be used due to its engine supportive characteristics (almost as similar as the conventional fossil fuels already being used in current IC engines) and its triple P factor like production, pollution, and price. Methanol can be produced by biomass and fossil fuels. It shows attractive emission control over CO, HC while the cost of production is almost 30% of the conventional fuels [2]. China has already started the blending of methanol with gasoline which is used as a fuel in their commercial vehicles. National Institution for Transforming India (NITI Aayog), an Indian government's policymaker body is thinking to reduce the import of crude oil up to US$100 Billion by 2030 by using Methanol as a fuel in their commercial engines [3] (Figs. 1 and 2).

3 Comparisons of Properties of Methanol with Other Fuels

Methanol exhibits many properties which make it a better option to be used in IC engines.

3.1 Physical Properties

Table 1 shows the physical properties of some of the important fuels like gasoline, methanol, ethanol, and methane to compare them well.

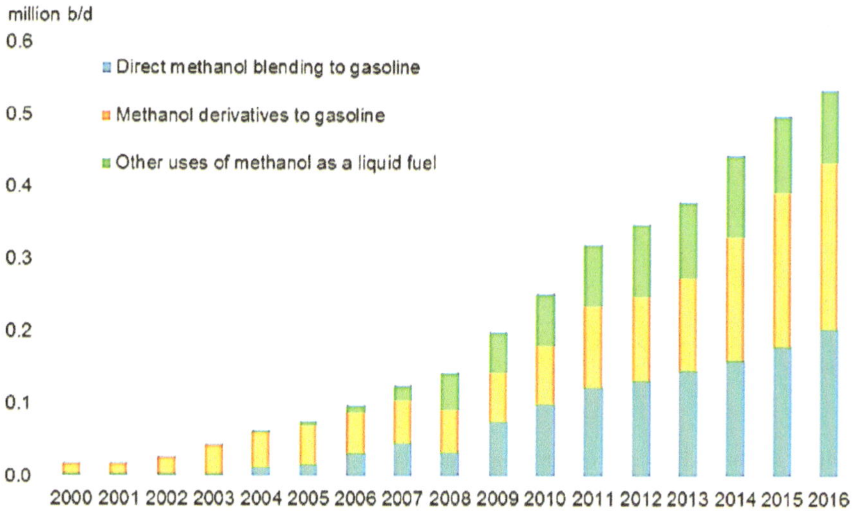

Fig. 1 Energy Information Administration [4]

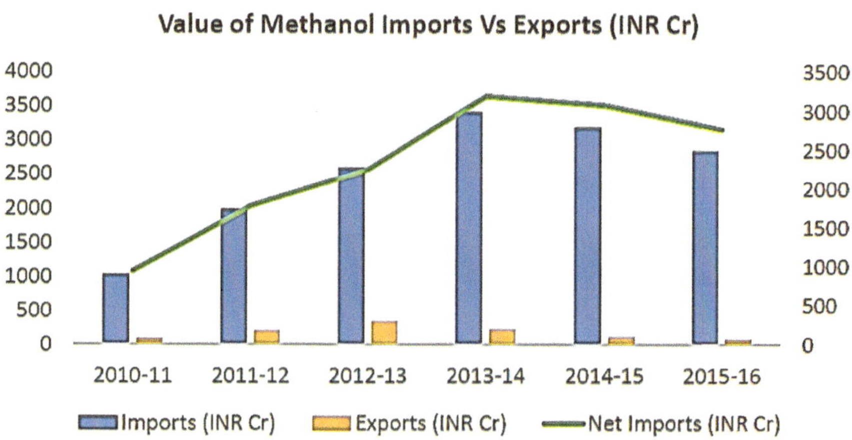

Fig. 2 Source—Ministry of Fertilizers and Petrochemicals [4]

3.2 *Chemical Properties*

Table 2 shows the chemical properties of some of the important fuels like gasoline, methanol, ethanol, and methane to compare them well.

Table 1 Physical Properties of some important fuels [5–8]

Property	Gasoline (typical)	Methanol	Ethanol	Methane
Chemical formula	Various	CH_3OH	C_2H_5OH	CH_4
Density (STP) [kg/m^3]	740	790	790	0.65
Vapour density (STP) [kg/m^3]	3.88	1.42	2.06	0.65
Boiling point at 1 bar [°C]	25–215	65	79	-161.5
Heat of vapourization [kJ/kg]	180–350	1100	838	510
Surface tension (20 °C) [mN/m]	21.6	22.1	22.3	na
Dynamic viscosity (20 °C) [mPas]	0.6	0.57	1.2	0.01

Table 2 Chemical Properties of some important fuels [5–8]

Property	Gasoline (typical)	Methanol	Ethanol	Methane
Chemical formula	Various	CH_3OH	C_2H_5OH	CH_4
Molecular weight [kg/kmol]	107.00	32.04	46.07	16.04
Oxygen Content by mass [%]	0.00	49.93	34.73	0.00
Carbon Content by mass [%]	86	37.48	52.14	74.87
Lower heating value [MJ/kg]	42.90	20.09	26.95	50.00
Higher heating value [MJ/kg]	48.00	22.88	29.85	55.00
Volumetric Energy Content [MJ/m^3]	31,746	15,871	21,291	32.50
Stoichiometric AFR [kg/kg]	14.70	6.50	9.00	17.65
Stoichiometric AFR [kmol/kmol]	54.49	7.22	14.36	9.81
Specific CO_2 emissions [g/MJ]	73.95	68.44	70.99	54.87
Specific CO_2 emissions relative to gasoline	1.00	0.93	0,96	0.74
Auto-ignition temperature [K]	465–743	738	698	813
Adiabatic Flame Temperature [K]	~2275	2143	2193	2225
Flash point [^{0}C]	-45	11	14	-188

3.3 *Mixture Properties*

Table 3 shows the mixture properties of various fuels to compare them well.

Table 3 Mixture properties of some important fuels [5–8]

Property	Gasoline (typical)	Methanol	Ethanol	Methane
Chemical formula	Various	CH_3OH	C_2H_5OH	CH_4
Energy per Unit Mass of Air [MJ/kg]	2.92	3.09	2.99	2.83
Energy per Unit Mass of mixture [MJ/kg]	2.73	2.68	2.70	2.68
Energy per Unit Mass of Air relative to gasoline	1.00	1.06	1.03	0.97
Energy per Unit Mole of Air [MJ/kmol]	84.24	89.22	86.44	87.76
Energy per Unit Mole of mixture [MJ/kmol]	82.72	78.36	80.81	74.20
Energy per Unit Mole of mixture relative to gasoline	1.00	0.95	0.98	0.90
Reid vapour pressure (at 37.8 °C) [psi]	7.00	4.60	2.30	2400
Stoichiometric mass fraction (in air)	0.06	0.13	0.10	0.05
Stoichiometric volume fraction (in air)	0.02	0.12	0.06	0.10
Heat of vapourization per unit mass of Air [kJ/kg]	24.10	170.90	93.60	28.89
Evaporation energy relative to gasoline	1.00	7.09	3.88	-
C_p/C_v for a stoichiometric mixture	na	1.37	1.38	1.39
Mole Ratio of Products to Reactants (incl. N2)	0.94	1.06	1.07	1.00
Ratio of Triatomic to Diatomic Products	0.35	0.53	0.44	0.40
Mass diffusivity in air [cm^2/s]	0.05	0.14	0.11	0.16
Lower flammability limit in air [vol%]	1.30	6.70	3.30	5.00
Higher flammability limit in air [vol%]	7.60	36.00	19.00	15.00
Lower flammability limit in air [λ]	160	1.81	1.91	1.99
Higher flammability limit in air [λ]	0.26	0.23	0.28	0.59
Lower flammability limit in air [φ]	0.62	0.55	0.52	0.50
Higher flammability limit in air [φ]	3.89	4.32	3.60	1.68
Minimum ignition energy in Air [mJ]	0.25	0.14	0.23	0.28
Quenching distance [mm]	~2.0	1.85	1.65	2.03

4 Merits and Demerits of Methanol as a Fuel

Methanol has some very good advantages over gasoline and some demerits as well as compared to gasoline. This section will depict the advantages and disadvantages of methanol with respect to gasoline as a fuel.

4.1 Merits

1. High Octane Number
2. Knock resistance
3. High efficiency
4. Smoke free combustion
5. Lower emission of NOx, CO, $CO_{2,}$ and HC
6. Higher latent heat of evaporation
7. Lesser exhaust temperature
8. Higher volatility
9. Competitive with demand and supply cycle

4.2 Demerits

1. Long ignition delay
2. Poor self-ignition point
3. Poor miscibility with diesel
4. Cold starting problem
5. Corrosive in nature
6. Poor lubrication property
7. Low viscosity

5 Methanol Production Methods

Methanol is one the most demanding chemical used in various industries and for purposes like textile, transportation, and production of hydrocarbons, olefins, and some aromatic compounds. In fact, methanol is the world's top 5 demanded chemical. Methanol is produced in two steps:

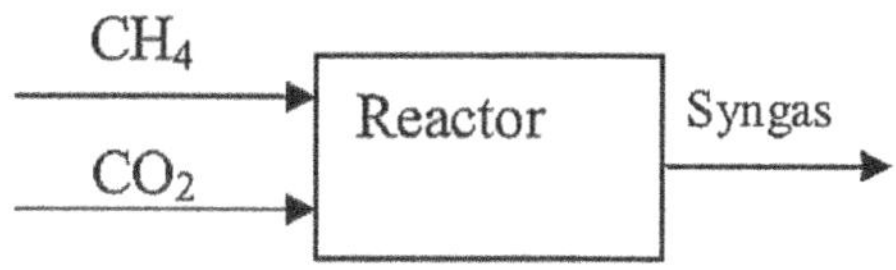

Fig. 3 Syngas production flow [15]

5.1 *Production of Synthetic Gas from Organic Feedstock*

Synthetic gas is also termed as Syngas. It is the combination of CO and H2 which can be produced by various feedstocks such as biogas, wood, biomass coal, etc. syngas synthetic by various methods through the catalytic conversion of feedstocks (Fig. 3).

There are various reforming methods to produce syngas from feedstock:

$$(Steam\ reforming\ reaction)\quad 2\mathrm{CH_4} + 3\,\mathrm{H_2O} \rightarrow \mathrm{CO} + \mathrm{CO_2} + 7\,\mathrm{H_2} \tag{1}$$

$$(Partial\ oxidation)\quad \mathrm{C} + 1/2\mathrm{O_2} < - > \mathrm{CO} \tag{2}$$

$$(Water-gas\ reaction)\quad \mathrm{C} + \mathrm{H_2O} < - > \mathrm{CO} + \mathrm{H_2} \tag{3}$$

$$(Water-gas\ shift\ reaction)\quad \mathrm{CO} + \mathrm{H_2O} < - > \mathrm{CO_2} + \mathrm{H_2} \tag{4}$$

5.2 *Methanol Synthesis from Syngas*

To produce methanol, syngas is further treated at 6–7 MPa and 200–250^{0}C in a closely packed reactor, since it is an exothermic reaction so catalytic selection is done to it in mind. The following are the reactions which depict the conversion of syngas into methanol [9, 10].

$$CO_2 + 3H_2 \leftrightarrow CH_3OH + H_2O \quad \left(-\Delta H_{298K,50\,\mathrm{bar}} = 40.9\,KJ/mol\right) \tag{5}$$

$$CO + 2H_2 \leftrightarrow CH_3OH \quad \left(-\Delta H_{298K,50\,\mathrm{bar}} = 90.7\,KJ/mol\right) \tag{6}$$

$$CO_2 + H_2 \leftrightarrow CO + H_2O \quad \left(-\Delta H_{298K,50\,\mathrm{bar}} = 49.8\,KJ/mol\right) \tag{7}$$

Selection of reactor and catalyst is done on the basis of reaction, amount of methanol production, and exothermic energy emits from reaction [10].

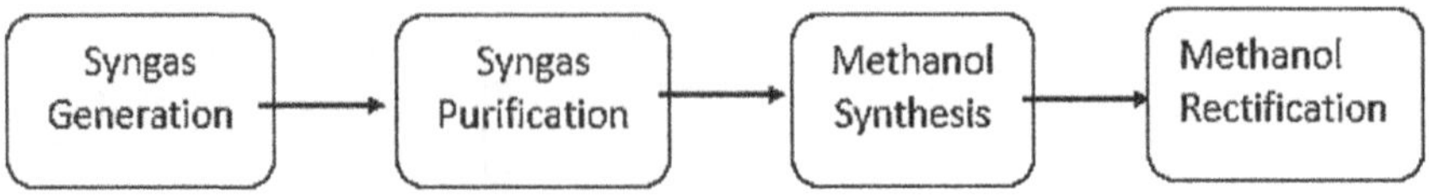

Fig. 4 Hierarchy to produce methanol from coal [11]

5.3 *Methanol from Coal Gasification*

Coal gasification is one of the old methods to produce methanol. It involves four processes as shown in the block diagram (Fig. 4):

Firstly, fine coal particles react at high pressure and temperature on the bed of a suitable catalyst bed to produce Syngas which, of course, has some impurities such as sulphur and nitrogen contentment. These impurities need to be taken off. To remove these impurities, produced syngas is passed through an amine solution. But as the result of this reaction, CO and H_2 availability reduced and that's why water shift reaction is needed to recover the CO and H2. As the result of the water shift reaction, methanol is produced but it needs further purification by scrubbing with amine [11].

Basic reaction for the production of methanol from coal gasification [12–14]

[Syngas Production]

$$3C + O_2 + H_2O \rightarrow 3CO + H_2 \quad \Delta H^0 = -142.22\,\text{kJ/mol} \tag{8}$$

[Water-gas shift reaction]

$$CO + H_2O \leftrightarrow CO_2 + H_2 \quad \Delta H^0 = -41.20\,\text{kJ/mol} \tag{9}$$

[Methanol synthesis reaction]

$$CO + 2H_2 \leftrightarrow CH_3OH \quad \Delta H^0 = -90.7\,\text{kJ/mol} \tag{10}$$

$$CO2 + 3H_2 \leftrightarrow CH_3OH + H_2O \quad \Delta H^0 = -49.5\,\text{kJ/mol} \tag{11}$$

5.4 *Methanol from Natural Gas*

Methanol from natural gas is produced in three steps; firstly, Syngas production; secondly, synthesis of methanol from syngas; and thirdly, rectification of methanol [11]. This process can be depicted by block diagram (Fig. 5):

Syngas can be produced by many methods such as the following:

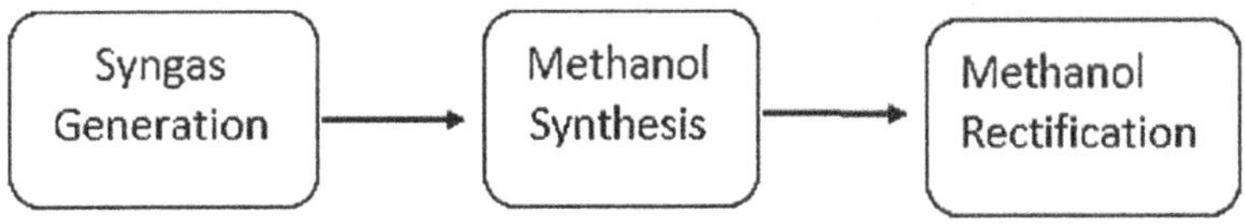

Fig. 5 Hierarchy to produce methanol from Natural [11]

5.4.1 Dry Reforming Reaction

This is an endothermic reaction that can take place at a very high temperature generally above 800^{0}C. That's why this is a slow reaction which is majorly affected by contact with reactant, temperature, and catalyst. The following is the DRR reaction [15].

$$CH_4 + CO_2 \rightarrow 2CO + 2H_2 \quad \left(\Delta H^0 = +260.62\,\text{kJ/mol}\right) \tag{12}$$

5.4.2 Steam Reforming Reaction (SRR)

SRR react efficiency with hydrogen yield. It is an endothermic reaction so it requires high energy input and there is some energy loss as well in this reaction. The production of hydrogen through SRR using methane is typically carried out at temperatures ranging from 700 to 900 °C and at pressures ranging from 20 to 40 bar [16]. When this reaction is used in industrial applications, energy demand is fulfiled by inter-stage heaters and heaters.

$$CH_4 + H_2O \rightarrow CO + 3H_2 \quad \left(\Delta\,H^0 = +226.81\,\text{kJ/mol}\right) \tag{13}$$

5.4.3 Water–Gas Shift Reactions (WGSR)

This reaction take place after the SRR reforming reaction in a two-stage reforming reactor model as shown in Fig. 6. CO and water molecules again react at a temperature ranging between 550 to 700 °C and form CO_2 and additional hydrogen [17].

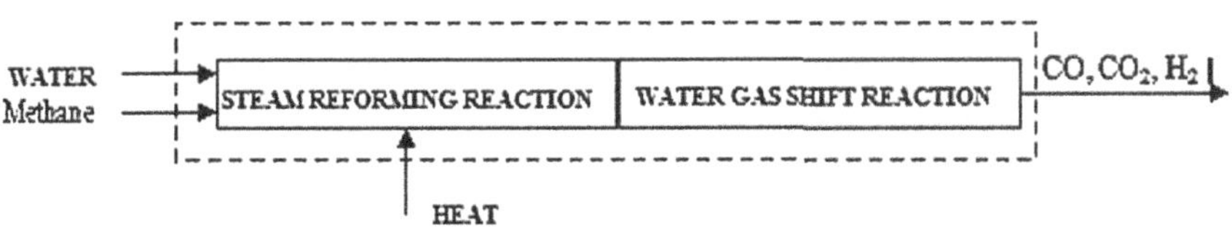

Fig. 6 Reactor model for steam reforming and water–gas shift reactions

$$CO + H_2O \rightarrow CO_2 + H_2 \quad (\Delta H^0 = -33.86\,kJ/mol) \tag{14}$$

5.4.4 Partial Oxidation (POX)

POX is an exothermic reforming reaction in which CH_4 is reacted with a limited supply of oxygen to do a partial oxidization reaction. The product of this reaction is CO and hydrogen. This is the self-sustain reaction due to its exothermic nature [18]. However, due to the exothermic nature of the POX process, the produced fuel has a lower calorific value than that of methane feed gas [19].

$$CH_4 + 0.5O_2 \rightarrow CO + 2H_2 \quad (\Delta H^0 = -22.63\,kJ/mol) \tag{15}$$

$$CH_4 + 2O_2 \rightarrow CO_2 + 2H_2O \quad (\Delta H^0 = -891.64\,kJ/mol) \tag{16}$$

Syngas synthesize into methanol when reacted at high pressure (8–10 MPa) and high temperature (250–300^{0}C) [20]. Methanol can be synthesized by hydrogenation of carbon monoxide and carbon dioxide [13, 21].

$$CO + 2H_2 \leftrightarrow CH_3OH \quad (\Delta H^0 = -90.7\,kJ/mol) \tag{17}$$

$$CO_2 + 3H_2 \leftrightarrow CH_3OH + H_2O \quad (\Delta H^0 = -49.5\,kJ/mol) \tag{18}$$

5.5 Methanol from Dry Milled Corn

The raw material to produce methanol from this process is dry milled corn. As per [37], a wind power-driven electrolysis process is being used which separates water into oxygen and hydrogen where oxygen is the byproduct and hydrogen feeds for the reaction through hydrogen compressor to react with CO_2 which is the byproduct coming out through the corn ethanol reactor. Along with the methanol, this synthesis reaction also produces flue gases, water, and steam that can be utilized in various productive works such as power production and electricity generation.

These are the reactions that are associated with this process [13]:

$$2H_2O \rightarrow 2H_2 + O_2 \quad (\Delta H_R^0 = +572\,kJ/mol) \tag{19}$$

$$CO_2 + 3H_2 \leftrightarrow CH_3OH + H_2O \quad (\Delta H^0_{(298K)} = -49.4\,kJ/mol) \tag{20}$$

$$O_2 + H_2 \leftrightarrow H_2O + CO \quad (\Delta H^0_{(298K)} = +41.12\,kJ/mol) \tag{21}$$

$$CO + 2H_2 \leftrightarrow CH_3OH \quad \left(\Delta H^0_{(298K)} = -90.55\,kJ/mol\right) \tag{22}$$

5.6 *Methanol from Wood*

This process is similar to the coal gasification process; the only change in feedstock in this process is wood instead of coal. The feedstock is first used to produce syngas and that syngas is further subjected for synthesis with methanol producing reactions; the requirement of hydrogen is maintained by water–gas shift reaction [22]

These are the reactions that are associated with this process:

Feedstock to Syngas conversion

$$3C + O_2 + H_2O \rightarrow 3CO + H_2 \quad \left(\Delta H^0 = -142.22\,kJ/mol\right) \tag{23}$$

$$CO + H_2O \leftrightarrow CO_2 + H_2 \quad \left(\Delta H^0 = -41.2\,kJ/mol\right) \tag{24}$$

Methanol from synthesis gas

$$CO + 2H_2 \leftrightarrow CH_3OH \quad \left(\Delta H^0 = -90.7\,kJ/mol\right) \tag{25}$$

$$CO_2 + 3H_2 \leftrightarrow CH_3OH + H_2O \quad \left(\Delta H^0 = -49.5\,kJ/mol\right) \tag{26}$$

5.7 *Methanol from Coal Power Plant*

Coal and hydrogen are the key feeds for this reaction. This type of reactor consists of a methanol reactor, distillation unit, electrolysis chamber, and CO_2 absorbing unit. Generally, a surplus amount of hydrogen is produced in this process which can be re-used in the power plant to increase the efficiency.

These are the reactions that are associated with this process [23]:

$$2H_2O \rightarrow 2H_2 + O_2 \quad \left(\Delta H^0_R = +572\,kJ/mol\right) \tag{27}$$

$$CO_2 + 3H_2 \leftrightarrow CH_3OH + H_2O \quad \left(\Delta H^0_R = -49.57\,kJ/mol\right) \tag{28}$$

$$CO_2 + H_2 \leftrightarrow H_2O + CO \quad \left(\Delta H^0_R = +41.27\,kJ/mol\right) \tag{29}$$

$$CO + 2H_2 \leftrightarrow CH_3OH \quad \left(\Delta H^0_R = -90.84\,kJ/mol\right) \tag{30}$$

5.8 *Methanol from Power plant's Flue Gases*

Methane, coal, carbon dioxide, and hydrogen are the main feeds for this reaction. Flue gases coming out from power plant like nitrogen, steam, oxygen, and carbon dioxide worked as reactants that are responsible to convert syngas into methanol. This reactor unit contains coal-fired thermal power plant, a tri-reforming unit, catalytic burner, heat recovery system, a syngas compressor, and a methanol reactor with a distillation chamber [24]:

These are the reactions that are associated with this process:

$$CH_4 + CO_2 \leftrightarrow 2CO + 2H_2 \quad \left(\Delta H^0 = +247.3\,\text{kJ/mol}\right) \tag{31}$$

$$CH_4 + H_2O \leftrightarrow CO + 3H_2 \quad \left(\Delta H^0 = +206.3\,\text{kJ/mol}\right) \tag{32}$$

$$CH_4 + 0.5O_2 \leftrightarrow CO + 2H_2 \quad \left(\Delta H^0 = -35.6\,\text{kJ/mol}\right) \tag{33}$$

6 Use of Methanol in IC Engine

Methanol can be used as a direct fuel cell or with levels of blending as per the requirements.

6.1 *Direct Methanol Fuel Cell Used in IC Engine*

The very first attempt was taken in 1960–70 to use methanol air fuel cell directly into the engine by electrolyte and sulphuric acid by using a catalyst; that catalyst is either noble metals such as platinum or non-noble metals such as metal phthalocyanines (MPC), metal tetra-azo-annulenes (MTAA), and metal tetraphenyl porphyrins (MTPP) [25]. Noble metal catalysts were a little bit effective but the cost is the major issue with it; on the other hand, non-noble metallic catalysts were not so much effective. Since the catalyst is the most important element for direct methanol fuel cell, the main problem is also non-availability of the most effective and economic catalyst [26]. In addition to catalyst, there are several problems which need to be solved before using direct methanol fuel cell such as the following:

- Chemical oxidation of methanol with electrolyte and catalyst
- Evaporation of methanol into the air
- Heat carry by fuel due to evaporation
- Lower efficiency due to evaporation
- Maintaining the level of water into the systems for the electrolyte

- CO_2 separation from the fuel

Unfortunately, none of the system has been developed yet which can work with direct methanol fuel cell but there is scope for the development and research in this area.

6.2 Methanol Used with Binary Blending

The blending of alcohols with fuels is simply termed as binary blending. Typically, this blending can be done at three levels as per [5, 26, 27].

- Low blend (percentage of methanol < 15 by volume)
- Mid blend (percentage of methanol from 15 to 30 by volume)
- High blend (percentage of methanol > 30 by volume)

In Israel, the M15 model has been adopted commercially while countries like USA and UK have adopted the M10 model [5, 28, 29]. Engine efficiency can be improved by the binary blending of methanol into the conventional engine fuel as it can increase the octane number.

6.2.1 Properties of Methanol Binary Blending

Vapour Pressure

When methanol blends with gasoline, alcohol chain length tends to increase. When gasoline is further added, their hydrogen bonds start to weaken so alcohol behaves like a low molecular mass component. Binary blending interaction between gasoline and methanol changes if this process further moves on [5, 30].

Distillation

Complete vapourization of fuel is very important in all running conditions of the engine. When more methanol is added into the gasoline, it starts lowering the volatility of fuel, and blended fuel starts acting as an azeotropic mixture. The azeotropic mixture is a constant boiling point mixture of two liquids whose proportions can't change by simple distillation. Due to this, the volatility of fuel is lowered. To deal with this scenario, engine parameter calibration is required [5, 31].

Flame Speed

As per the experiments done by Sileghem et al. [5, 32] to investigate the optimal blending for the optimized flame speed of burning inside the cylinder, the investigation suggested that burning velocity is not linear and various mixing rules are used to get acceptable results. Good results are obtained by Hirasawa et al. by using the energy faction mixing rule.

Delay Period

The delay period is different for gasoline and methanol. Delay in the function of temperature for both fuels is suggested by Yates et al. [33]. Sileghem et al. [32] used the empirical relations developed by Yates et al. to calculate the delay period of blending of methanol and gasoline by using the energy fraction mixing rule.

Density

The surrounding molecules' volume defines the volume of nearby molecules volume; this means the density of blend is decided by the volume of molecules of blended molecules. Since density plays an important role in the burning of fuel, hence the density of blend must be evaluated by using turner et. al investigations.

6.3 *Methanol Used with Ternary Blending*

Blending of alcohol is not limited to binary; in fact, two alcohols can be blend with fuel, and that type of blending is termed as ternary blending; generally, Gasoline, methanol, and ethanol are used in ternary blending which is usually termed as GEM. This is being observed experimentally that cold stating with GEM is comparatively difficult because ethanol is responsible to so that as ethanol has poor cold starting property [5, 28]. The ternary blend is also a good option to be used in IC engine as they are economical. Fewer changes are required in the engine to be operated with the ternary blend. Pure gasoline can be used to run the engine on contemporary to this. Optimum GEM blending is necessary for year-round working as ethanol is harder to run at low temperature [5].

6.3.1 Properties of Methanol Ternary Blending

Vapour Pressure

Vapour pressure of methanol blend fuel is more than that of ethanol blend fuel so if the methanol is increased in GEM, the vapour pressure of the ternary blend also increases which is responsible to enhance evaporative emissions. The kinetic energy of evaporation accelerates when vapour pressure increases [5, 34, 35].

Distillation

E85 blend without methanol exhibits low distortion, i.e. low volatility. Methanol has different kinds of characteristics. Azeotropic behaviour of GEM tends to change when the percentage of methanol increases in GEM. The cold starting problem tends to decrease when the blend of alcohol increases into the mixture [28].

Octane Number

At the iso-stoichiometric blending, the mixture exhibits nearly identical lower heating value and almost constant octane number by turner et.al. [28, 36].

Auto-Ignition

It is being observed experimentally that high concentration alcohols show iso-stoichiometric behaviour in spite of combustion method and engine types (SI or CI) [27]. A similar type of results was shown in another investigation that shows non-linearity behaviour [38].

At low temperature, the blend is unstable, i.e. fuel and alcohols separated in phases so ethanol is used to make this mixture more stable, which actually worked as a co-solvent [39].

6.4 Use of Methanol in SI Engine

Methanol is conventionally considered as SI engine's fuel because it shows suitable properties to run in SI engine; high auto-ignition temperature and high latent heat of vapourization are few of them. Methanol can be used directly or with the level of blends while making few modifications to the engine or engine parameters. Here are some important attributes of methanol while using as a fuel in an SI engine:

- Methanol has handsome oxygen content so that low stoichiometric air is required, i.e. low air–fuel ratio required for the combustion which leads to an increase in better combustion.
- Methanol has an almost similar value of physical properties such as viscosity, thermal diffusivity, and sound speed into the mixture that of gasoline mixture.
- Methanol consists of hydrogen bonding that's why methanol has a high latent heat of vapourization.
- Methanol shows high auto-ignition temperature so that it can be used with a higher compression ratio.
- Since methanol promotes a cooling effect inside the engine cylinder due to evaporation. This cooling effect is caused by the heat taken away with evaporated fumes which promotes the cooling effect and increases volumetric efficiency.
- Due to higher auto-ignition temperature, knocking tends to decrease.
- Exhaust temperature in the case of methanol is quite lower than that of gasoline.
- Methanol produced lesser NO_x due to lower adiabatic flame temperature inside the engine cylinder [5, 6].
- CO can sustain at a higher temperature and lack of oxygen promotes to produce CO [45]. The exhaust temperature of methanol is less as compared to gasoline and since methanol carries more oxygen molecules, it results in lesser production of CO.
- Methanol helps in complete combustion. This means it produces lesser unburnt particles, lesser hydrocarbons, and lesser UHC as compared to gasoline.

6.5 Use of Methanol in CI Engine

Since methanol has a high auto-ignition temperature, so it cannot be directly used in CI engines, so some modifications are required while running CI engine with methanol. Those modifications are done in engine construction as well as in engine performance parameters.

- Methanol has almost double auto-ignition temperature than mineral diesel, which shows a positive indication as to be used at a higher compression ratio.
- Methanol has good ON while diesel has good CN, so it is better to use methanol in blending along with methanol by using some solvents since diesel and methanol do not mix well with each other.
- Methanol–diesel engine is generally fuelled by duel injected system, i.e. methanol and diesel both injected into the engine cylinder separately.
- Methanol has a significant amount of oxygen, so methanol-fuelled diesel engine exhausts lower emission as compared to pure diesel.
- Methanol has a high latent heat of vapourization which leads CI engine to produce higher brake thermal efficiency, power output, and volumetric efficiency.

7 Engine Performance and Emission

Various experiments have been done to determine the performance evaluation among Gasoline, methanol, and ethanol.

7.1 *Engine Performance and Emission with Pure Fuels*

7.1.1 Engine Performance

Some of the experiments are done by Balki et.al [40] over a single cylinder, four-stroke, air-cooled engine petrol engine with an original compression ratio of 8.5. The compression ratio of the used engine was varied by different cylinder heads and by providing extra sealing with proper arrangements. Table 4 shows the various properties of different fuels used for the experiments.

When experiments done by Balki et. al. [40] with the gasoline, methanol, and ethanol with the blend, it was observed that brake-specific fuel consumption (BSFC) tends to increase as the CR increases. As per the Table 5, BSFC for methanol is almost double than of gasoline in the same CR. The reason for this behaviour of

Table 4 Properties of fuels used [40]

Property	Gasoline	Ethanol	Methanol
Chemical formula	-	C_2H_5OH	CH_3OH
Molar C/H ratio	0.44–0.50	0.33	0.25
Density (g/cm^3 at 20 ^{0}C)	0.72–0.76	0.790	0.792
Latent heating value (kJ/kg)	44,300	26,900	20,100
Stoichiometric air/fuel ratio	14.6	9.0	6.47
Auto-ignition temperature (^{0}C)	228–470	363	455
Boiling point (^{0}C)	27–225	78.30	64.5
Heat of vapourization (kJ/kg)	349	923	1178

Table 5 Variation BSFC at different CRs [40]

CR	BSFC (g/kWh)		
	Gasoline	Ethanol	methanol
8.0	341.03	473.29	672.75
8.5	278.78	453.72	600.34
9.0	273.31	443.99	594.87
9.5	302.2	442.94	590.83

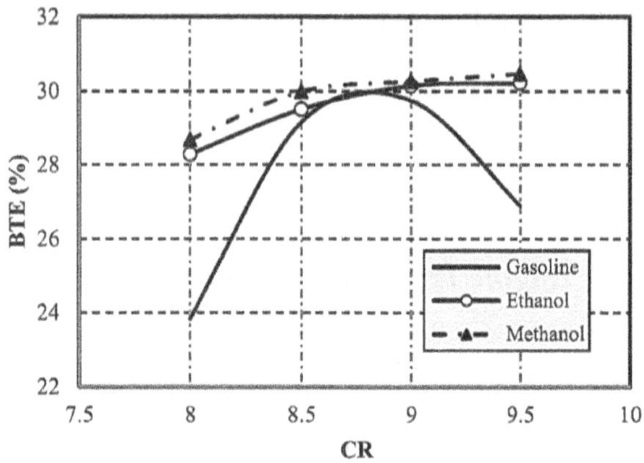

Fig. 7 Variation in BTE at different CRs [40]

methanol and ethanol is that the heating value of gasoline is very much higher than that of alcohol fuel.

Engine cylinder experiences more pressure when increasing CR, thus more brake mean effective pressure and hence more brake thermal efficiency. But when CR further increases, BTE tends to decrease as shown in Fig. 7. This is due to an increase in surface-to-volume ratio and higher friction and pumping loss. As shown in Fig. 7, it is clear that there is a drop in BTE for methanol and ethanol at 9.5 CR. The value of BTE is more than that of gasoline.

7.1.2 Engine Exhaust Emission

As per the experiments done by Balki et. Al, Fig. 8 shows the various graphs which depict the engine exhaust emissions.

Alcohol fuels have a higher oxygen level than gasoline so their combustion efficiency is better than gasoline. Combustion efficiency can be measured by the emission of unburnt hydrocarbons and CO emission. As the CR increases, the UHC and CO levels decrease. When CR further increases, there is an increase in UHC and CO level due to an increase in surface-to-volume ratio in the combustion chamber. It is being experimentally calculated that there is a drop of about 22.79% and 28.22% of UHC in ethanol and methanol as compared to gasoline. While a drop in CO is about 14.99% and 29.37% on average in ethanol and methanol as compared to gasoline. Ethanol and methanol have lower hearting value than gasoline so a less high temperature will be achieved in the combustion chamber which is responsible to produce lower NO_x and CO_2. There is an increase in its value as the CR increases but after 9.5 CR NO_x and O_2 slightly increase due to an increase in surface-to-volume ratio of combustion ratio.

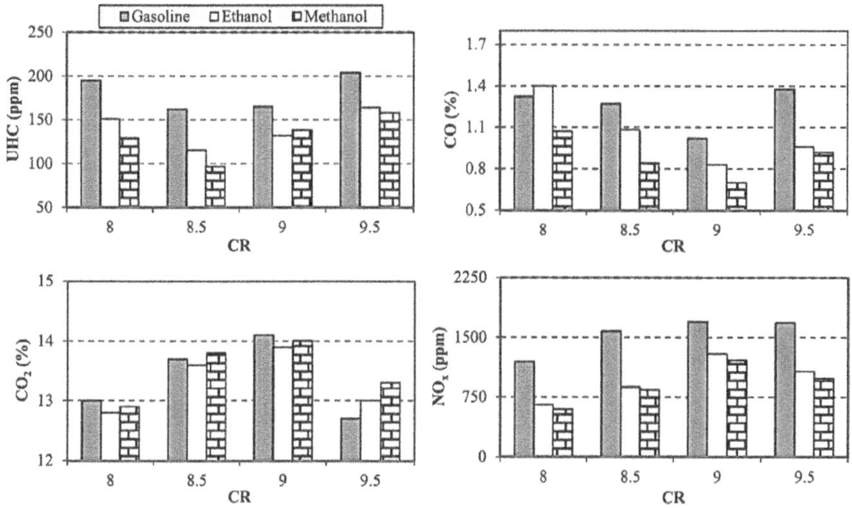

Fig. 8 Variation in engine exhaust emission at different CRs [40]

7.2 *Engine Performance and Emission with Blending of Fuels*

7.2.1 Engine Performance

Agarwal et. al. [41] have done numerous experiments to predict the engine behaviour and exhaust emission of gasoline and alcohol blends. As per this study, experiments were done by him over a four-cylinder, four-stroke, water-cooled, MPFI SI engine (Maruti Suzuki; Zen). This engine was subjected to run over a range of speed and blends. Engine speed ranges were 1500, 2500, and 3500 rpm. The used fuel was pure gasoline, M10, and M20 blend.

Figure 9 shows the trends of BSFC against the engine speed for different fuels. As per the graph, performance lines drop because methanol has higher oxygen content as compared to gasoline. Since methanol has higher oxygen content due to this, its heating value is lower because oxygen does not affect heating value significantly. While gasoline has almost double the heating value as compared to methanol as per table 6.

Figure 10 shows the variation of BTE against different loads and speeds. This fashion of graph is achieved because methanol has a higher latent heat of vapourization. Due to higher latent heat of vapourization, there is a significant loss of heat inside the engine when the engine worked with blends and pure alcohol.

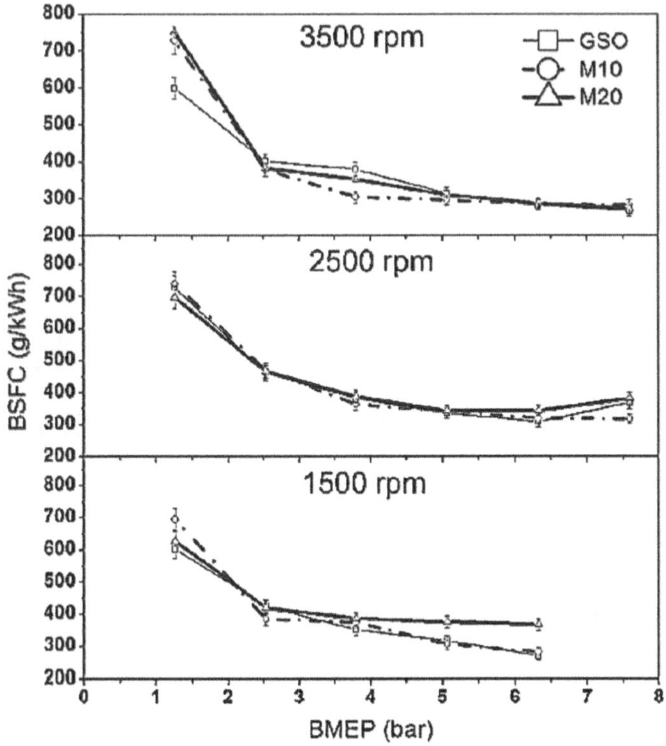

Fig. 9 BSFC at different engine loads and speeds [41]

Table 6 Properties of fuels [41]

Fuel property	Methanol	Gasoline	Diesel
Formula	CH_3OH	$C_nH_{1.87n}$	$C_nH_{1.7n}$
Molecular weight	32.04	100–110	200
Research octane no (RON)	106	92–98	–
Density, kg/L	0.792	0.72–0.78	0.81–0.89
Boiling point/range, °C	65	27–225	188–343
Specific heat (Vapour) kJ/kg · K	1.72	~1.7	~1.7
Viscosity, cSt @40 °C	0.59	0.37–0.44	2.6–4.1
Latent heat of vapourization, kJ/kg	1178	349	233
Higher heating value, MJ/kg	22.7	47.3	43.8
Lower heating value, MJ/kg	20.0	44.0	41.0
Lower flammability limits, Vol.%	7.3	1.4	1.0

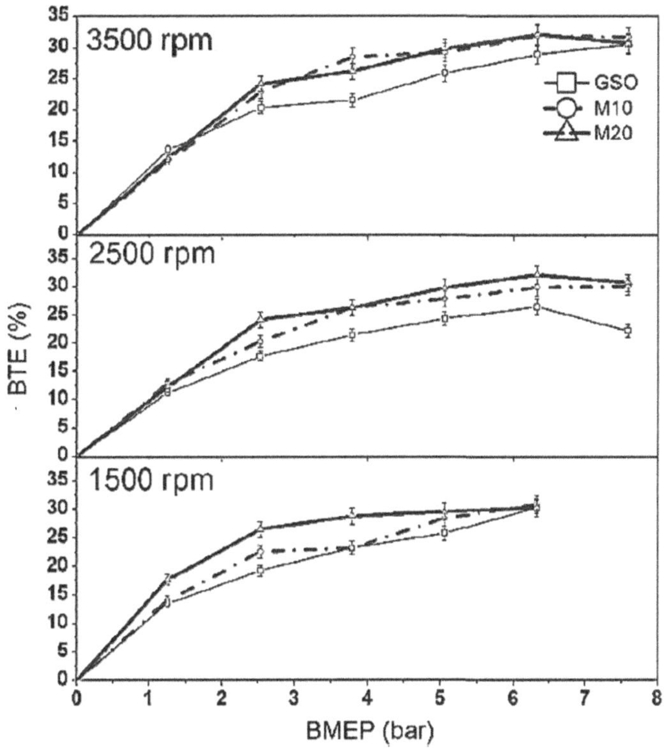

Fig. 10 BTE at different engine loads and speeds [41]

7.2.2 Engine Exhaust Emission

Figure 11 shows the HC and CO exhaust from the engine at different engine loads and speeds. At lower engine load, the mixture was too lean to burn and it was difficult to sustain the flame by the mixture which leads to produce higher HC emission. At higher engine load, the engine has less oxygen which leads to produce higher HC emissions. Furthermore, alcohol has a higher latent heat of vapourization. Due to higher latent heat of vapourization, engine cylinder temperature cools down which causes difficulty in combustion and misfire, which leads to produce more HC. At higher engine load, the temperature inside the engine is relatively higher which leads to produce lesser CO emission. Simultaneously at the higher engine load, CO emission increases due to less availability of oxygen. At lower engine speed, due to higher combustion temperature, CO emission decreased with engine load. It was observed that at the same engine speed, CO emission was less for alcohol blends due to the presence of high oxygen content.

Figure 12 shows the variation of NO and smoke opacity at different loads and speeds. NO is mainly formed due to higher engine temperature and oxygen availability during combustion. NO emission increases with engine load. At lower load

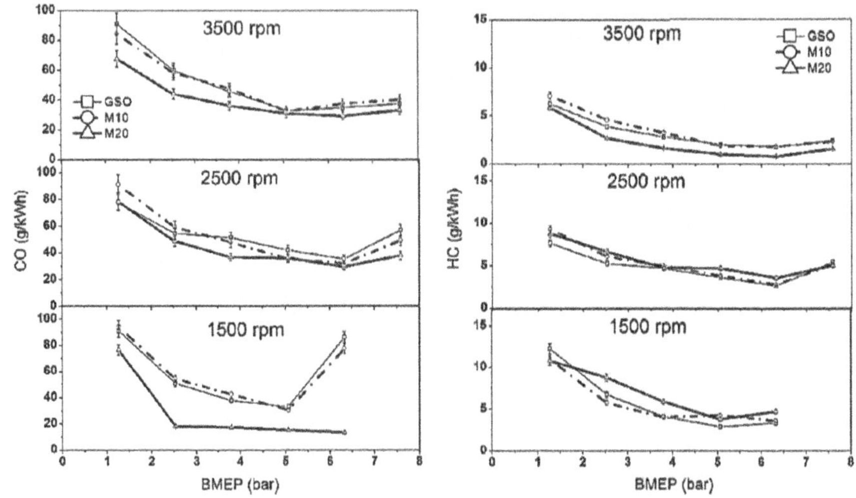

Fig. 11 HC and CO emission at different engine loads and speeds [41]

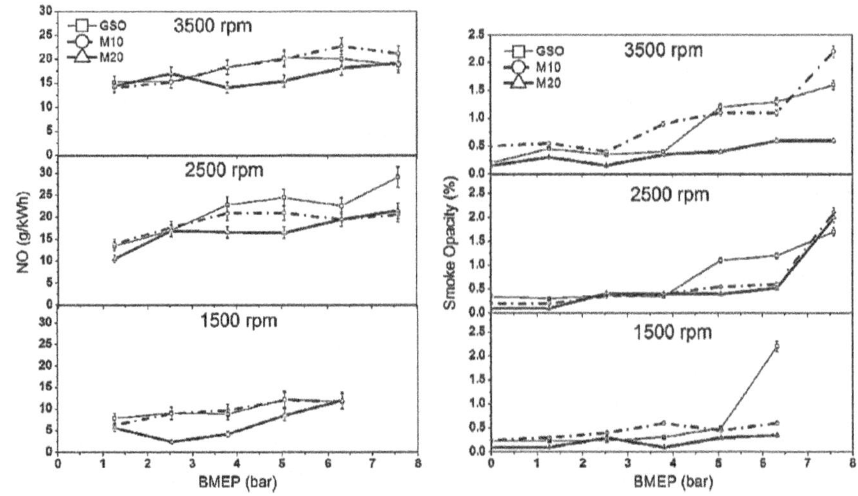

Fig. 12 Smoke opacity and NO emission at different engine loads and speeds [41]

on the engine, lower engine temperature dominates over the high oxygen availability which results in lower NO emissions. Gasoline blends have lower NO emission because of the higher latent heat of vapourization of methanol, which results in lower peak pressure and temperature inside the engine. This tends to produce lower NO emission. Gasoline produces higher smoke opacity at a higher engine load. While methanol and its blends produce lower smoke opacity due to higher oxygen content as Fig. 12 suggests.

8 Impact of Methanol on Various Concerns

8.1 Impact of Methanol on Health

Methanol is responsible for helping in digestion in the human body which indicates that some amount of methanol is always present in our body; in fact, some traces of methanol is also present in almost every fruit in diluted form but when the amount of methanol increased up to 100 ml into human blood, it can cause death. Methanol often faced criticism to be used in the engine due to its toxicity. It can cause blindness, neurological disorders, and death when it is ingested more than a certain amount. It is too difficult to detect by smelling when it has a concentration of more than 2000 ppm [42].

8.2 Impact of Methanol on Environment

Methanol is water soluble, and as a result, it is not possible to distinguish its presence in water physically. It is diffused to the atmosphere at a significant rate than any other chemical in the same category. Methanol biodegrades much faster than hydrocarbons which results in short lifetimes.

8.3 Impact of Methanol on Fuel Handling

Even a small amount of methanol blend with gasoline changes vapour pressure and evaporated emissions. Due to an increase in vapour pressure, a large volume of storage tanks is required to accumulate evaporative emissions.

8.4 Impact of Methanol on Safety

Safety is one of the reasons why methanol is attracting researcher to be used it in the engine in addition to its performance and emission ability. Methanol has a high auto-ignition temperature, radiates less heat, can be extinguished by water, and burns with smoke.

8.5 Impact of Methanol on Engine Material

Since methanol is corrosive in nature, so few modifications are required in engine fuel systems while using methanol as a fuel in the engine. Coating of various additives helps to protect the engine and its components from corrosion caused by methanol. Various metal matrix composites of aluminium can enable engine corrosive resistive [43, 44].

9 Conclusion

In this chapter, a comprehensive review of methanol as a fuel in IC engine is presented. The main objective of this chapter is to review methanol as an alternative to gasoline and diesel in commercial IC engine. Comparison of various properties of methanol with ongoing fuels supports its candidature to be used as an alternative to gasoline and diesel, which depicts that methanol can be used directly or with levels of blending into the engine. Of course, few modifications are required in the engine. Although methanol is best suitable to be used in SI engine due to its excellent properties like high auto-ignition temperature, high octane number, higher latent heat, impressive emission control, and low cost with minimum modification in the engine, it can also work with diesel engines. Of course, it's quite challenging than SI engine but there is a huge scope for that too. It is being observed that binary blend of methanol is performing well in gasoline engines; countries like Israel, China, USA, UK, and other European countries are already running commercial vehicles with various blending ratios like M15, M10, and so on. Methanol production is discussed; there are many methods and processes available to produce methanol from various feeds like wood, biomass, biogas, syngas, natural gas, and coal. Syngas is working as a middle reactant for the production of methanol as methanol is synthesized by the syngas commercially. Methanol production is even more appreciable as it can be synthesized from greenhouse gases such as water steam/vapour, carbon dioxide, and carbon monoxide. Methanol exhibits many impacts on health as it is toxic in nature, on the environment as it is easily diffused in the air at atmospheric temperature, on safety due to its flammability, on engine material compatibility as it is corrosive, and handling as it has higher vapour pressure. It has been seen that methanol has more benefits on using it as a fuel in IC engine compared to its disadvantages.

References

1. Central intelligence agency. https://www.cia.gov/library/publications/the-world-factbook/geos/xx.html
2. Kim S, Dale BE (2005) Environmental aspects of ethanol derived from no-tilled corn grain: nonrenewable energy consumption and greenhouse gas emissions. Biomass Bioenerg

28(5):475–489
3. Methanol Economy (2018) NITI Aayog working on roadmap for India onWorld Environment Day. https://pib.nic.in/newsite/PrintRelease.aspx?relid=179785
4. https://www.niti.gov.in/writereaddata/files/document_publication/Article%20on%20Methanol%20Economy_Website.pdf
5. Verhelst S, Turner JW, Sileghem L, Vancoillie J (2019) Methanol as a fuel for internal combustion engines. Prog Energy Combust Sci 70:43–88
6. Vancoillie J (2013) Modeling the combustion of light alcohols in spark-ignition engines (Doctoral dissertation, Ghent University)
7. Vancoillie J, Verhelst S (2010) Modeling the combustion of light alcohols in SI engines: a preliminary study. In: FISITA 2010 World Automotive Congress. International Federation of Automotive Engineering Societies
8. De Cuyper T, Demuynck J, Broekaert S, De Paepe M, Verhelst S (2016) Heat transfer in premixed spark ignition engines part II: Systematic analysis of the heat transfer phenomena. Energy 116:851–860
9. Aasberg-Petersen K, Nielsen CS, Dybkjær I, Perregaard J (2008) Large scale methanol production from natural gas. Haldor Topsoe, p 22
10. van Sint Annaland MM, Gallucci IFF, de Groot IMT, Spallina IVV (2017) Techno-economic analysis of methanol production using chemical looping reformin
11. Valera H, Agarwal AK (2019) Methanol as an alternative fuel for diesel engines. In: Methanol and the alternate fuel economy. Springer, Singapore, pp 9–33
12. Trop P, Anicic B, Goricanec D (2014) Production of methanol from a mixture of torrefied biomass and coal. Energy 77:125–132
13. Matzen M, Alhajji M, Demirel Y (2015) Chemical storage of wind energy by renewable methanol production: Feasibility analysis using a multi-criteria decision matrix. Energy 93:343–353
14. Trudewind CA, Schreiber A, Haumann D (2014) Photocatalytic methanol and methane production using captured CO2 from coal-fired power plants. Part I–a Life Cycle Assessment. J Clean Prod 70:27–37
15. Rathod VP, Bansal P, Bhale PV (2015) Analytical and experimental investigations for hydrogen rich syngas production by biogas reforming processes. Energy Procedia 75:728–733
16. Ryi SK, Park JS, Kim DK, Kim TH, Kim SH (2009) Methane steam reforming with a novel catalytic nickel membrane for effective hydrogen production. J Membr Sci 339(1–2):189–194
17. Tsolakis A, Golunski SE (2006) Sensitivity of process efficiency to reaction routes in exhaust-gas reforming of diesel fuel. Chem Eng J 117(2):131–136
18. Tomishige K, Matsuo Y, Sekine Y, Fujimoto K (2001) Effective methane reforming with CO_2 and O_2 under pressurized condition using NiO–MgO and fluidized bed reactor. Catal Commun 2(1):11–15
19. Brown LF (2001) A comparative study of fuels for on-board hydrogen production for fuel-cell-powered automobiles. Int J Hydrogen Energy 26(4):381–397
20. Klier K (1982) Methanol synthesis. In: Advances in catalysis, vol. 31. Academic Press, pp 243–313
21. Brynolf S, Fridell E, Andersson K (2014) Environmental assessment of marine fuels: liquefied natural gas, liquefied biogas, methanol and bio-methanol. J Clean Prod 74:86–95
22. Andersson J, Lundgren J, Marklund M (2014) Methanol production via pressurized entrained flow biomass gasification–Techno-economic comparison of integrated vs. stand-alone production. Biomass Bioenerg 64:256–268
23. Räuchle K, Plass L, Wernicke HJ, Bertau M (2016) Methanol for renewable energy storage and utilization. Energy Technol 4(1):193–200
24. Minutillo M, Perna A (2010) A novel approach for treatment of CO_2 from fossil fired power plants. Part B: the energy suitability of integrated tri-reforming power plants (ITRPPs) for methanol production. Int J Hydrogen Energy 35(13):7012–7020
25. Binder H, Köhling A, Sandstede G (1972) From electrocatalysis to fuel cells. SANSTEDE (ed) Seattle, pp 15–31

26. McNicol BD, Rand DAJ, Williams KR (1999) Direct methanol–air fuel cells for road transportation. J Power Sources 83(1–2):15–31
27. Turner JW, Pearson RJ, McGregor MA, Ramsay JM, Dekker E, Iosefa B, ... ac Bergström K (2012) GEM ternary blends: testing iso-stoichiometric mixtures of gasoline, ethanol and methanol in a production flex-fuel vehicle fitted with a physical alcohol sensor (No. 2012–01–1279). SAE Technical Paper
28. Turner JW, Pearson RJ, Bell A, de Goede S, Woolard C (2012) Iso-stoichiometric ternary blends of gasoline, ethanol and methanol: investigations into exhaust emissions, blend properties and octane numbers. SAE Int J Fuels Lubric 5(3):945–967
29. The Standards Institution of Israel. Israel M15 press release. https://methanolfuels.wpengine.com/wpcontent/uploads/2013/05/Israel-M15-Press-Release-8 2016.pdf
30. Australian GEM fuel program.https://methanolfuels.org/wpcontent/ uploads/2013/05/Grant-Lukey-Coogee-Energy-Australia.pdf (2013)
31. Bourhis G, Solari JP, Dauphin R, De Francqueville L (2016) Fuel properties and engine injection configuration effects on the octane on demand concept for a dual-fuel turbocharged spark ignition engine (No. 2016–01–2307). SAE Technical Paper
32. Sileghem L, Coppens A, Casier B, Vancoillie J, Verhelst S (2014) Performance and emissions of iso-stoichiometric ternary GEM blends on a production SI engine. Fuel 117:286–293
33. Yates A, Bell A, Swarts A (2010) Insights relating to the autoignition characteristics of alcohol fuels. Fuel 89(1):83–93
34. Turner JW, Pearson RJ, Dekker E, Iosefa B, Johansson K, Ac Bergström K (2013) Extending the role of alcohols as transport fuels using iso-stoichiometric ternary blends of gasoline, ethanol and methanol. Appl Energy 102:72–86
35. Turner J, Griffith W (2012) Processing of fuel and recirculated exhaust gas. UK Patent GB2484495
36. Turner D, Xu H, Cracknell RF, Natarajan V, Chen X (2011) Combustion performance of bio-ethanol at various blend ratios in a gasoline direct injection engine. Fuel 90(5):1999–2006
37. Waqas M, Naser N, Sarathy M, Feijs J, Morganti K, Nyrenstedt G, Johansson B (2017). Auto-ignition of iso-stoichiometric blends of gasoline-ethanol-methanol (GEM) in SI, HCCI and CI combustion modes
38. Anderson JE, Leone TG, Shelby MH, Wallington TJ, Bizub JJ, Foster M, ..., Polovina D (2012) Octane numbers of ethanol-gasoline blends: measurements and novel estimation method from molar composition (No. 2012–01–1274). SAE Technical Paper
39. Qi DH, Liu SQ, Zhang CH, Bian YZ (2005) Properties, performance, and emissions of methanol-gasoline blends in a spark ignition engine. Proc Inst Mech Eng Part D: J Autom Eng 219(3):405–412
40. Balki MK, Sayin C (2014) The effect of compression ratio on the performance, emissions and combustion of an SI (spark ignition) engine fueled with pure ethanol, methanol and unleaded gasoline. Energy 71:194–201
41. Agarwal AK, Karare H, Dhar A (2014) Combustion, performance, emissions and particulate characterization of a methanol–gasoline blend (gasohol) fuelled medium duty spark ignition transportation engine. Fuel Process Technol 121:16–24
42. Bromberg L, Cheng WK (2010) Methanol as an alternative transportation fuel in the US: Options for sustainable and/or energy-secure transportation. Sloan Automotive Laboratory, Massachusetts Institute of Technology, Cambridge, MA
43. Bansal P, Upadhyay L (2013) Experimental investigations to study tool wear during turning of alumina reinforced aluminium composite. Proc Eng 51:818–827
44. Bansal P, Upadhyay L (2016) Effect of turning parameters on tool wear, surface roughness and metal removal rate of alumina reinforced aluminum composite. Proc Technol 23:304–310
45. Bansal P, Pandya A, Ahuja A (2016, December) Experimental performance evaluation on SI engine with gasoline and liquefied petroleum gas as fuel. In: Techno-Societal 2016, International conference on advanced technologies for societal applications. Springer, Cham, pp 183–191

Methanol as a Low-Cost Alternative Fuel for the Reduction of Emissions

Boris Chudnovsky, Alex Lazebnikov, Mordechai Reshef, and Ilya Chatskiy

Abbreviations

HFO	Heavy fuel oil
MEOH	Methanol
GT	Gas turbine
SSH	Secondary superheater
RH	Reheater
TT7	Gas turbine exhaust temperature
CHP	Combined heat and power generation
HWB	Hot water boiler
HRSG	Heat recovery steam generator
CC	Combined cycle

Nomenclature

CO	Carbon monoxide
NOx	Nitrogen oxide
SO_2	Sulphur dioxide
CH_2O	Formaldehyde

B. Chudnovsky (✉) · A. Lazebnikov · M. Reshef · I. Chatskiy
Generation Project Division, Electric Corporation and Dor Chemicals (Joint Venture), Haifa, Israel
e-mail: borisz@iec.co.il

A. K. Agarwal et al. (eds.), *Methanol*, Energy, Environment, and Sustainability,
https://doi.org/10.1007/978-981-16-1224-4_3

1 Introduction

The main strategies for reducing NOx emission are: (1) Modification of the combustion process to control the fuel and air mixing and reduce flame temperatures; (2) Post-combustion treatment of the flue gases to remove NOx. Here, we shall discuss the first form of NOx reduction only. The approach to NOx reduction can include combustion tuning that provides a modification of the combustion process: (1) control fuel and air mixing, (2) reduction of flame temperature, (3) use of low NOx burners, (4) staged combustion, (5) gas reburning technology, and (6) alternative fuel combustion. In this chapter, we shall discuss our experience with the combustion of an alternative fuel—methanol and minor tuning to inexpensively reduce NOx emissions. Since methanol combustion occurs with lower combustion temperature than with fuel oil combustion and since it contains no sulphur and ash, the outcome is that there are lower NOx emissions, no SO_2 emissions, and very low particulates emissions. It is applied to types of combustors.

The concept of using methanol as a fuel received much attention because methanol is the most versatile and clean burning fuel of all fuels derived from non-petroleum energy sources such as natural gas, coal, biomass, and carbon dioxide. Using CO_2 for methanol production leads to the reduction of greenhouse gas emissions so that methanol can actually be considered as an enviro fuel. Methanol is also safer, cheaper, and easier to ship than liquefied natural gas which is very important where transportation of large amounts of fuel to remote areas is required while other fuels are not available.

IEC (Israel Electric Corporation) and Dor Chemicals present results for specific utility boilers, gas turbines, small industrial boilers, etc. that show the control of NOx, SO_2, and particulate emission by combustion of methanol and its mixing with fuel oil. By using an appropriate combination of methanol and fuel oil, a desirable emission requirement can be achieved while optimizing the costs.

The results presented here are from a 140 MWe unit consisting of a CE (Combustion Engineering) tangential fired pressurized boiler and a Pratt and Whitney FT4C TWIN PAC 50 MWe GT. In these studies, we have tested co-firing of methanol with heavy fuel oil in utility boilers and full methanol firing in gas turbines. We found that the use of methanol as an alternative fuel enabled us to meet the commonly accepted emission for NOx by using co-firing with heavy fuel oil in the boiler. We also found that using methanol solely firing the gas turbine has drastically reduced NOx emission. Besides that, methanol firing leads to a considerable reduction of particulates and SO_2 emissions. We have performed optimization of boiler operation conditions for various fractions of methanol and heavy fuel oil co-firing.

The firing of synthetic methanol to produce clean energy is becoming highly desirable as it decreases all kinds of emissions, including CO_2 emissions. Using methanol for electricity generation leads to a reduction of CO_2 formation. In this chapter, we also evaluate the performance and emissions of small co-generation units based on micro turbines and engines burned methanol with different cycle configurations. Given the fact that due to the increase in renewable energy introduction into the grid distributed energy facilities are becoming very significant. Hence, the development of small CHP projects burning synthetic clean methanol with low emissions is challenging and should be promoted. We believe that the conclusions of the present experience are general and can be applied to other utility boilers and gas turbines.

2 Technological Background

Introducing a new technology for energy generation is a major task, and when electricity is concerned, it is particularly a complex and lengthy process. Factors that influenced energy production are: (1) Substantial capital costs since investors seek a high rate of return on their investments. (2) Environmental regulations are driven by 'green' and environmental power groups that have become a major influencing force on legislators as it is a hot political issue. (3) Electricity prices are controlled by utility regulators aimed at reducing prices. (4) Ongoing demand for electricity and rapid growth in many parts of the world that expect cheap electricity. (5) It is a challenge to control the grid with as much as possible renewable energy sources throughout days and seasons of the year including the shortest days when solar energy inputs are lowest.

Methanol burning projects in power generation utilities have been attempted throughout the years for demonstration and commercial purposes. Methanol burning was done at the beginning of the '70s [1]; this attempt was done for demonstration purposes only. In 1974, a test was performed by [1] with a 20 MWe gas turbine. The methanol was burned as a liquid fuel. The measured NOx emission was 74% less than that was obtained when diesel oil was burned. Measured CO emissions values were close to diesel oil burning [1–4].

From 1984–1985, GE has performed methanol burning tests of heavy-duty gas turbine combustors MS6001B. The methanol was burned as a liquid, without and also with water injection. The tests demonstrated that methanol fuel can be successfully burned without requiring major modifications of the combustor. NOx emissions were approximately the same as burned no.2 distillate with water injection. Using methanol for electricity supply leads to the reduction of CO_2 and will conform to the targets of the Paris Protocol.

Liner metal temperatures and dynamic pressures do not get affected considerably by methanol combustion and GE performance were in the acceptable range. The obtained results show that in comparison with distillate fuel methanol burning leads to significant emission reduction with a slightly higher unit capacity and lower heat rate.

Methanol may be successful used in diesel engines. As follows from the publication [5] a large Ro-Pax ship, Stena Germanica, was converted to methanol burning. This project was executed by Stena with engine manufacturer Wärtsilä's assistance and a methanol producer Methanex. Since the methanol fuel produces no sulphur emissions and low value of nitrogen oxide emissions, it is compliant with current emissions reduction regulation. The conversion of engine to methanol burning required just the addition of methanol pumps, fuel rail, and injectors re-design.

Recent gas turbine experience of methanol burning was done by Dor Chemicals and Israel Electric Corporation [6]. Dor Chemicals in cooperation with the Israel Electric Corporation in 2012 and by the end of 2013 has converted a twin pack gas turbine unit (2 engines) to use methanol and as result, enabling a substantial reduction in emissions. This has enabled the gas turbine to operate for many hours per day without restrictions. This unit has become operational in 2014.

Methanol can be produced from several renewable feedstocks that are available in large quantities and most important from CO_2. The latter can lead to a combination of extracting unwanted exhaust gas for methanol production. Methanol's advantage in fuel source mix is that it is related to as a renewable fuel. Methanol as a backup combustion fuel may be produced in a solar central receiver (SCR) and used in the system designed for 24/7 electricity generation during the winter days of the year when the solar energy input is limited. Possible solutions of power to fuel applications are presented below.

The process for methanol synthesis based on water electrolysis and carbon dioxide extracted from the coal-fired boiler is as follows [7]:

Water electrolysis—H_2 and O_2 are produced by using electrical energy. Hydrogen, obtained during electrolysis process, is a source for methanol synthesis, and produced oxygen is used for external users.

Amine-based CCS (Carbon Capture Storage) is connected to coal power plant exhaust ducts, and CO_2 is extracted for cleaning the flue gas and for the methanol synthesis. Methanol reactor synthesis of CH_3OH from H_2 and CO_2 is shown in Fig. 1.

The methodology developed by NewCO_2Fuels (NCF) is presented below [8]. In this method, solid oxide cells are used in electrolysis mode at a temperature of about 800 °C for the dissociation of CO_2 to CO and O_2.

Simultaneously, the same device can dissociate water (H_2O) to H_2 and O2. The dissociation of both CO_2 and H_2O at 800 °C is endothermal. NCF developed a heating mechanism, based on a working fluid flow, to supply the needed thermal energy. This working fluid is heated by concentrated solar energy; its outflow from the dissociation

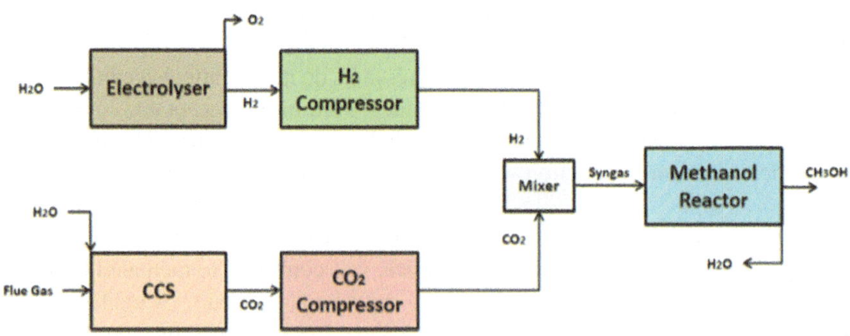

Fig. 1 The simplified plant layout [7]

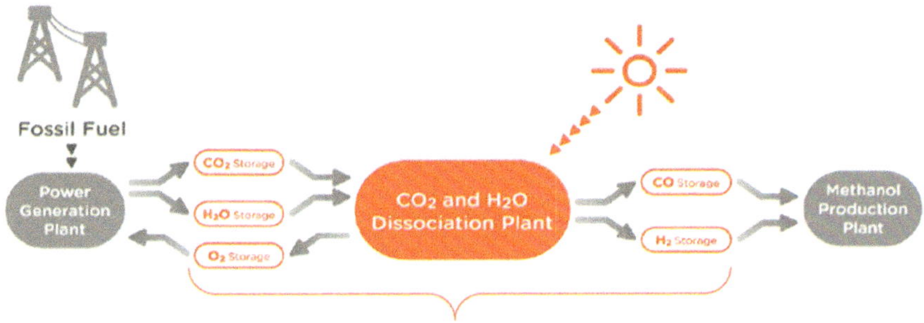

Fig. 2 Concept of novel technology of fuel production from CO_2 proposed by NewCO$_2$Fuels [8]

process is at a sufficient temperature to produce steam for energizing a Rankine cycle, which produces the electricity needed for the dissociation process. Thus, the proposed NCF dissociation process is powered by concentrated solar energy.

The CO, or the mixture of CO and H_2 (called Syngas), can then be used as gaseous fuel (e.g. in power plants) or converted to liquid fuel (e.g. methanol), which can be stored, transported, and used in motor vehicles and electricity generation facilities. The oxygen produced in the process can be used in the combustion of the clean fuel, for example in oxy-fuel combustion in power plants (Fig. 2).

As follows from the MHPS [9], methanol synthesis may be based on captured CO_2 from fossil fuel power plant and water treatment electrolysis. The produced renewable fuel may be used for electricity generation and/or for vehicles.

3 Methodology of Combustion Process Prediction

3.1 Gas Turbines

To predict the combustion behaviour of methanol firing and co-firing with fuel oil and their consequent impact on performance and pollutant emission from utility boilers and gas turbines, we use a methodology and software that combines experimental work and computational simulations.

The simulations were performed in advance to the retrofit of the combustor of an FT4C TWIN PAC 50 MWe GT designed by Pratt and Whitney. The detailed method of combustor simulations (thermal and emissions) is described in the works [10, 11]. The special software code for combustor simulations was developed. For NOx simulation, Chemkin software was applied. For cycle simulation, Thermoflow software was used.

The method consists of two consecutive stages. The first stage includes liner hydraulic, heat balance, heat transfer, and metal temperature calculation.

The target of hydraulic calculation is the determination of air flow through all parts of the combustor, air and flue gas velocities, excess air coefficient, and pressure drop in each element of the combustor. The calculation is based on an iteration method. The first assumption for the hydraulic calculation is based on the proportionality of air flow distribution to the cross-sectional area of each element of the combustor. At the next stages of calculations, the friction factor (flow coefficient) is taken into consideration. The final iteration is based on the following condition:

$$100 * \left|\left(G_{gl} - G_{air}\right)/G_{air}\right| \leq 3.0\% \tag{1}$$

where G_{gl}—flue gas flow at last liner section and G_{air}—liner air flow

The equation for excess air determination is

$$\alpha = \left(G_g - B_{fuel})/L_0 * B_{fuel}\right), \tag{2}$$

where G_g—flue gas flow and B_{fuel}—fuel flow.

Fuel burn-out is calculated based on the following equation:

$$\xi_2 = \xi_0 * X_0^{\ b * P^{\bullet} / \alpha}, \tag{3}$$

where $\xi_0 = 0.99$ (experimental constant); X_0—relative length ($X_0 = x/L$);$P^{\bullet}$—pressure ratio ($P^{\bullet} = P_{comb}/P_0$); b—experimental factor that depends on liner design; α—flame zone excess air.

Using burn-out equation and excess air at each liner section, mass average flue gas temperature through liner length is calculated. Flue gas components are determined based on fuel burning stoichiometric reactions.

In order to estimate methanol firing on combustor reliable operation, liner metal temperature is calculated for both fuels: LFO and methanol.

Liner metal temperature is calculated based on the following equation:

$$\sigma * \varepsilon_F^{Eff} * \left(T_{FlameEff}^4 - T_L^4\right) + \alpha_1^c * (T_m - T_L) = \alpha_2^c * \phi * (T_L - T_2) + \sigma * \varepsilon_R * \left(T_L^4 - T_{shell}^4\right), \tag{4}$$

where σ—Stefan–Boltzmann constant; ε_F^{Eff}—flame effective emissivity; $T_{FlameEff}$—flame effective temperature; T_L—liner metal temperature; α_1^c—liner inside convective heat transfer factor; T_m—boundary layer temperature (is a function of cooling air temperature); α_2^c—liner outside (skin) heat transfer factor; T_2—combustion air temperature; ε_R—liner relative emissivity factor; T_{shell}—combustor sleeve inside temperature.

Combustor sleeve inside temperature is calculated as

$$T_{shell} = T_2 + (T_L - T_2)/\left(1 + 0.04 * (\rho_2 * W_2)^{0.8}\right), \tag{5}$$

where W_2—combustion air velocity between liner and sleeve; ρ_2—combustion air density.

The second stage NOx emission calculation. The emission calculation is based on the generally recognized mechanism of the gas-phase reaction, occurring during the combustion process, [12–15]. The mechanism was confirmed by a comparison of results from detailed kinetic calculations with experimental data. The considered mechanism includes 234 elementary reactions with recommendations for the rate coefficients, among them the gas-phase reaction mechanism for thermal, prompt and nitrogen formation. The mechanism takes into account the chemical kinetics of the combustion process and is realized mathematically as a system of 234 ordinary stiff differential equations.

However, the application of this purely kinetic mechanism to the real combustion chamber, in which diffusion processes take place in parallel with kinetic ones, requires additional consideration. In line with this approach, we have divided these reactions into three following groups: hydrocarbon oxidation (1); non-thermal NOx formation (2); thermal NOx formation (3). The first group reactions (stage) include combustion of hydrocarbons and preliminary mixing between fuel, air, and combustion products. The initialization of the second group reactions is run by intermediate products of the first group. The third group is unlimited by diffusion. In accordance with the above, the model was supplemented with a time delay τ for each reaction group. We define τ as the time taken to prepare the reactants for reaction in each of the corresponding groups. In doing so, we get $\tau_1 = \tau_2 = \tau_d$ and $\tau_3 = 0$ and these values are taken into account during the integration of the model equations. It should be noted that diffusion time τ_d depends on the longitudinal coordinate. Hence, the problem has been simplified to the determination of the kinetic time τ_k which is the upper integration limit of the model kinetic equation, equal to $\tau_k = \tau_s - \tau_d$, where τ_s—is residence time. Current time t and longitudinal coordinate x are connected by the simple relation $t = x/v(x)$, where $v(x)$ is the liner velocity of combustion products in the combustor. Furthermore, it was assumed that the combustor volume consists of N consecutive connected together elementary sub volumes. Each of these sub volumes is considered as a well-stirred reactor. For each of this reactor, we can write:

$$\xi_1(x) = \xi_1(j) = \sum_{j}^{N_j} B_j/B_0 = \sum_{j=1}^{N_j} \int_{\tau_j}^{\tau_{j+1}} f\big(B_j, C_{ji}, T\big)dt, \tag{6}$$

where C_{ji}—is a concentration of the i reactant, $i = 1,2,.0.234$; B_j—fuel concentration in j reactor; B_0—total flow rate of fuel. Using $C_{ji}(x)$ by means of the integration model, one can find the completeness of combustion, ξ_1.

In order to find the unknown value of the delay time $\tau_k(x)$, we have to invoke an additional Eq. (6). Setting ξ_1 equal to ξ_2, one can get an equation that may be used to determine τ_k.

The accuracy of the model was checked by comparing predicted results to experimental data obtained during tests and depicted in Fig. 3. Tests were carried out on different combustion chambers, including test rigs and full-scale combustors. As the

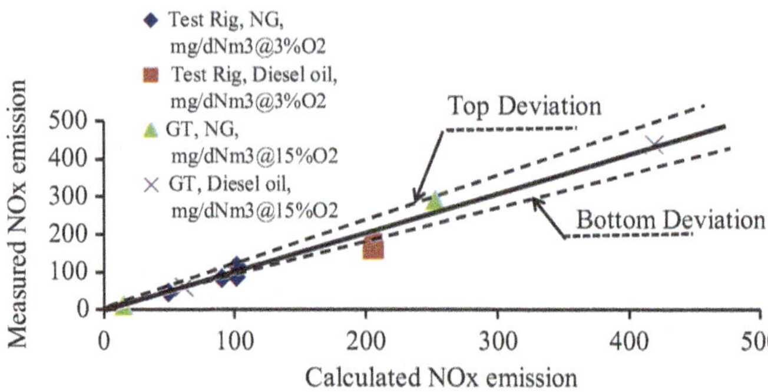

Fig. 3 Comparison between measured and calculated NOx emission (mg/dNm^3)

comparison shows, the root mean deviation between calculated and measured data was less than 13% for a mode range of parameters. (It should be noted that the model was checked for different GT sizes and designs; the obtained maximum 13% deviation may be defined as a good approximation and this comparison confirms that the model is adequately simulating the real process).

3.2 Boiler

For the analysis of the operational parameters for methanol and heavy fuel oil co-firing in a boiler designed for fuel oil, we use the standard method for furnace heat transfer calculations [16, 17]. Based on this method, we consider the following heat transfer mechanism, where the furnace heat transfer depends on furnace wall fouling, emissivity, and radiation properties of the flue gases.

The incident heat flux q_{inc} of the water walls is

$$q_{inc} = q_{wnet}/\varepsilon_w + \sigma_o T_w^4, \tag{7}$$

where ε_w is spectral emissivity, σ_0—Stefan–Boltzmann constant, q_{wnet} is net (absorbed) radiation heat flux, and T_w—water wall temperature with the contamination, calculated by

$$T_w = T_{fl} + (\varepsilon\, q_{net}), \tag{8}$$

where T_{fl} is water wall medium temperature and ε is waterwall fouling factor.

Therefore, incident radiation fluxes are equal to

$$q_{inc} = q_{wnet}/\varepsilon_w + \sigma_0\, [T_{fl} + \varepsilon\, q_{wnet}]^4. \tag{9}$$

Based on performed tests it was found that methanol and fuel oil co-firing are very similar to fuel oil firing; however, the radiation properties of flue gases formed during co-firing are slightly different from oil firing. Flame emissivity factor for co-firing compared with fuel oil firing is reduced and furnace exit temperature is increased accordingly, in spite of the fact that methanol combustion temperature is lower than that during fuel oil firing. The flame emissivity depends on the radiation of three atomic gases, ash, and coke particles. Especially, for liquid and gaseous fuel, it also depends on luminous flame distribution in the furnace, named as factor m [17]. It was taken into consideration that methanol flame is almost transparent, indicating the presence of entirely 100% of 3 atomic gases (no particles emitting luminous flame) and as a result, the m factor was assumed equal to zero at the furnace exit. For fuel oil, this factor was taken according to a recommendation in [17]. For a blend of methanol and liquid fuels, the factor m was calculated according to the methanol heat fraction. In order to achieve more accurate simulation results, including NOx emission calculation, the furnace was divided into four zones, and each zone was calculated separately.

The factor m is expressed by the following equations:

For LFO and methanol mix,

$$m_{(LFO+MEOH)} = q * (0.7 - 0.15 * n) + 0.333 - 0.0833 * n, \tag{10}$$

where q is LFO mass fraction and n is furnace zone number.

For HFO and methanol mix,

$$m_{(HFO+MEOH)} = q * (0.934 - 0.188 * n) + 0.333 - 0.0833 * n, \tag{11}$$

where q is HFO mass fraction and n is furnace zone number.

For integral furnace calculation for methanol and its mix with fuel oil, the factor m is varied from zero to its value recommended in [17] depending on the heat fraction of methanol in the fuel blend. The validation of the methodology was performed by comparison of furnace exit temperature (FEGT) executed by reverse heat balance calculation with direct furnace calculation according to [17] using the above-mentioned methodology. During the boiler trial run, the main boiler parameters were measured and flue gas temperature distribution through the boiler was estimated. Besides that, the pressure parts metal temperature was evaluated for performed test using measured steam temperature and flue gas temperature obtained from simulation results. The above-mentioned methodology is able to predict boiler behaviour at a hundred percent of methanol burning.

4 Experience of Methanol Burning in Gas Turbines

4.1 Gas Turbines Short-Term Operation

Israel Electric and Dor Chemical converted a gas turbine unit of Pratt and Whitney, an aero-derivative FT4C-1, from diesel oil to methanol burning. A gas turbine consists of a multistage axial flow compressor, a can-annular combustion section, and a two-stage axial flow reaction turbine as can be seen in Fig. 4 (left). The can-annular combustion section is shown in Fig. 4 (right). At the first step of the conversion of the gas turbine to methanol burning, only modification of the fuel supply system was performed [18]. The existing gas turbine firing system, high-pressure fuel pump, and control valve remained without any change. Due to the fact that the heating caloric value of methanol is about half of the diesel oil, the gas turbine capacity was limited accordingly. A temporary methanol tank was installed nearby the gas turbine with an associated forwarding pump (Fig. 5, left). The new methanol pipe line and its connection to existing diesel oil pipes are shown in Fig. 5 (right). In order to perform continuous measurement of the gas turbine parameters, temporary performance display (Fig. 6, left) and emission continuous monitoring were installed (Fig. 6, right). The general results of GT tests are shown below. It can be seen from Fig. 7 (top) that load increasing leads to a rise in exhaust temperature. The methanol firing causes an exhaust temperature reduction in comparison to light fuel oil firing due to lower methanol adiabatic temperature at the same air flow control. Comparison of gas turbine exhaust spread for methanol and diesel oil burning is shown in Fig. 7 (bottom).

The main goal of the performed work was to compare GT emissions performance between light fuel oil and methanol firing. The obtained results are presented in Fig. 8. As can be seen from the presented data, the tests were performed at a wide load range up to 23 MWe unit load during light fuel oil firing and approximately up to 16 MWe load during methanol firing. Further load increasing at methanol firing could not be reached due to fuel flow capacity limitation caused by the fuel modulating valve, the pump, and the nozzle's size. The performed tests clearly show that methanol firing leads to a considerable NOx reduction for all loads range. Figure 8 (top) depicts

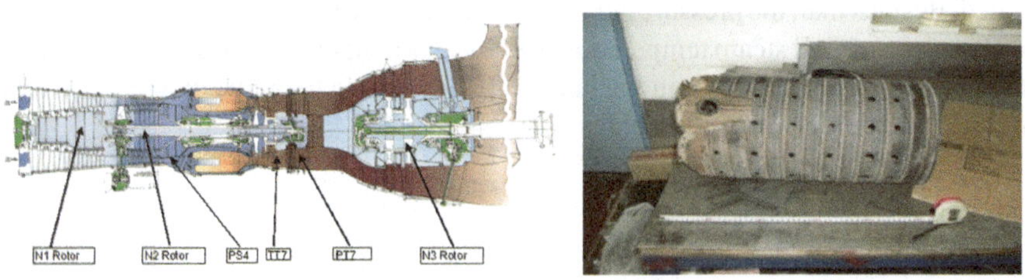

Fig. 4 Schematics of a 50 MWe P and W Gas Turbine (left) and a can-annular combustion section (right)

Fig. 5 Temporary methanol tank with forwarding pump (left) and methanol pipe line connection to existing diesel oil pipes (right)

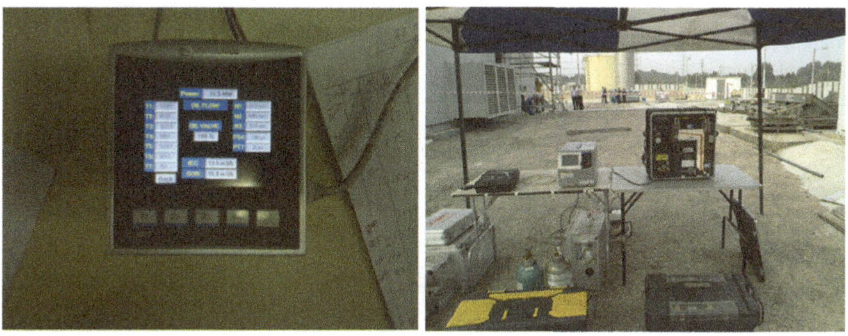

Fig. 6 Temporary GT performance display (left) and emission continuous monitoring (right)

the results of NOx emission measurements obtained during the tests. Comparison of CO emissions (Fig. 8, bottom) during all performed tests shows that its value is higher at methanol firing, especially at low load. Probably, the main reason for this phenomenon is lower combustion temperature and higher excess air during methanol firing.

However, load increasing leads to CO reduction and it may be assumed that at full load, CO emission during methanol firing will be close to light oil firing. One of the possible ways for CO reduction is adjusting fuel and air control during methanol firing in order to achieve the excess air similar to light fuel oil burning. Particulates emission as a function of GT load is shown in Fig. 9. As follows from the obtained results, methanol firing significantly reduces unburned carbon in flue gases. As was expected, during methanol firing, SO_2 emission is zero.

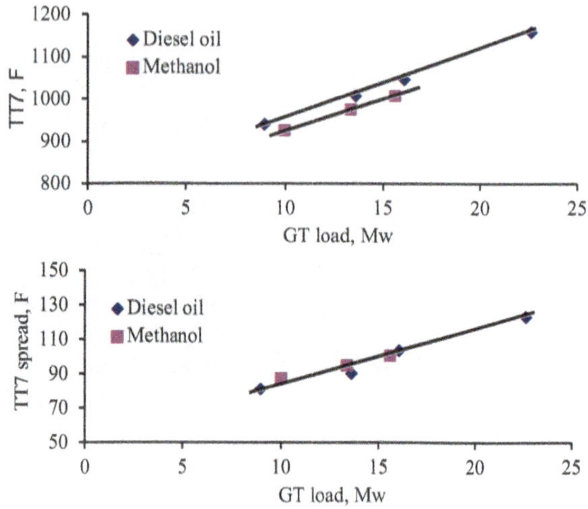

Fig. 7 Average TT7 temperature (top) and exhaust temperature spread (bottom) as a function of GT load

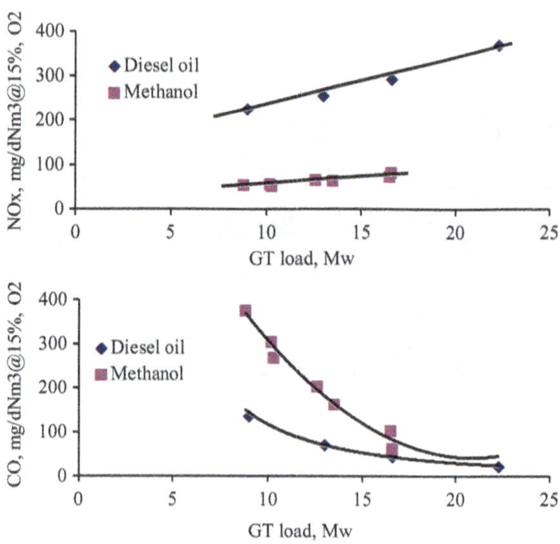

Fig. 8 Comparison of measured NOx emission (top) and CO emission (bottom) for diesel oil and methanol firing

4.2 Gas Turbines Long-Term Methanol Conversion

Encouraged by the very good results of the short-term test, it was decided to go forward for a long-term test which could pave the way to demonstrate technical and commercial feasibility.

The first aim was to restore the output capacity that this engine can provide. This capacity was deprived because of the low heating value of the methanol (practically half of the light fuel oil). It was therefore needed to enable the machine to allow for double flow amounts. To do that, it was needed to release all the bottle-necks in

Fig. 9 Comparison of measured particulates emission for diesel oil and methanol firing

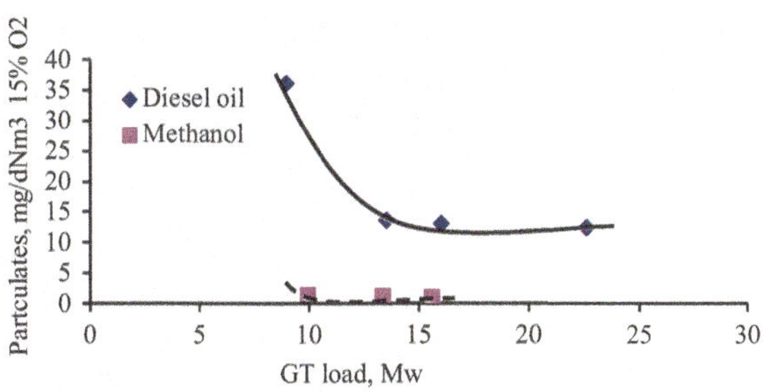

the fuel supply and injection system. The team's first intention was to request the OEM—WGPW—Wood Group Pratt and Whitney for higher capacity equipment, i.e. a higher capacity modulating valve and a higher capacity high-pressure shaft-driven fuel injecting pump. This requested OEM equipment was not available. It was, therefore, necessary to search for a different concept [19].

The novel concept was to replace the original shaft-driven high-pressure pump and a modulating valve with an electrical pump installed on an off-engine basement skid and driven by a VSD—Variable Speed Drive. The VSD is controlled from a digital fuel controller which is replacing the previous fuel control system which governed the fuel amounts injected into the engine via the modulating valve. On top of that, two more bottle-necks had to be released on the machine. The P and D—Pressure and Dump valve and the fuel nozzles. The P and D valve was found to allow for double the flow by only replacing its screen with a coarser screen. Yet, the fuel nozzles had to be replaced from the original Excello type (Fig. 10 left) into the Delevan high flow nozzles (Fig. 10, right).

These nozzles were originally designed to allow for double the flow for water injection lowering NOx applications (WLN—Wet Low NOx).

In order to simplify the conversion of the unit to work with methanol, the gas turbine's sequencer was left in the system and a controller was designed to work in conjunction with it. The original startup and shut down procedures were left untouched as the system was originally designed. To do so, the system had to work with diesel fuel for these phases of work. On top of making the conversion simpler, it has served for the unit to have a dual fuel system for better redundancy.

The pumps' skid was therefore designed to have two branches; one for the light fuel and one for the methanol. This enabled to start the engine with a normal startup procedure and then at any state after that (before synchronization or after that at any load) to switch gradually from diesel oil burning into methanol burning. The pumps and valve skids are shown in Fig. 11 (left) and 11 (right).

On top of the main converted system, some BOP (balance of plant) changes were required. These included the methanol supplying and storage systems and the specialized firefighting system.

The methanol supplying system consisted of a new unloading piping system with two pumps, a conversion of a light fuel oil storage tank of 2000 m^3, and two forwarding pumps to bring the methanol to the injection skid. A floating roof was

Fig. 10 Excello fuel nozzles (left) and Delevan fuel nozzles (right)

installed inside the storage tank in order to inhibit the evaporation of methanol which is very volatile.

The fire detection systems were based on infrared detectors since the flame temperature of methanol is lower than light fuel oil flame temperature. Also, the flame is invisible and an infrared video camera was purchased to assist in detecting the source of fire in case anything is ignited. The fire extinguishing system was designed to use AR—Alcohol Resistant—foams that can stay active in the presence of methanol.

Prior to the long-term methanol conversion, a combustor simulation was performed using the above-mentioned developed methodology based on the simulation results validation of the short-term conversion trial run performance data. The results of the simulation for the GT performance and emission for methanol and light fuel oil firing is shown in Figs. 12 and 13.

Comparison of calculated flue gas temperature distribution shows that methanol firing caused flame temperature reduction (Fig. 12) that leads to decreased NOx formation. Predicted NOx emission formation is presented in Fig. 13. As can be seen from the performed simulation, methanol firing is associated with a very considerable NOx reduction. The calibration of the calculation model was carried out for light fuel oil firing. The measured NOx value during light oil operation is in good agreement

Fig. 11 Injection pumps skid-design (left) and injection pumps and stop valve skids-installed (right)

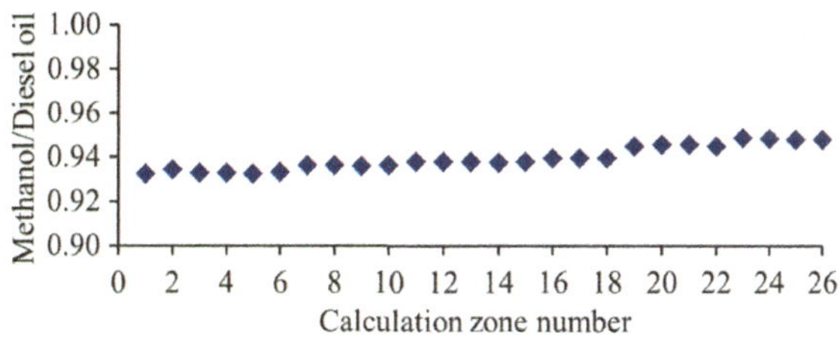

Fig. 12 Relative flame temperature distribution

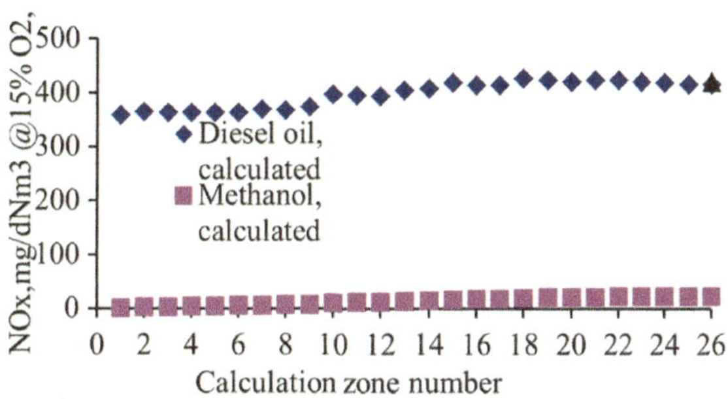

Fig. 13 NOx emission through liner length (bottom)

with calculated results (Fig. 13). Lower combustion temperature during methanol firing leads to liner wall temperature reduction (Fig. 14). The general results of GT during long-term tests are presented in Fig. 15. It can be seen from Fig. 15 (top) that load increasing leads to a rise in exhaust temperature.

The methanol firing causes a reduction in the exhaust temperature in comparison with light fuel oil.

This is due to lower methanol adiabatic temperature at the same air flow control. GT heat rate as a function of load is shown in Fig. 15 (bottom). As can be concluded from the measured data, the heat rate during methanol firing is almost similar to diesel oil firing at the same load range and in good agreement with the calculated value for both fuels.

The obtained results for emission measurements are presented in Figs. 16, 17, 18 and 19. Figure 16 (top) depicts the results of NOx emission measurements obtained during the tests. As can be seen from the presented data, the tests were performed at a wide load range of up to 23 MWe unit load during light fuel oil firing and methanol firing. Based on the presented data, it may be concluded that methanol firing leads to

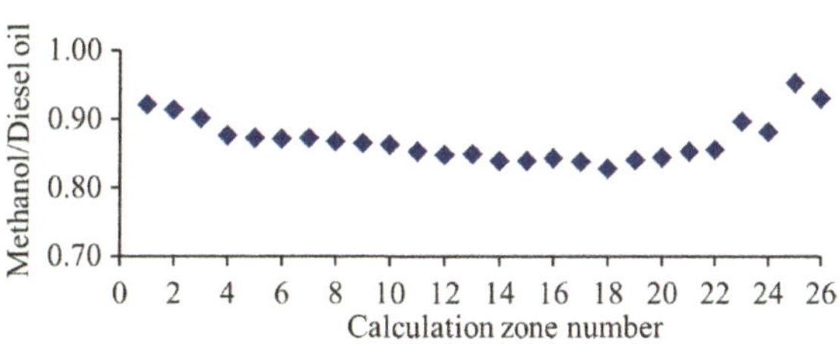

Fig. 14 Relative liner wall metal temperature distribution through liner length

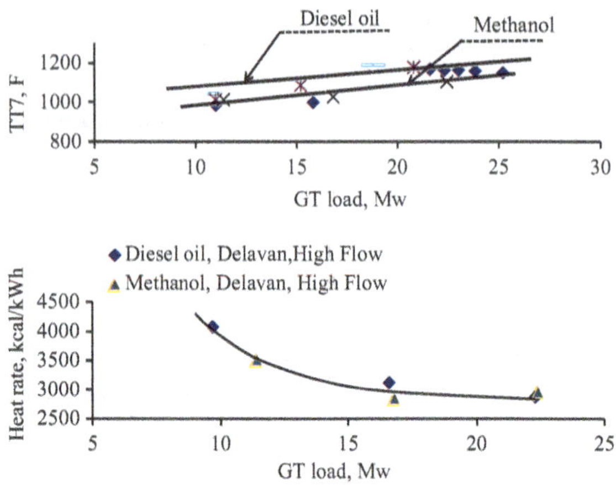

Fig. 15 Comparison of measured TT7 temperature (top) and heat rate (bottom) for diesel oil and methanol firing

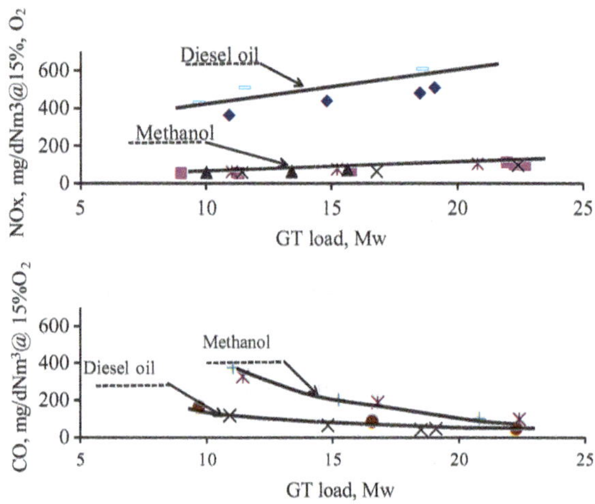

Fig. 16 Comparison of measured NOx emission (top) and CO emission (bottom) for diesel oil and methanol firing

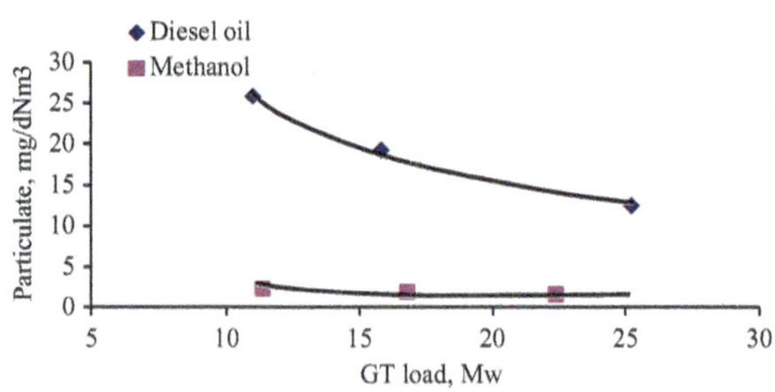

Fig. 17 Comparison of measured particulates emission (referred oxygen 15%) for diesel oil and methanol firing

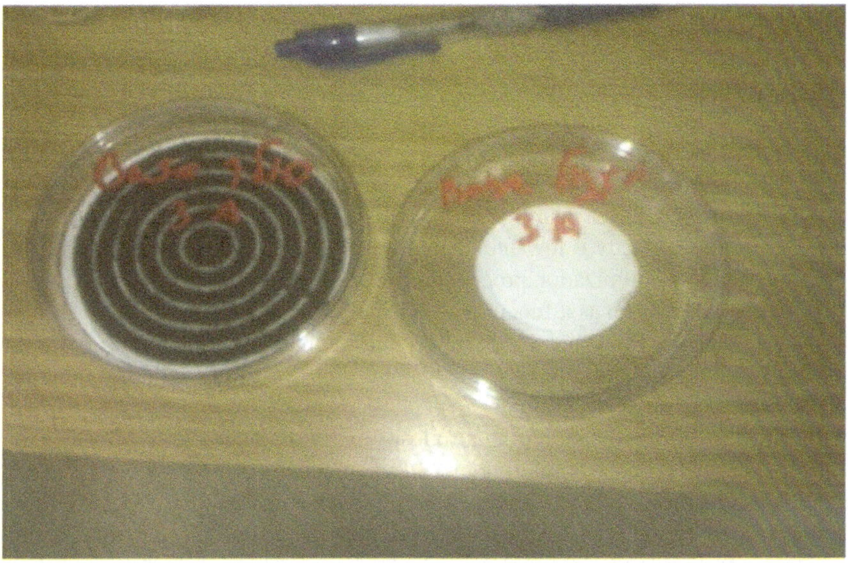

Fig. 18 Comparison of filters for diesel oil and methanol firing

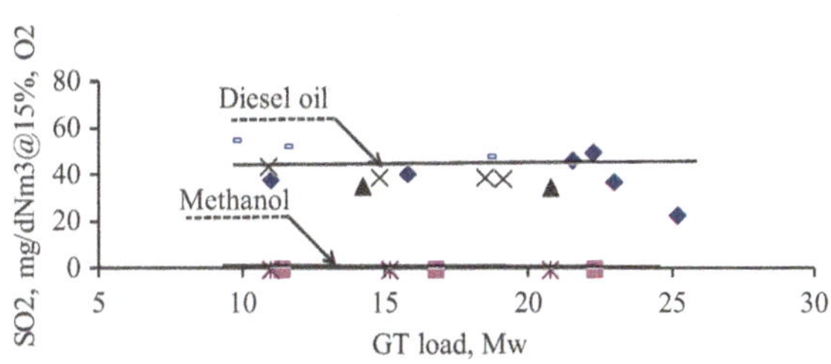

Fig. 19 Comparison of measured SO_2 emission for diesel oil and methanol firing

a considerable NOx reduction for all performed tests. Comparison of CO emissions (Fig. 16, bottom) during all performed tests shows that its value is slightly higher during methanol firing, especially at low load. One of the reasons for CO increase is high exhaust temperature spread both for oil#2 and methanol firing. In order to improve exhaust flue gases' spread, a P & D valve that was malfunctioning was replaced. As can be seen from the performed measurements, exhaust spread was reduced and as a result, due to the removal of 'cold' spots, CO emissions were also decreased.

The other reason for this higher CO emission is lower combustion temperature and higher excess air during methanol firing. It is seen that load increasing lowers this phenomenon and leads to CO reduction. One of the possible ways for CO reduction is adjusting fuel and air control during methanol firing in order to achieve excess air similar to that of fuel oil burning. It should be note, that for the methanol firing test, an old gas turbine, not equipped with IGV (inlet guide vane), was used and therefore, it was not possible to adjust air quantities. Air reduction may be achieved only by plugging some of the air ports of the combustion liner. This kind of adjustment requires a special investigation for both methanol and light fuel oil firing. In addition, high flow Delevan nozzles (the only available P & W nozzles with high flow rate required for methanol firing) were installed instead of the original

Excello nozzles are optimized for light fuel oil and water injection and not for methanol firing. From measured data, it is seen that CO emission for the original Excello atomizer is lower than that of high flow Delevan nozzles. However, these Excello nozzles' capacity is low and, therefore, not suitable for methanol firing. We shall continue this optimization process during the long-term trial run.

Particulates emission as a function of GT load is shown in Fig. 17. The display of the difference in filter dirt for diesel oil and methanol is shown in Fig. 18. As follows from the obtained results, methanol firing drastically reduces unburned carbon formed during combustion.

As was expected, methanol firing also leads to SO_2 reduction to practically zero. Comparison of SO_2 emissions during fuel oil and methanol firing is shown in Fig. 19. Based on all of these obtained results, it may be concluded that the presented long-term operation of the GT conversion to methanol firing is able to provide the full original gas turbine load capacity while reducing harmful emission to the atmosphere in a very significant manner.

4.3 Methanol and Diesel Oil Mixture Burning in Gas Turbine

We have also analyzed and tested the concept of using methanol and diesel oil mixtures [20]. Results of a full-scale test of methanol firing mixed with fuel oil at an FT4C TWIN PAC 50 MWe GT are presented below. These show the control of NOx, SO_2, and particulate emission by using various combinations of methanol and fuel oil fractions. It was found that increasing the heat fraction of methanol in the fuel mixture has caused a drastic reduction of NOx, SO_2, and particulates emissions. The general results of the tests are shown in Figs. 20, 21, 22, 23, 24 and 25. It can be seen from Fig. 20 (top) that load increasing leads to a rise in exhaust temperature as expected and methanol heat fraction rise leads to an exhaust temperature reduction. The exhaust temperature spread is shown in Fig. 20 (bottom). As follows from the obtained results, a higher methanol heat fraction leads to a slight reduction of the exhaust spread.

GT heat rate as a function of load is shown in Fig. 21 (top). As can be concluded from the measured data, heat rate depends only on GT load and does not depend on the rate of methanol fraction. Figure 21 (bottom) depicts results of NOx emission measurements obtained during the tests. As can be seen from the presented data, the tests were performed at load ranges from 13 to 21 MWe while methanol heat fraction changed from 0 to 100%. Based on the received data, it may be concluded that methanol firing leads to significant NOx reduction for all performed tests.

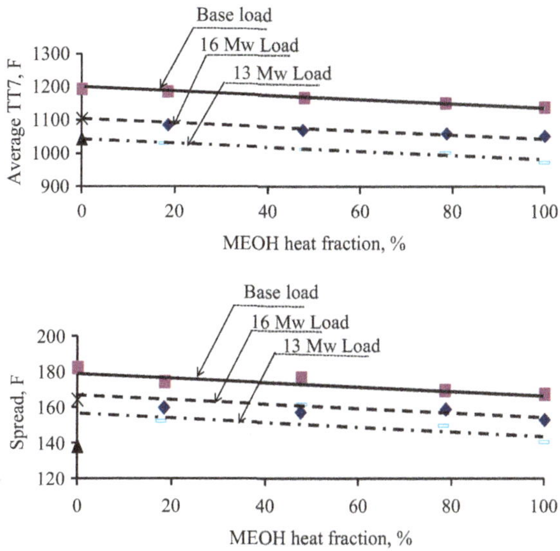

Fig. 20 Average exhaust temperature TT7 (top) and spread (bottom) as a function of methanol heat fraction

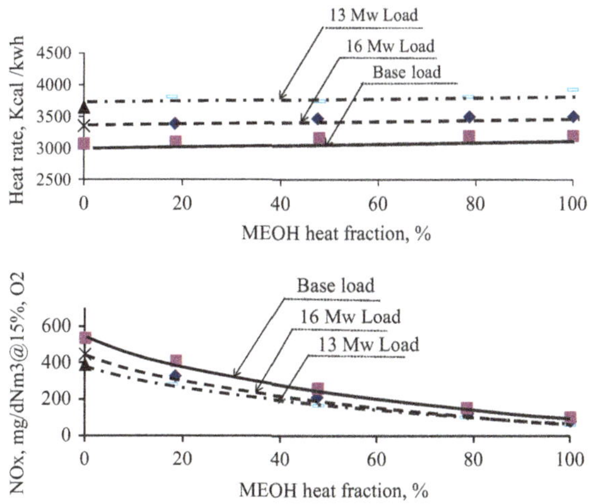

Fig. 21 GT heat rate (top) and NOx emission (bottom) as a function of methanol heat fraction

Comparison of CO emissions (Fig. 22, top) obtained from all performed tests shows that its value is increased during load reduction and with a higher heat fraction of methanol in the fuel mixture.

Oxygen concentration in exhaust flue gas as a function of GT load and methanol heat fraction is shown in Fig. 22 (bottom). As can be seen from the figure, methanol firing is correlated with higher excess oxygen and it may lead to combustion over-cooling and as a result to increased CO emissions. Generalization of the obtained data shows that the main reason for CO increase is high excess air and, as a result,

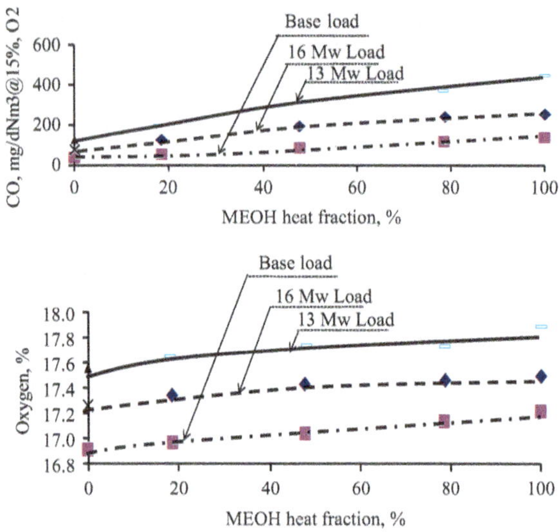

Fig. 22 CO emission (top) and exhaust oxygen concentration (bottom) as a function of methanol heat fraction

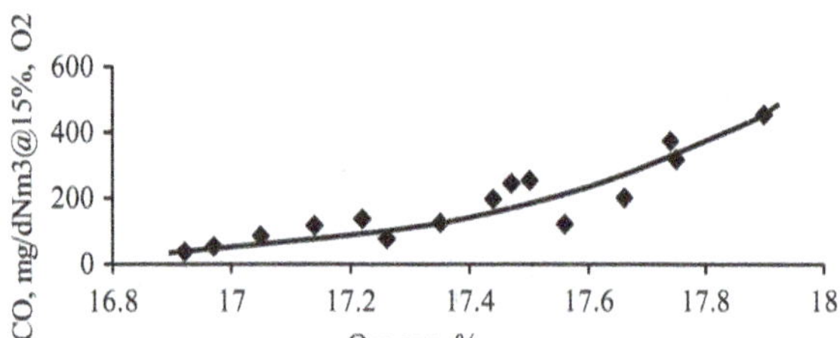

Fig. 23 CO emission as a function of oxygen concentration

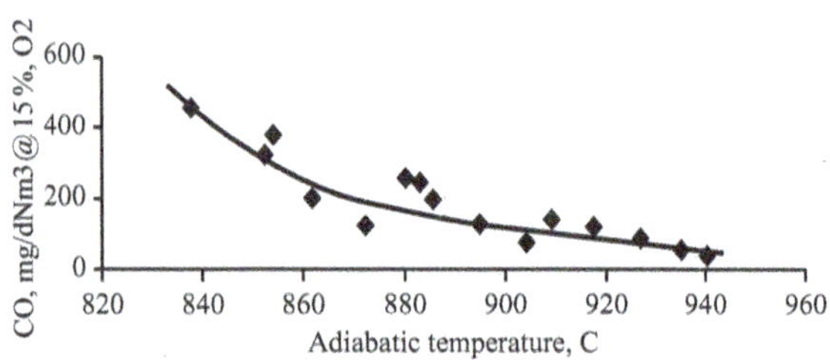

Fig. 24 CO emission as a function of adiabatic flame temperature

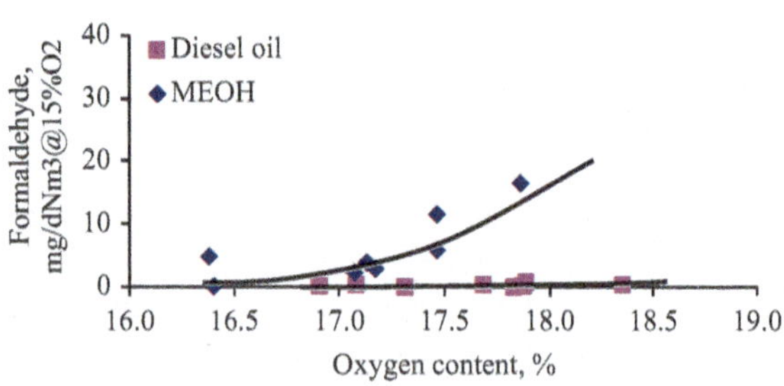

Fig. 25 Formaldehyde emission as a function of oxygen concentration

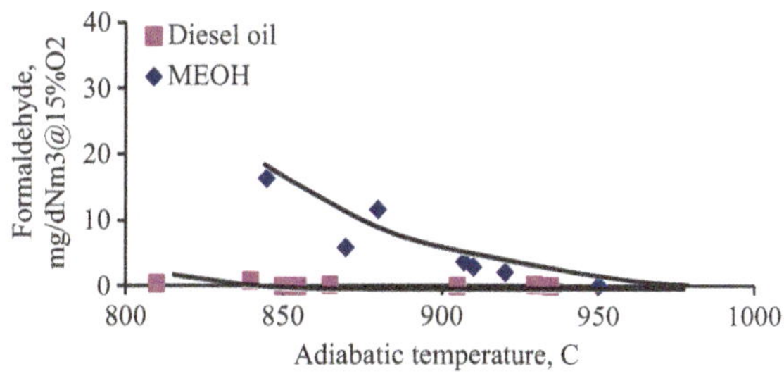

Fig. 26 Formaldehyde emission as a function of adiabatic flame temperature

lower combustion temperature, especially during load reduction and when methanol heat fraction is increased (Figs. 23 and 24).

The same phenomenon is measured for formaldehyde formation. Decreasing excess oxygen (increasing of combustion temperature) leads to a reduction of formaldehyde formation (Figs. 25 and 26).

A similar effect of combustion air increase on CO and formaldehyde formation was obtained in the work of Levy's team [21] where decreasing excess air (increasing combustion temperature) has significantly reduced CO formation. Another confirmation of the influence of excess air reduction on the improvement of combustion characteristics is presented in the work of Deuker [22]. This paper discusses Siemens' experience in the reduction of CO on their 'F' generation gas turbines. This was done by several stages of upgrading Siemens' GT combustion performance by reducing air flow to the burners. As learned from the paper, the main driver for the increase of CO is the decrease of flame temperature.

Constant load but higher flame temperature can only be achieved by reducing the air mass flow through the burners. Hence, every measure that reduces the air mass flow to the burners while keeping the load constant will increase flame temperature and, as a consequence, result in a reduction of CO. Siemens' 'F' upgrade GT has succeeded in reducing 30% of the CO by the reduction of the combustion air.

It should be noted that NOx emission when methanol is burned remained low across all load ranges (Fig. 27). Based on the above, we can conclude that a possible way for CO reduction is the adjustment of fuel and air ratio in order to reduce the excess air, and thus slightly increase air temperature at the GT inlet which will lead to an increase in flame temperature. As was mentioned above, NOx emission when methanol is firing remained low throughout a wide load range. Hence, a slight

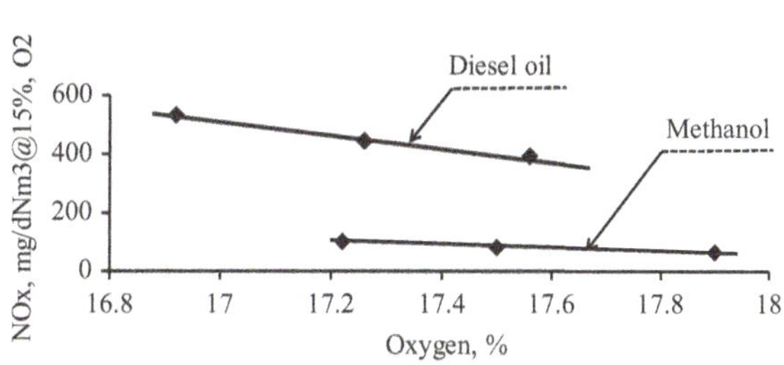

Fig. 27 . NOx emission as a function of oxygen concentration

increase of flame temperature will not adversely impact the lowering of the NOx emission.

4.4 Evaluation of CO and Formaldehyde Reduction in Gas Turbine

In order to estimate formaldehyde and CO formation, we performed simulations and special tests using installed bleed valves for combustion air control. Opening and closing bleed valves allowed us to control air flow through combustors and analyze its influence on gas turbine performance and emissions. For the methanol flame simulation, we have used a methodology that combines experimental work and computational simulations as described in the following references [11, 13, 14, 19]. The emission calculation was based on the comprehensive kinetic mechanism of gas-phase reaction for CO, CH_2O and CH_3OH combustion, as described in [23]. We have used the standard CHEMKIN transport package for the kinetic simulation. The simulation results of gas turbine emission for methanol at different loads are shown in Figs. 28, 29 and 30.

As can be seen from the performed simulation, load reduction leads to flue gas temperature decrease which leads to the growth of CO and formaldehyde formation. The simulation results (Fig. 28) are in good agreement with measured data, especially for CO formation. Formaldehyde measured data is higher than that obtained from the simulation; however, the tendency of flame temperature influence remained the

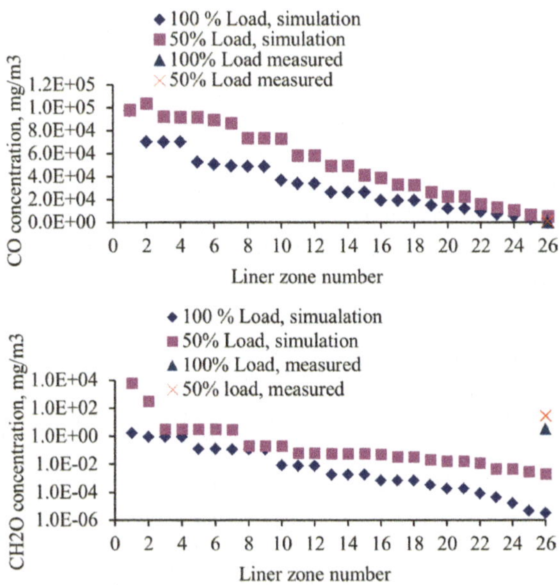

Fig. 28 CO (top) and formaldehyde (bottom) emissions distribution through liner length and its comparison with the measured value

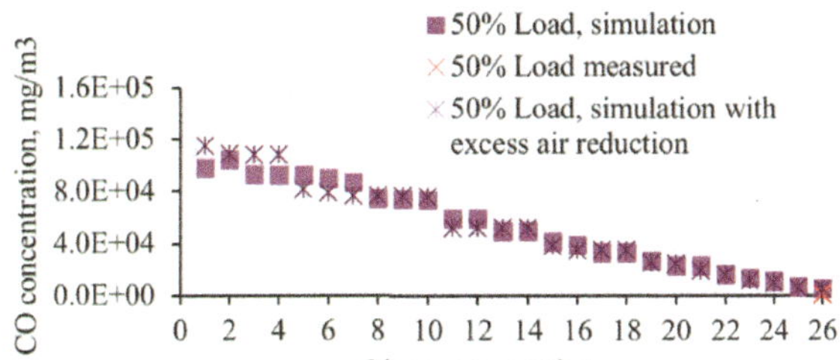

Fig. 29 Comparison of simulation at 50% load of CO emissions distribution through liner length with actual and reduced excess oxygen

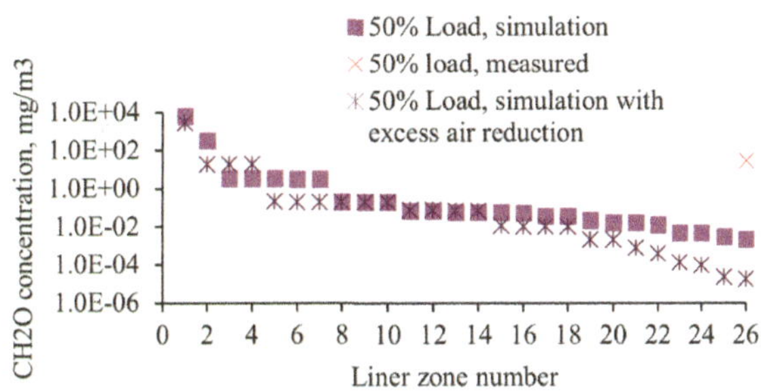

Fig. 30 Comparison of simulation at 50% load of formaldehyde emissions distribution through liner length with actual and reduced excess oxygen

same according to the calculations' results. This is due to the fact that at partial load, the formaldehyde and CO formation significantly increase, probably, because of combustor excess air increase and, as a result, combustion temperature decreases. We have performed a simulation with reduced air flow to the burners. The calculation results for 50% load with different air flows are shown in Figs. 29 and 30. As follows from the simulation's results, reducing the air mass flow to the combustor while keeping the load constant will increase flame temperature and, as a consequence, result in a reduction of CO and formaldehyde. Based on obtained results, we have performed a series of tests with control of air mass flow to the combustor. For air flow control, we have used three bleed valves that enabled us to change air flow to the combustor. The results of performed measurements are shown below in Figs. 31 and 32.

As follows from the obtained results, increasing combustion air leads to CO and formaldehyde rising (Figs. 32, 33) and NOx (Fig. 34) emission reduction.

The main reason for that phenomenon is the reduction of the combustion temperature. CO, formaldehyde, and NOx emission as a function of TT7 (exhaust) temperature is shown in Figs. 32, 33 and 34. As can be seen from the presented data (Figs. 32 and 33), CO and formaldehyde are significantly reduced with temperature increase. At the same time, NOx emission is only slightly increased as a function of the temperature. The reason for the moderate increase in NOx is the low combustion temperature of methanol (Fig. 34). This temperature is lower than the characteristic temperature required for significant NOx formation according to the thermal NOx emission formation mechanism. Therefore, based on simulations and measured results, it may

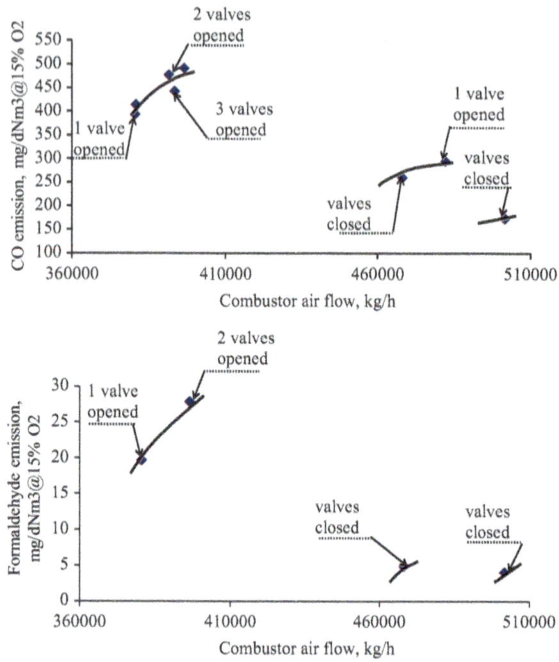

Fig. 31 CO (top) and formaldehyde (bottom) emissions at the exhaust as a function of combustor air flow

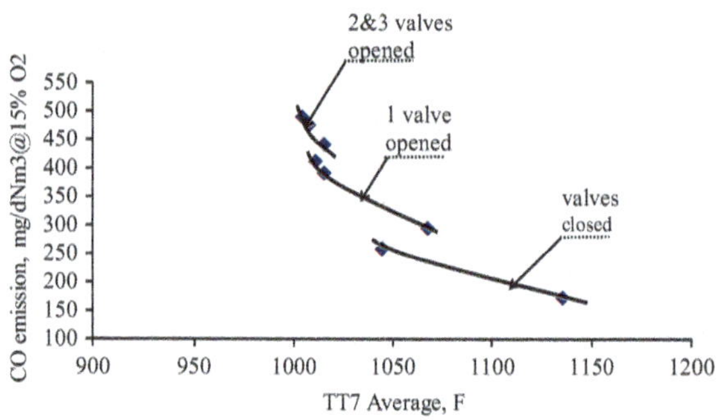

Fig. 32 CO emissions as a function of TT7 exhaust temperature

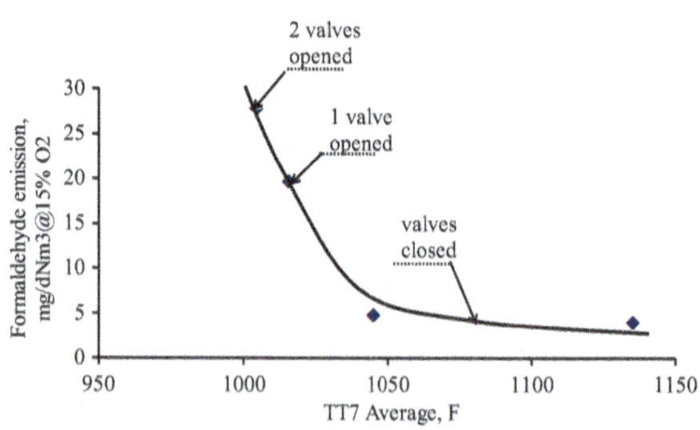

Fig. 33 Formaldehyde emissions as a function of TT7 exhaust temperature

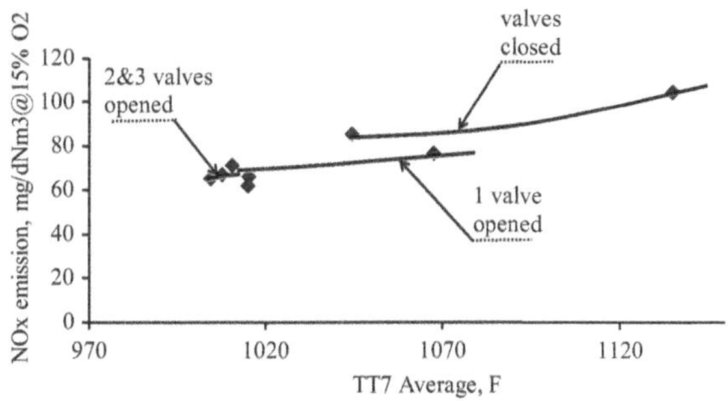

Fig. 34 NOx emissions as a function of TT7 exhaust temperature

be concluded that air flow control to the combustor, especially, to burners is very important for CO and formaldehyde reduction.

5 Experience of Methanol Burning at Utility Boilers

5.1 Large Utility Boilers

The large utility boiler that was converted to a combined heavy fuel oil and methanol burning is a 140 MWe tangential firing unit. The description of this project is presented in work done at the Haifa Power utility in 2012 [18]. The 140 MWe unit is equipped with a boiler designed by (CE) Combustion Engineering (Fig. 35, top). The boiler is equipped with twelve Low NOx burners (Fig. 35 bottom), which are located on three levels and are tangentially fired within the furnace, with control over the distribution of the secondary air along the burner. A fraction of the secondary air is fed through the closed-coupled overfire air ports, installed inside each burner. This configuration enabled us to control the stoichiometric ratio in the combustion zone. The 140 MWe unit boiler was tested for different methanol fractions of the total boiler heat capacity (from 33% to 50% heat) at different boiler loads. Generalization of the obtained results shows that boilers' performance during methanol co-firing is similar to heavy fuel oil firing. The comparison of the boiler's emission with heavy fuel oil firing and methanol co-firing is shown below.

In order to carry out methanol co-firing tests, the following modifications were performed:

- Installation of a new pump suitable for methanol supply of

  4-20 barg and mass flow rate of 28 t/h.

- Installation of a methanol tank.
- A 2.5" return line and control valve for methanol flow control.
- The existing 2" and 1" pipes with valves were replaced by larger diameters: 4" and 1.5", respectively.

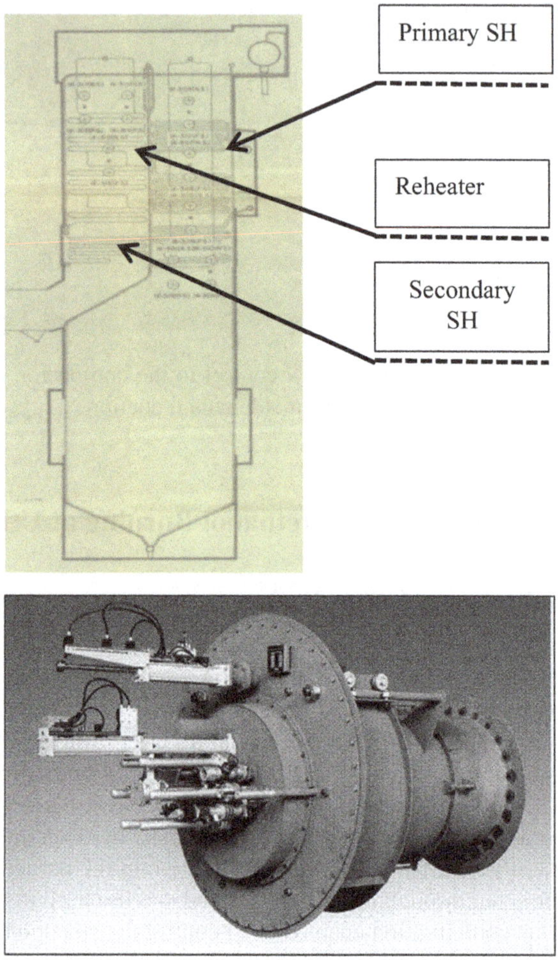

Fig. 35 Schematic of 140 MWe unit tangential firing boiler (top) and burner general view (bottom)

- Installation of two gate valves: one gate valve disconnects the light fuel oil line and the other disconnects the new methanol line.
- Replacement of HFO atomizer with methanol atomizer in burners.
- Replacement of the existing flame detectors with detectors consistent with methanol firing detection.
- Control system software modification (BMS modification).

The methanol tank and pump installation are shown in Fig. 36 (top). The display of methanol supply control is shown in Fig. 36 (bottom). The comparison of boiler emission is shown in Figs. 37 and 38. As can be seen from the obtained data, the tests were performed at a wide load range up to 110 MWe unit load. Further load increasing was limited due to the boiler's insufficient conditions prior to planned maintenance. The different methanol heat fractions and burner arrangements were

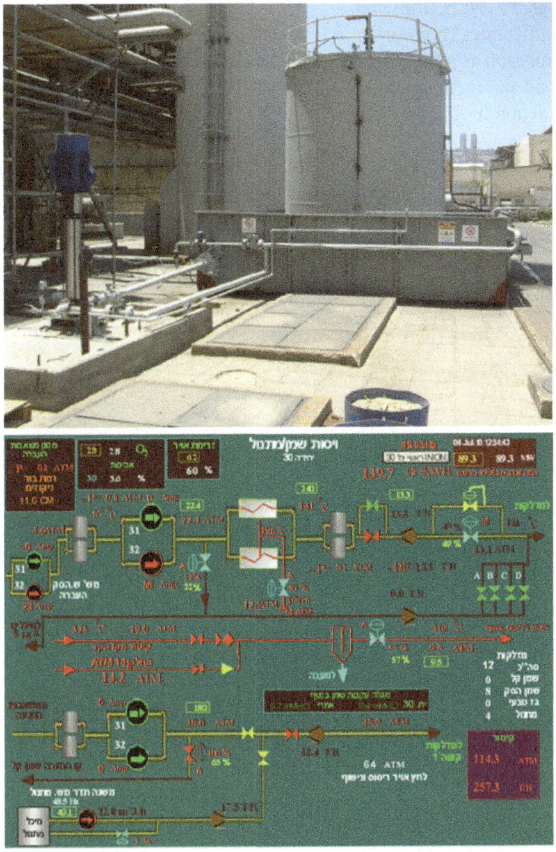

Fig. 36 Methanol pump and tank installation (top) and display of methanol supply control (bottom)

evaluated during the tests. Based on the presented data, it may be concluded that methanol and heavy fuel oil co-firing leads to considerable NOx reduction for all performed tests. Increasing methanol fraction has caused NOx reduction as can be seen in Fig. 37, top. CO emissions during all performed tests were negligible. Particulates emission as a function of boiler's load is shown in Fig. 37, bottom. As follows from the obtained results, methanol and heavy fuel oil co-firing drastically reduce unburned carbon formation during combustion. As was expected, methanol co-firing also leads to SO_2 reduction. Comparison of SO_2 emissions during heavy fuel oil firing and methanol co-firing at 50% heat fraction is shown in Fig. 38.

During the performed study, we have also investigated methanol co-firing impact on furnace heat transfer based on the methodology described above in this chapter. Furnace heat absorption at different boiler loads and methanol heat fractions is shown in Fig. 39

Oil firing operation leads to an increase in furnace heat absorption while methanol co-firing operation causes a decrease in furnace heat absorption. This is due to the different radiant heat transfer properties of the flue gases. Based on the performed

Fig. 37 NOx (top) and particulates (bottom) emission as a function of unit load and methanol heat fraction

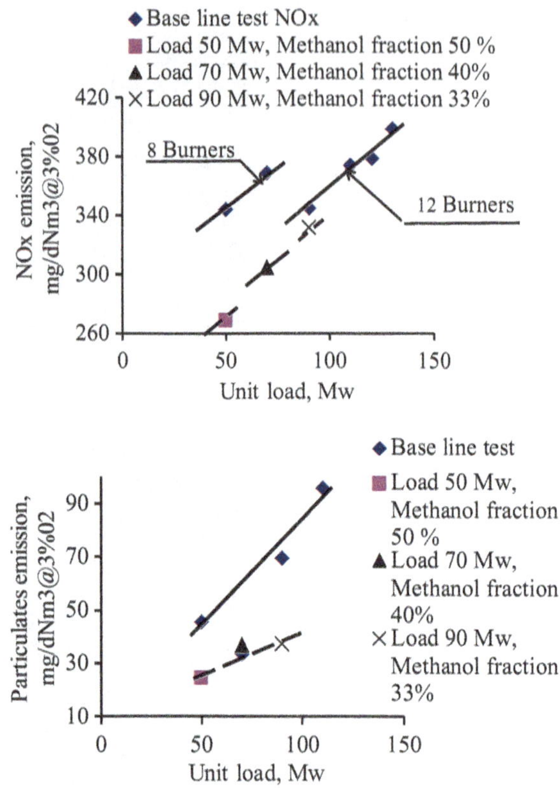

Fig. 38 SO_2 emission as a function of unit load and methanol heat fraction

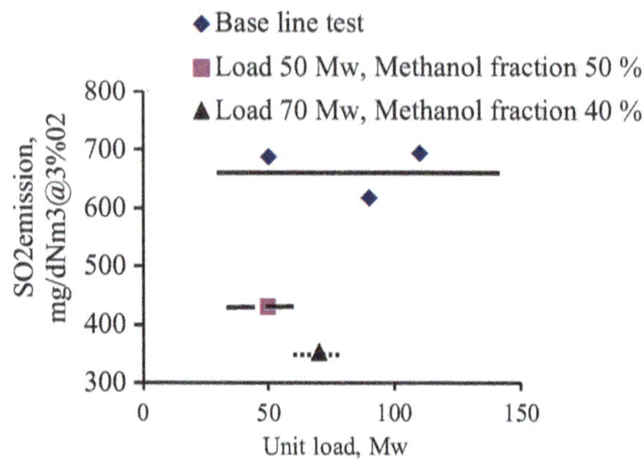

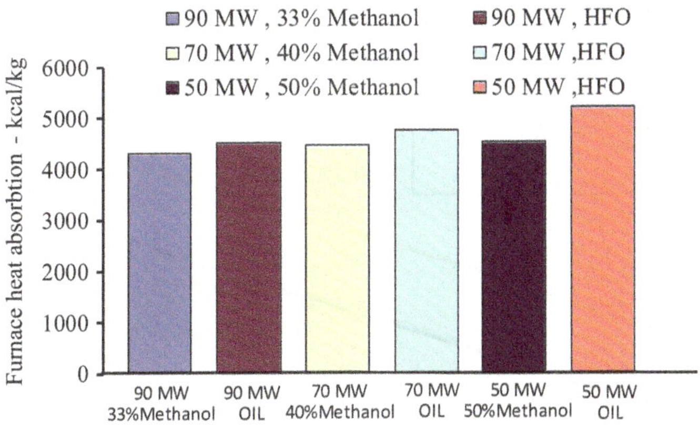

Fig. 39 Furnace heat absorption for different boiler operation modes

simulation, the reason for different radiant heat transfer properties is the lack of char and ash formation during methanol firing and, as result, lower flame emissivity. It was confirmed by visual inspection of the initial flame zone. When methanol is fired, the flame is seen to be almost transparent while the heavy fuel oil flame is very well seen, most probably, because of the char particles that considerably increase the flame emissivity. As a result, during the co-firing mode of operation, furnace heat absorption is slightly decreased and furnace exit flue gas temperature (FEGT) is increased. Based on the measured Secondary Superheater, Primary Superheater, and Reheater steam outlet temperature and using obtained flue gas temperature distribution through pressure parts, midwall tube metal temperature were calculated for heavy fuel oil firing and methanol co-firing with heavy fuel oil.

The results of the calculations for a 90 MWe unit load are presented in Figs. 40, 41 and 42. Comparison of the outlet tube metal temperature with the allowable metal temperature shows that for both modes of operations—heavy fuel oil firing and methanol co-firing—current metal temperatures are lower than the allowable temperatures. Hence, in spite of the increasing FEGT temperature during methanol co-firing, the pressure parts metal temperature remains in the safety temperature zone and the boiler can operate without any restriction. The results presented here clearly show that with minor low-cost fuel system retrofit, methanol co-firing leads to significant NOx, SO_2, and particulates emission reduction while maintaining similar facility's performance and characteristics. It may be concluded that our experience gives useful results for design and/or optimization purposes.

5.2 *Small Industrial Boilers*

In addition to the experience gained in burning methanol in a large utility boiler, DOR Chemicals have acted and developed capabilities in converting an industrial

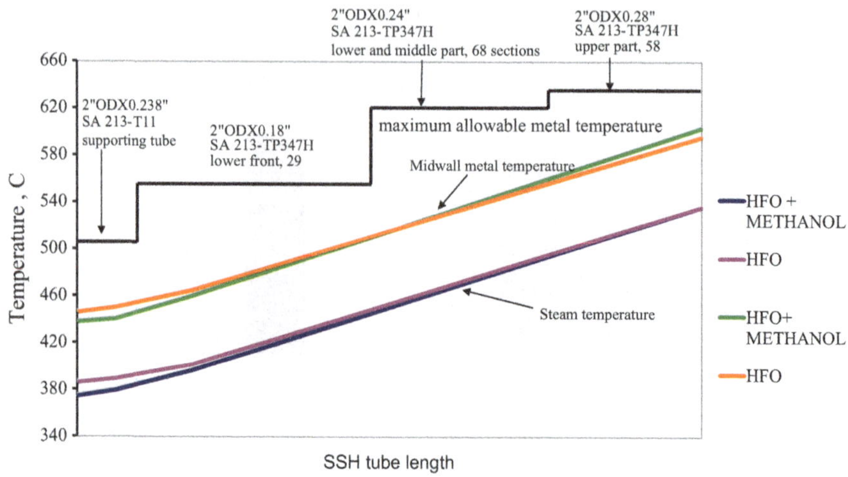

Fig. 40 Secondary superheater tube metal temperature for heavy fuel oil and methanol co-firing 90 MWe load

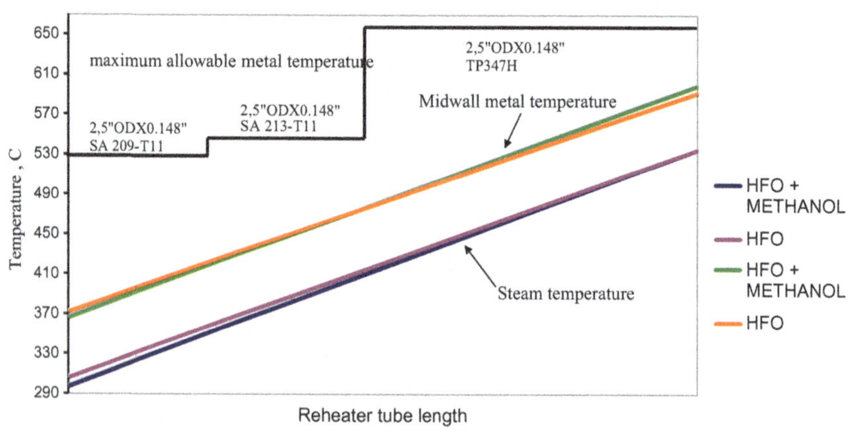

Fig. 41 Reheater tube metal temperature for heavy fuel oil and methanol co-firing 90 MWe load

boiler to methanol firing [24]. A fire tube DESA boiler with 10 t/h steam capacity designed by the German Company 'Standardkessel' (Fig. 43, top) was selected for that. Three types of tests were done: (1) methanol burned separately, (2) methanol burned in co-firing with heavy fuel oil, and (3) methanol burned in co-firing with light fuel oil. It was found that using methanol as an alternative fuel enabled a drastic reduction in NOx, SO_2, and particulates emissions. In order to mix methanol with other liquid fuels, we have used a special patented mixing device ('Fuel Activation Device'—FAD) developed by Turbulent Energy Inc. This device was used to prepare

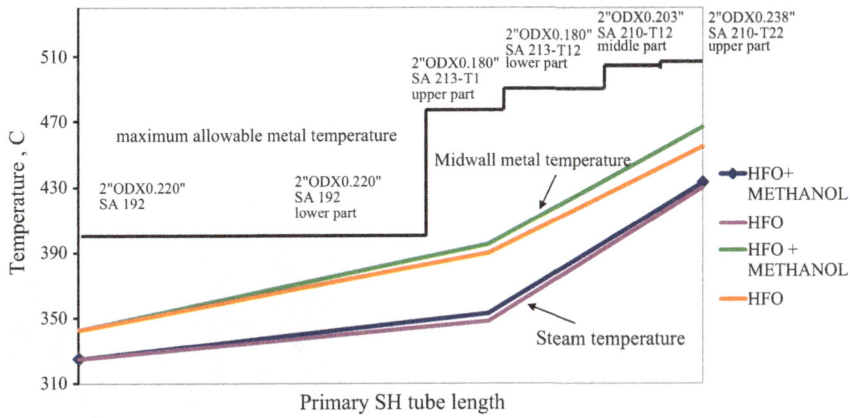

Fig. 42 Primary superheater tube metal temperature for heavy fuel oil and methanol co-firing 90 MWe load

premixed or in-line blends [25, 26]. We have performed optimization of boiler operation conditions for different heat fractions of methanol and fuel oil (light/heavy) co-firing.

The boiler performance parameters are as follows:

- Rated capacity 10 t/h
- Design pressure 15.4 barg
- Live steam temperature Saturated
- Feed water temperature 100 C

Steam temperature, as it is saturated, is uncontrolled. The boiler is equipped with a Saacke burner. The burner is shown in Fig. 43 (bottom). The preparation of mixtures of different fuels ran in the tests were, as mentioned, performed by using the FAD. That made it possible to provide homogenous and stable mixings of two or more liquids. This enabled the improvement of the combustion process and reduction of emissions while optimizing the fuel blend used in accordance with the required allowed emissions at a minimum cost. There were two different stages of the tests. The first stage included the use of pre-prepared mixtures of different fuels in a container by using the FAD. The second stage has used an in-line supply of fuels through the FAD where the fuels were mixed in real time. It should be noted that in both cases of blend supply no special logistics or additional storage or pumps were necessary. Figure 44 shows the on site arrangement required to support either premix (top) or in-line (bottom) fuel flow via the FAD. The boiler tests were performed with the original atomizing system and as a result of increasing the methanol fraction, due to its lower heat value and the limited flow capacity of the nozzles, boiler capacity was reduced. It should be noted that neither the boiler nor the combustion system control was changed prior to these tests.

The first series of tests were performed using fuel from the containers, where blends were prepared in advance by using the FAD. The results presented below

Fig. 43 DESA boiler (top) and Saace burner (bottom) general view

represent HFO and its mix with methanol. It should be noted that during the firing of these blends, an 'Octaro' additive was added to the blend in order to ensure acceptable viscosity of unheated HFO. Figure 45 (top) depicts boiler efficiency versus boiler exit excess air for HFO. As can be seen from the presented data, boiler efficiency is slightly increased for blend burning. The same series of tests were carried out for LFO and its mixtures with methanol using FAD premixed fuel blends fed directly from containers. Test results are shown below. For LFO and its blends, the tendency is slightly changed. Increasing the methanol fraction in the mixture slightly increased boiler exit temperature and as a result, there is decreased boiler efficiency for a maximum and partial load. The obtained results are shown in Fig. 45 (bottom). During the tests, boiler emissions performance was measured for methanol blend with both HFO (Fig. 46) and LFO (Fig. 47). As can be seen from the data, increasing methanol fraction significantly reduces NOx emissions in HFO with a methanol blend. In the

Fig. 44 Off-line fuel blend supplying arrangement using the container (top) and in-line fuel blend supplying arrangement using Turbulent Energy FAD device (bottom)

case of LFO#2 and the methanol blend, the dependence is distinctly according to the increase in the methanol fraction. It should be noted that NOx formation is based on two main mechanisms—thermal NOx formation and fuel-bound nitrogen. Thermal formation strongly depends on the flame temperature, especially at the initial flame zone. As indicated above, increasing the methanol fraction in the blend causes the increasing furnace combustion temperature.

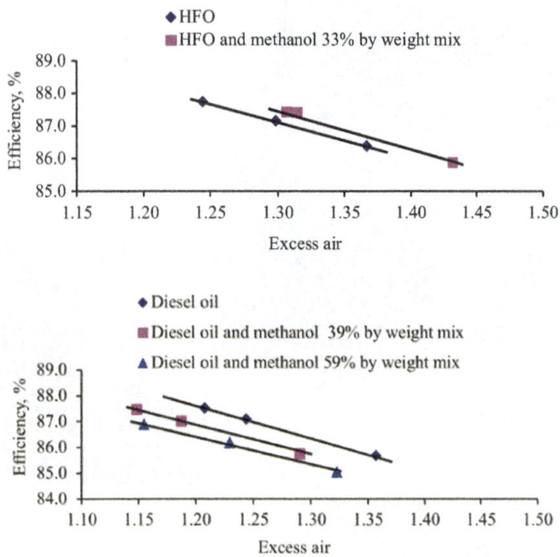

Fig. 45 Boiler efficiency when burning blend of HFO and methanol (top) and LFO#2 and methanol (bottom) as a function of excess air

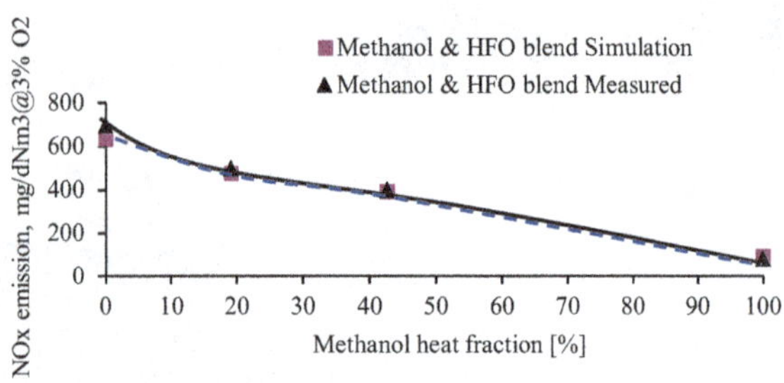

Fig. 46 NOx emission for a blend of HFO and methanol as a function of methanol heat fraction

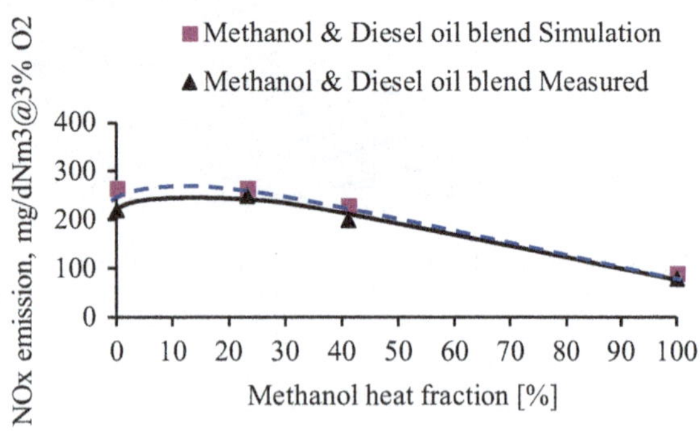

Fig. 47 NOx emission for a blend of LFO and methanol as a function of methanol heat fraction

The main reason for this phenomenon is the different radiant heat transfer properties. Due to a lack of ash and coke in the flue gases mix, the flame emissivity is reduced and as a result, the flame temperature is increased in comparison to unmixed LFO or HFO, especially at low methanol fractions. Higher flame temperatures lead to increased NOx emissions in accordance with the thermal NOx emission formation mechanism. However, the combustion temperature of methanol is lower than that of LFO and HFO and further increasing the methanol fraction does not change the flame temperature significantly, in spite of the lower emissivity in the initial combustion zone and as a result, the thermal NOx formation remains at an almost constant rate. Here, the significant NOx formation mechanism is the second mechanism of NOx formation which is based on fuel-bound nitrogen. Replacing any heat fraction of HFO and LFO with methanol leads to a reduction of fuel-bound nitrogen in the blend. As a result, NOx produced by the second mechanism should be reduced. Therefore, the total NOx formation depends on a combination of these two different mechanisms. It should be noted that fuel-bound nitrogen content is significantly different in LFO as compared to HFO. In LFO, nitrogen content is 10–15 times lower than in HFO. Based on the performed simulation, the contribution of the bound nitrogen content reduction on NOx formation at lower methanol fractions (when methanol and LFO are fired) is lower than the impact of the temperature increase in the initial flame zone on thermal NOx formation. As a result, the total NOx emissions are increased at low methanol fractions. Further increase in the methanol fraction reduces the fuel bound nitrogen and together with the lowering of temperature due to more methanol combustion (and less thermal NOx), NOx emissions are reduced.

Due to the fact that nitrogen content in HFO is significant, even a low methanol fraction leads to a reduction of fuel-bound nitrogen and results in NOx emissions reduction. This is probably the reason that we obtained a linear effect of NOx emissions reduction versus methanol heat fraction for HFO. The obtained test results are compatible with the performed simulation using the methodology described in the works done by [24, 27].

The existing firing system influence other emissions in addition to the influence on NOx as follows: (1) The system provides a complete burn-out of both HFO and LFO and its mix with methanol that is confirmed by CO emissions reduction to zero at all ranges of boiler operation. (2) SO_2 emissions are reduced according to methanol heat fraction in the mix as methanol does not contain any sulphur in it.

In the second series of tests, mixing was performed in line using the FAD. For this purpose, the FAD was installed in the firing system inlet and different fuel pipe lines were connected directly to the FAD. As follows from the performed tests, when LFO and methanol are burned utilizing the Turbulent FAD as in-line mixing, it leads to a more efficient boiler operation.

The obtained results show that burning blends of fuel oil and methanol lead to a significant reduction in emissions in industrial boilers. NOx emissions were reduced by approximately more than 50–80% depending on liquid fuel type in the blend and the fraction of methanol in the industrial boiler. It should also be noted that mixing of liquid fuel with methanol reduces SO_2 emissions in accordance with the methanol heat fraction. Methanol burning leads to more than a 10% reduction in CO_2 emission.

6 Methanol as a Renewable Fuel for Energy Storage

Methanol can be produced from various raw materials in different production processes. One possible process can be by using CO_2 as a raw material. If CO_2 is extracted from the flue gas of a conventional utility and then transferred into a methanol production process, then two combined advantages could be gained: (1) the polluted flue gas can be cleaned and thus release less CO_2 emission to the environment, and (2) clean synthetic and non-harmful methanol can be produced and be stored as a renewable energy source, to be used at times when energy is required while other renewable energies are absent.

6.1 Performance Evaluation

Methanol as a renewable fuel is produced by using renewable electricity, carbon dioxide (extracted from flue gases), and hydrogen (produced by electrolysis process). Methanol, the simplest alcohol, can be used as long-term chemical energy storage. Below we present several evaluations of small co-generation plants that may be used as distributed energy-producing and burning methanol as a renewable fuel. For this purpose, we have run simulations for different configurations of co-generation plants. Two engines were used for the evaluation: Capstone gas turbine and OTTO engine.

For a co-generation plant based on a Capstone gas turbine, the following alternatives were run:

- Full steam cycle—HRSG with a steam turbine.
- HRSG for industrial steam production.
- HWB (Heat recovery hot water boiler) for water supply.
- For a co-generation plant based on an OTTO engine, the followings options were executed:
- HRSG for industrial steam and hot water production.
- HWB (Heat recovery hot water boiler) for water supply.

As GT and engines which were supposed to carry out the performance simulation for a co-generation plant, different sizes of Capstone GT units with an output power ranging from 65 to 1000 kWe and engines ranging from 100 to 250 kWe (converted by DOR chemicals to methanol burning) were selected [28]. To predict the co-generation plant performance, we had run different cycle configurations using commercial Thermoflow Inc. software code. Emission reduction data was estimated by application of the knowledge obtained from the long-term operation of the converted to methanol gas turbine and from literature related to micro turbines emission performance.

Some simulation results for different cycle configurations and engines' loads are presented below (Figs. 48, 49, 50 and 51). As follows from the simulations' results, the unit heat rate is increased with unit capacity rising. Comparison of co-generation plants using waste heat for steam production shows that increasing engine

Fig. 48 CHP based on a Capstone 200 kWe GT with a HWB

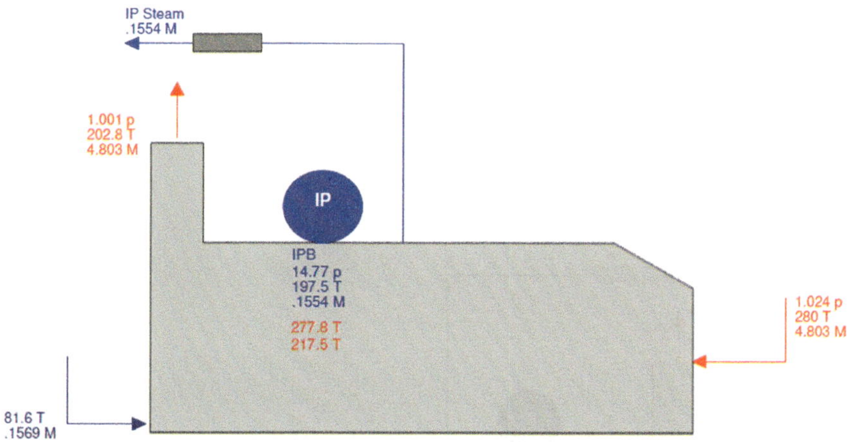

Fig. 49 CHP based on a Capstone 200 kWe GT with HRSG application

capacity leads to a rise in steam production so that higher capacity co-generation plants are suitable for the industry with high steam demand. Co-generation plants for hot water production were found more effective for CHP (based on gas turbines with low exhaust temperature) than for steam generation applications. The application of engines with high exhaust temperature shows

that it may be effective for both applications: CHP for industrial steam generation and hot water production. However, a comparison of HRSG and HWB capacity shows that CHP based on a GT could generate more capacity than that estimated for an OTTO engine. The simulation data are summarized in Table 1 and Figs. 52 and 53. NOx emission performance for a Capstone GT [28] presented in Fig. 54 was estimated. As follows from [29], NOx emission performance for Capstone GT equipped with a diffusion combustor is equal to 95 ppmvd@15%O_2 (195 mg/dNm3@15% O_2) at full load. Load reduction leads to NOx emission decrease. The authors have also

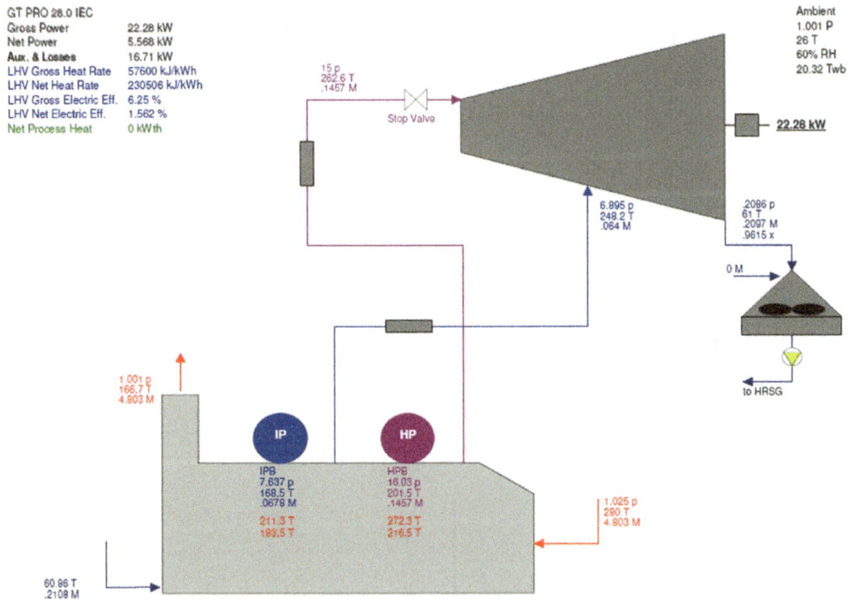

Fig. 50 Combined cycle based on Capstone GT 200 kWe

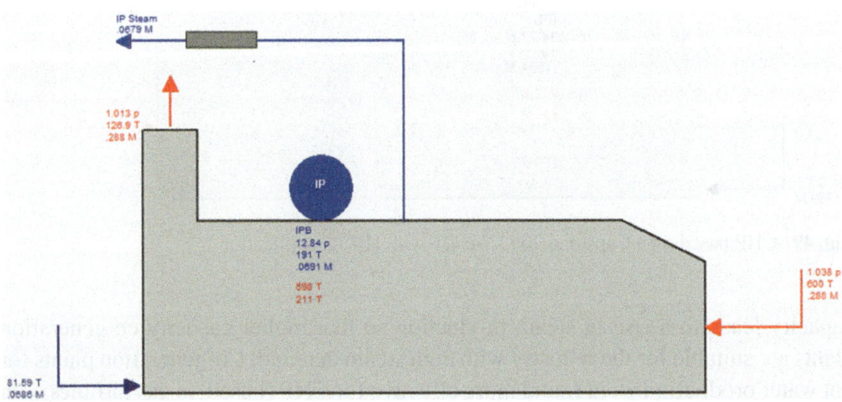

Fig. 51 CHP based on OTTO engine 100 kWe with HRSG application

performed some tests for different fuels burning at the same combustor. As follows from the obtained results, NOx emission significantly depends on flame adiabatic temperature (this fact is also confirmed by other literature data). Knowing methanol's adiabatic flame temperature may allow us to estimate the NOx emissions in this type of GT when burning methanol. Generalization of GT emission performance and using methanol flame adiabatic temperature estimated, NOx emission is approximately (see the dashed line in 6.1.7) 35 ppmvd@15% O_2 (72 mg/dNm3@15% O_2) at full load.

Table 1 CHP performance characteristic

Engine characteristics	GT						OTTO Engine			
	1		2		3		4		5	
Nominal Electricity capacity (kW)	65	65	200	200	1000	1000				
Net Electricity capacity (kW)	61	61	190	190	950	950	100	100	250	250
Methanol Heating value, kcal/kg	4777									
Methanol flow, kg/h	41.7	41.7	115.8	115.8	579	579	30.0	30	75.0	75.0
Type of CHP	HWB	HRSG	HWB	HRSG	HWB	HRSG	HWB	HRSG	HWB	HRSG
Heat output (kW equivalent)	119.2	67.2	264	110.8	1292	553	50.2	45.1	125.7	106
Total CHP efficiency, %	77.8	52.4	72.1	47.5	71.3	47.5	90.2	85.9	90.2	84.4
Water/steam capacity, kg/h	1250	83.1	2780	155.4	1390	776	541	65	1352	153

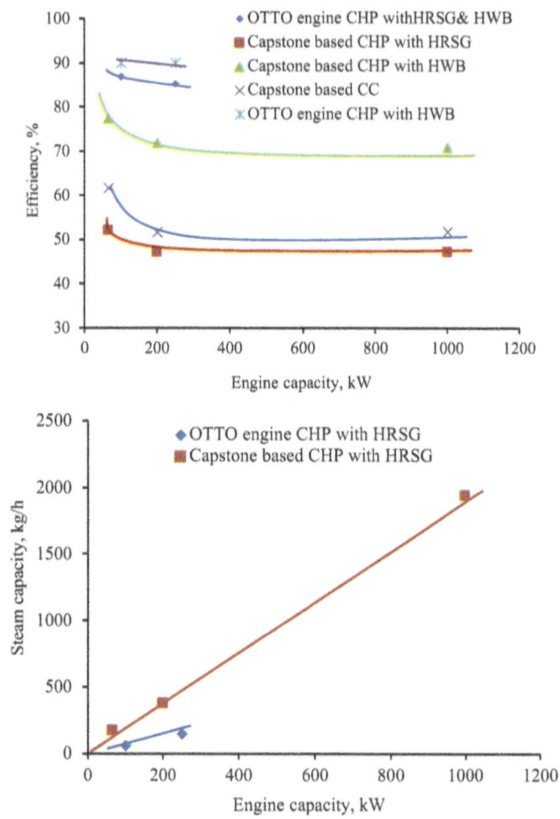

Fig. 52 Comparison of CHP efficiency (top) and HRSG steam capacity (bottom) for different cycle configurations

The estimation of NOx emission based on methanol adiabatic flame temperature and data from [29] is in good agreement with measurements obtained from the FT4C TWIN PAC 50 MWe GT that was converted for the long-term methanol firing. Based on the existing experience of this mentioned GT, the estimated particulates emission should be lower than 5 mg/dNm3@15% O_2. Due to the fact that methanol is synthesized from captured CO_2, carbon dioxide emission should be equal to zero and the lack of sulphur in methanol leads also to zero SO_2 emission.

6.2 *Cost Estimation*

Based on the performed simulations done for different cycle configurations, electricity generation cost can be estimated for different methanol price scenarios. Using an economic feasibility study of methanol production by synthesis from captured CO_2 [7] was established. It was done for three different scenarios of methanol prices depending on electricity generation costs for methanol synthesis. As follows from

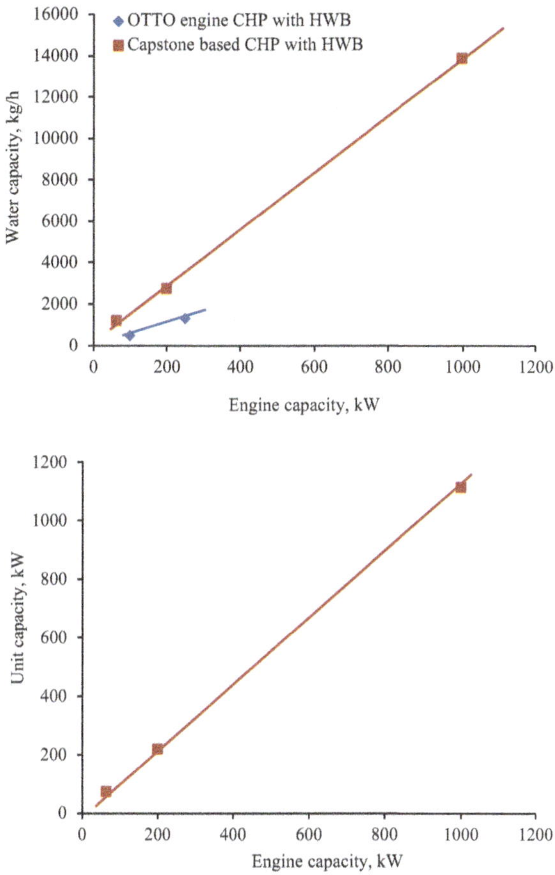

Fig. 53 Comparison of CHP HWB capacity for different cycle configurations (top) and combined cycle capacity based on Capstone GT (bottom)

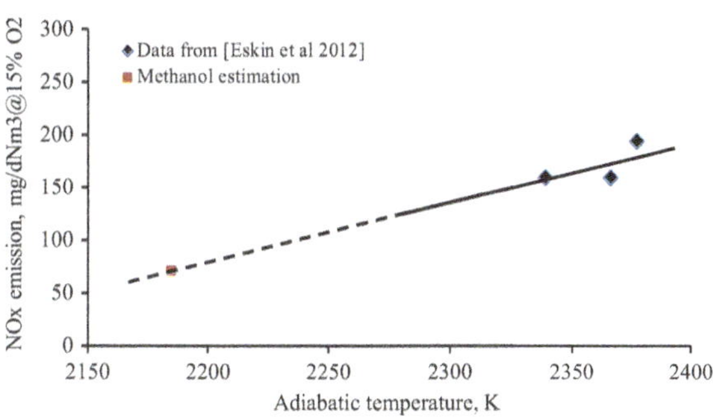

Fig. 54 NOx emission performance

Table 2 for each alternative, the annual benefit obtained from the sale of methanol and oxygen (accomplished with methanol production) was analyzed while the annual costs based on electrical energy purchasing from the grid (at each country's market price) and from the fixed annual cost were taken into account. The methanol costs as a function of electricity prices are shown in Fig. 55. Based on the above-mentioned scenarios of methanol costs and methanol consumptions in the investigated CHP cycles, electricity prices were estimated. Figure 56 is presenting the estimated electricity price for providing hot water from a hot water boiler (HWB) in a combined co-generation plant (CHP) scenario. The estimated installed cost of the evaluated CHP as indicated [28] versus its capacity is illustrated in Fig. 57 (top). A comparison of different energy storage costs is shown in Fig. 57 (bottom). The three different electricity production prices for methanol depend on methanol production cost as shown in Table 2.

As follows from the presented data [30], methanol-based energy storage cost is comparable with other storage types and in some scenarios even lower. Therefore, the power to fuel for methanol synthesis from CO_2 and hydrogen might be a promising technology that should be developed in the near future. The production of methanol allows an alternative option for long-distance energy transportation.

Power-to-liquids offers a possibility for the large-scale production of liquid fuel from H_2O, CO_2, and renewable energy, like wind, solar, geothermal, or hydropower. There are further options for liquid fuel production by direct conversion of sunlight. The solar–thermochemical option is producing syngas, a hydrogen and carbon monoxide mixture, directly from concentrated solar radiation. It is based on two-step redox cycles. Direct syngas production from concentrated solar radiation and subsequent synthesis of Fischer–Tropsch fuels. It has been demonstrated on a laboratory scale only. Based on the current methanol combustion experience, the use of methanol for electricity generation will require very little change in existing equipment and will not create technological or environmental complications. It also will not impact significantly on capital and operation costs. Using methanol for electricity generation would lead to CO_2 emission reduction and assist in meeting the targets of the Kyoto Protocol (United Nations, 1998) and Paris protocol (2015).

7 Conclusion

We have showed that one can successfully predict the performance and pollutant emissions of methanol firing from full-scale utility boilers and gas turbines. The prediction was based on a methodology that combined experimental work with computational simulations. Based on predicted results, we have performed a retrofit of a utility boiler and a gas turbine for methanol firing and then carried out series of tests for methanol firing and co-firing with fuel oil. The performed experiments tested different facilities loads, methanol fractions, and operation modes. The results presented here clearly show that with a minor low-cost fuel system, retrofit methanol

Table 2 Annual revenues, costs, methanol production costs, and pay back periods (PBP) are calculated for Italian, German, and Chinese scenarios [7]

	Italy @ 53.95 €/MWh		Germany @ 33.31€/MWh		China @ 10€/MWh	
	Revenues	Cost	Revenues	Cost	Revenues	Cost
Methanol	21,185,200.00 €		21,185,200.00 €		21,185,200.00 €	
Oxygen	12,410,406.00 €		12,410,406.00 €		12,410,406.00 €	
Electrical Energy		30,158,947.73 €		18,620,844.28 €		5,590,166.40 €
Fixed Annual Cost		4,285,859.25 €		4,285,859.25 €		12,410,406.00 €
Methanol production cost	650 €/tonne		433 €/tonne		186 €/tonne	
PBP@400 €/ton	>20 years		6 years		3 years	

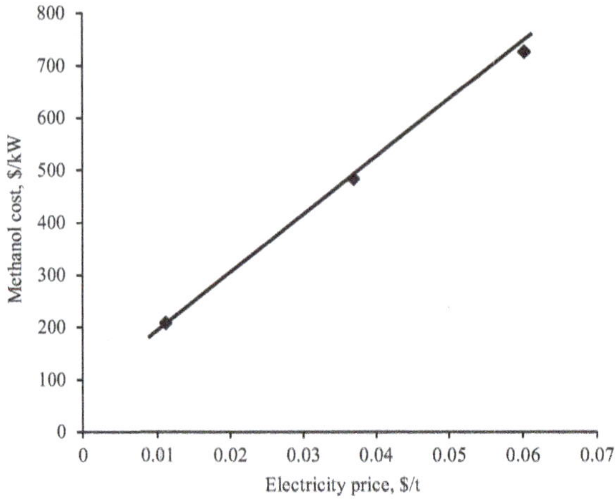

Fig. 55 Methanol price as a function of electricity cost (top, [7]) and electricity price for CHP with HWB scenario (bottom)

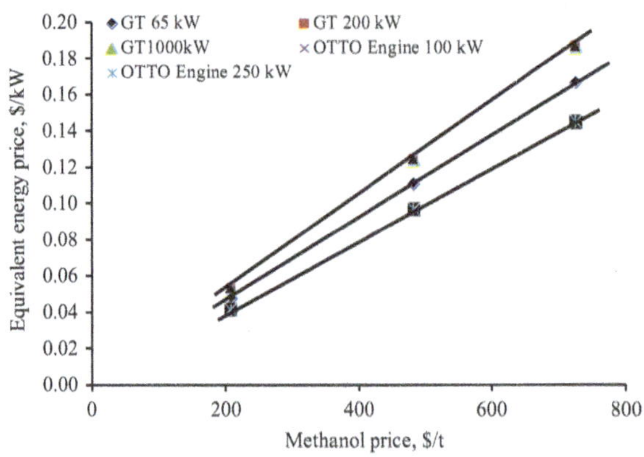

Fig. 56 Methanol price as a function of electricity cost (top, [7]) and electricity price for CHP with HWB scenario (bottom)

firing leads to significant NOx, SO_2, and particulates emission reduction while maintaining similar facilities' performance characteristics. We may conclude that our experience gives useful results both prior to firing and/or for optimization purposes. Renewable methanol may be used as a fuel or chemical feedstock and can serve as a long-term chemical energy storage. We believe that the conclusions of the present work are general and can be applied to other boilers and gas turbines as well.

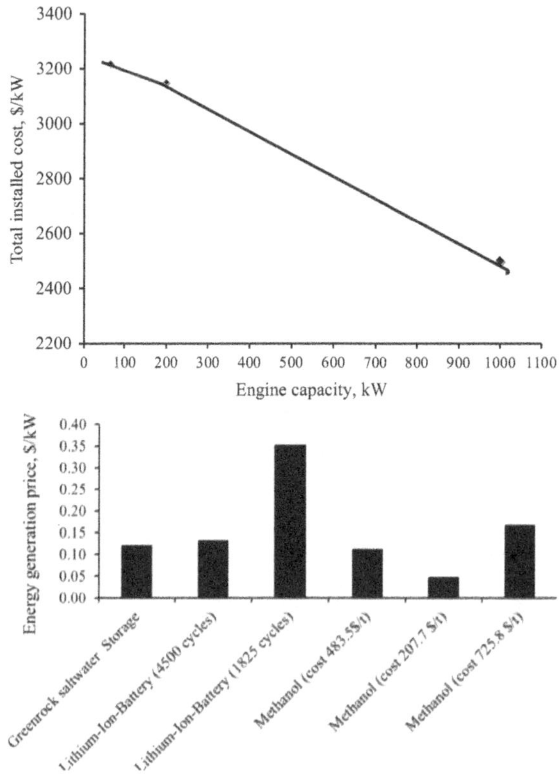

Fig. 57 Total installed cost per engine (top, [28]) and electricity price for different storage types (bottom)

References

1. Test and evaluation of methanol in a gas turbine system. Southern California Edison Company, EPRI Report AP-1712, February 1981
2. A special report: burning tomorrow's fuels. Power, S14-S15, February 1979
3. Methanol. Clean coal stationary engine demonstration project. Executive Summary. California Energy Commission, Report P500-86-004, February 1986
4. Methanol. power generation—Demonstration test starts for a power source at peak demand. Japanese High-Technology Monitor, 5 April 1993
5. Day WH (2016) Methanol fuel in commercial operation on land and sea. Longview Energy Associates, LLS, Gas Turbine Words, November–December 2016
6. Chudnovsky B, Reshef M, Keren M, Baitel S (2015) Successfully implementation of methanol firing at 50 MWe gas turbine for long term operation. In: Proceedings of the 20th International conference on power & energy ASME POWER & ENERGY 2015, June 28–July 2, 2015, San Diego, California, USA
7. Bellotti D, Rivarolo M, Magistri L (2018) Economic feasibility of methanol synthesis as a method for CO_2 reduction and energy storage. In: 10th International conference on applied energy (ICAE 2018), 22–25 August 2018, Hong Kong, China
8. Newsletter New CO_2 Fuels. Issue 1, September 2012. www.newCO$_2$fuels.co.il

9. Buddeberg T, Bergins C, Schmidt S, Fell HJ (2017) Production of fuels from hydropower and carbon di-oxide from organic waste in Norway. In: 4th International engine congress, Baden-Baden
10. Chudnovsky B, Talanker A (1996) Preliminary gas turbine combustor thermodynamic calculation for converting to dry low NOx (DLN) operation and system. Test Report, April 1996
11. Sudarev A, Antonovsky V (1985) Gas turbine combustion chamber. Heat Transfer, Mashinostroenie, Leningrad, 272 pp
12. Levy Y, Gany A, Goldman Y, Erenburg V, Sherbaum V, Ovcharenko V, Rosentsvit L, Chudnovsky B, Herszage A, Talanker A (2010) Increasing operational stability in low NOx GT combustor by a pilot flame. GT2010-22785. In: Proceeding of ASME Turbo Expo 2010, Glasgow, UK
13. Miller J, Bowman G (1989) Mechanism and modeling of nitrogen chemistry in combustion. In: Progress in energy combustion, science 1989, vol 15, pp 287–338
14. Chudnovsky B et al (2006) IEC experience in NOx emission control. In: 22nd Annual symposium of the Israeli Section of the Combustion Institute
15. Borodyanski G, Chudnovsky B, Nekhamkin Y et al (1997) NOx and CO_2 emission reduction when reformed fuel burning. In: Proceeding of the third conference on combustion and emission control, Bath, UK, 11–12 July 1997
16. Chudnovsky B, Talanker A (2004) Effect of bituminous coal properties on heat transfer characteristic in the boiler furnaces. In: Proceedings of the ASME international mechanical engineering conference, Anaheim, California, USA, November (CDROM)
17. Blokh A (1984) Heat transfer in steam boilers. Originally published by Energoatomizadat, Leningrad, USSR, as Teploobmen v Topkakh Parovykh Kotlov with revision for the English edition
18. Hain Y, Chudnovsky B et al (2012) Methanol fuel as low cost alternative for emission reduction in gas turbines and utility boilers. In: Proceedings of the 2012 20th International conference on nuclear engineering collocated with the ASME 2012 power conference ICONE20-POWER 2012 July 30–August 3, 2012, Anaheim, California, USA
19. Chudnovsky B, Reshef M et al (2015) Successful implementation of methanol firing at 50 MWe Gas Turbine for Long Term Operation. In: Proceedings of ASME POWER 2015, June 28 July 2, San Diego, USA
20. Chudnovsky B, Reshef M, Talanker A (2017) Evaluation of methanol and light fuel oil blends firing at a 50 MWe gas turbine. In: Proceedings of ASME POWER 2017, 28.06-30.06.17, Charlotte, NC, USA
21. Levy Y, Erenburg V, Roizman A, Sherbaum V, Ovcharenko V (2016) Experimental comparison of methanol and kerosene combustion. In: Proceedings of 29th annual symposium of Israeli Section of the Combustion Institute, Tel Aviv, Israel
22. Deuker E, Koenig M, Moiller M, Slad J, Streb H (2016) SGT5- 4000F gas turbine and combined cycle power plant evaluation reflecting the changing market requirements. Siemens AG, Energy Sector Germany and Siemens Energy, Inc, USA, Message
23. Li J, Zhao Z, Kazakov A, Chaos M, Dryer FL, Scire JL (2006) A comprehensive kinetic mechanism for CO, CH_2O, and CH_3OH combustion". Published on Line in Wiley InterScience, October 2006
24. Chudnovsky B, Livshits D, Baitel S (2016) Burning methanol and its blends attractive alternative for emission reduction. In: Proceedings of the 21st ASME International conference on power & energy, Charlotte, NC, USA, June 2016
25. Livshits D et al (2014) Fuel composite, devicefor producing thereof and system of use. United States Patent, 8,715,378, May 6, 2014
26. Livshits D et al (2014) Method of dynamic mixing of fluids. United States Patent, 8,746,965, 10 June 2014
27. Pavlov V, Shteiner I (1984) Optimization of combustion of liquid and gaseous fuels in industrial facilities, published by Energoatomizadat. USSR, Leningrad, p 120

28. Catalog of CHP Technologies. Section 5. Technology characterization- Microturbines. US EPA, Combined Heat and Power Partnership, March 2015
29. Eskin L, Holton M, Turner B, Jiklik R, Klassen M, Roby R (2012) Long term demonstration of a Lean, Premixed, Prevaporized system for gas turbines. In: Proceedings of the 2012 20th International conference on nuclear engineering collocated with the ASME 2012 Power Conference ICONE20-POWER2012 July 30–August 3, 2012, Anaheim, California, USA
30. Cost Comparison of Battery Storage - BlueSky Energy I Stromspeicher Batterien.html

Energy Analysis of Methanol Synthesis via Reverse Water-Gas Shift Reactor

Asad A. Zaidi, Muhammad Nihal Naseer, and Tahir Abdul Hussain Ratlamwala

Nomenclature

m	Mass flow rate
P	Pressure
T	Temperature
h	Specific Enthalpy
s	Specific Entropy
v	Specific Volume
ΔH	Enthalpy of reaction

1 Introduction

In this era of industrialization, global energy demand is increasing day by day owing to economic and population growth [1–4]. The stats show that in 2020 population growth rate is 1.05% (lower than 1.08, 1.10, and 1.12 in 2019, 2018, and 2017, respectively) [5]. More than 81 million people are becoming part of this world every year [5]. This growing population is the main contributor to the increment in energy demands [6–9]. This energy demand is expected to increase by 21% over the span of the next two decades and hit the margin of about 17,000 Mtoe/year by 2040 [10]. In order to meet this demand, the world is majorly relying on fossil fuel incineration as depicted in Fig. 1. It is eminent that in 2019 fossil fuels dominated the energy market by contributing 88.58% [11].

Fossil fuel burning overlaps with excessive release of greenhouse gases, majorly CO_2 emission. This CO_2 accumulates in the atmosphere and results in global

A. A. Zaidi (✉) · M. N. Naseer · T. A. H. Ratlamwala
National University of Sciences & Technology (NUST), Sector H-12, 44000 Islamabad, Pakistan
e-mail: asadali@pnec.nust.edu.pk

A. K. Agarwal et al. (eds.), *Methanol*, Energy, Environment, and Sustainability,
https://doi.org/10.1007/978-981-16-1224-4_4

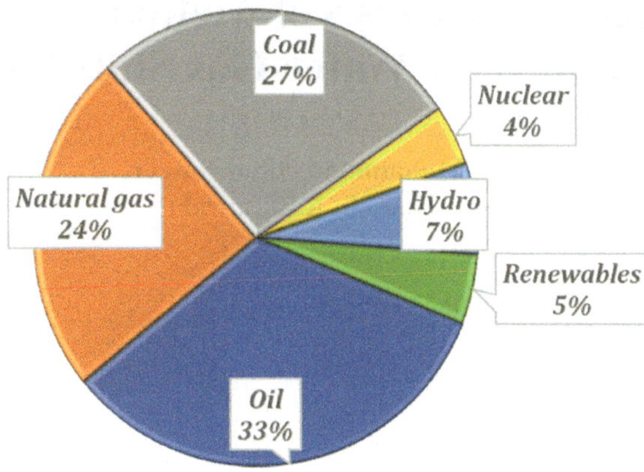

Fig. 1 Contribution of various sources in energy generation for the globe in 2019

warming. No doubt, there are other greenhouse gases (GHGs) also in the atmosphere, such as NO_x and CH_4, that are causing global warming but the presence of CO_2 is more vulnerable [12]. The vulnerability of various GHGs can be accessed from their radioactive forcing (RF) factor. A high value of RF represents a higher capability to absorb solar radiations and cause more global warming. From Fig. 2 [13, 14], it is inferred that CO_2 has the highest RF value that makes it devastating for environmental sustainability [14]. In addition, a comparatively long-life span of CO_2 in the atmosphere is also exacerbating. 40% of the total quantity of emitted CO_2 resides in the atmosphere for 100 years, 20% after 1000 years, and 10% after 10,000 years [15].

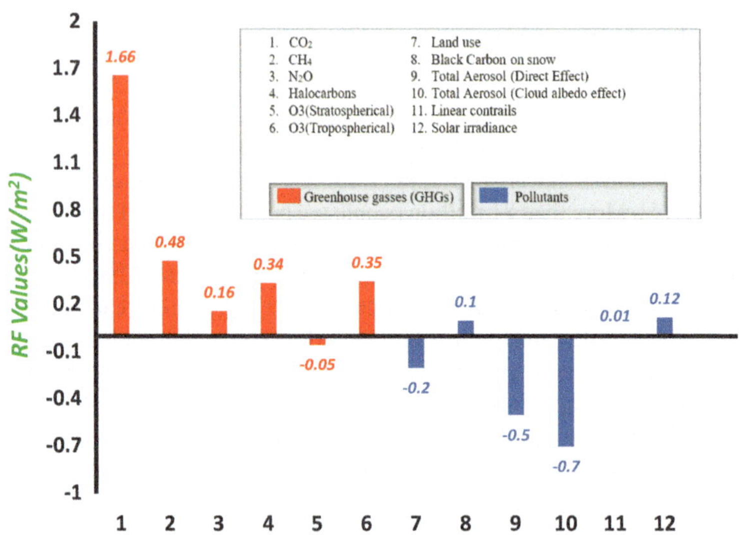

Fig. 2 Radioactive forcing (RF) value of various greenhouse gases and pollutants

In spite of being detrimental, CO_2 is still released at an enormously high rate. The energy sector is one of the major contributors to CO_2 emission. In 2018, CO_2 from the energy sector made a new record of historically highest emission, i.e. increase by 1.7%. This created unexpected weather conditions in various regions and increased energy demand for heating and cooling by 2310.83 million tonnes of oil equivalents [16]. In 2019, the overall concentration of CO_2 crossed the margin of 400 ppm, this concentration is the highest concentration over more than 800,000 years. The per emission is recorded 36 billion tonne and on per capita base, concentration is increased by more than 100 folds [17, 18]. Out of this total released carbon dioxide, the energy production sector is responsible for 4000 Mtonne of CO_2 in 2019 [17, 18]. Hence, it can be concluded that in order to survive on the earth, achieve a sustainable environment by reducing global warming, and to diminish unexpected energy rises due to global warming, there is a dying need for clean and green fuel [19–21].

Methanol being a clean fuel promises a sustainable environment in the future [1, 20, 21]. Methanol is currently used in a number of applications, Fig. 3, with an annual production of more than 90 million metric tonne [22]. Methanol can be dated back to the period of Adam and Eve because methanol synthesis is as old as fire. It has been termed as wood alcohol as Egyptians obtained it by the pyrolysis of wood. But it was first isolated in 1661 by Robert Boyle who termed it as 'spirit box'. Another major milestone in the methanol research was covered in 1812 when methyl alcohols and ethyl alcohols were differentiated. Wood alcohol has been extensively used in World War I but the first thermochemical methanol production process was reported in 1923. The very first commercial synthetic methanol was patented in 1961 by ICI. These days methanol can be produced of 97% better quality than in 1961. It could be inferred from this fact that since the day of identification, the production and utilization of methanol have always been a pivot to researchers. Moreover, the word

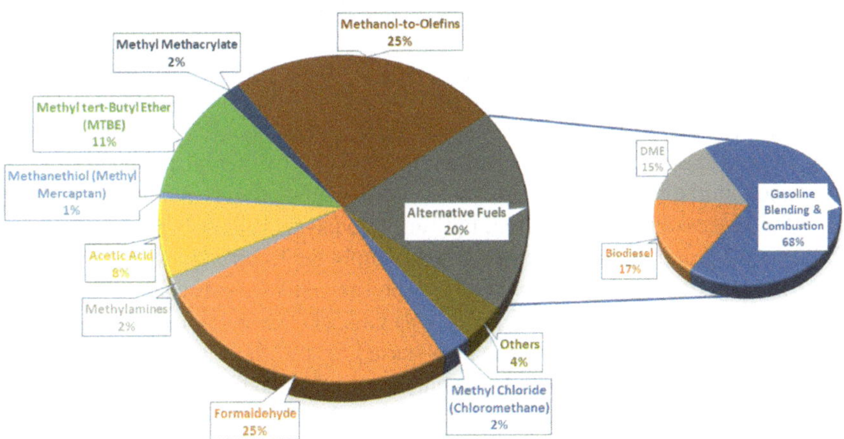

Fig. 3 Methanol Demand of globe in 2019 [25]

'Methanol' finds its root in Greek, derived from two words '*methy*' (wine) and '*hŷlē*' (wood) which literally means wood wine. [22–24].

A lot of attention has been given to the synthesis of methanol. Methanol is assumed to be one of the most cost-effective and viable options for the fulfilment of energy requirements in the coming future [26, 27]. As faced in every process, the fundamental problem is the development of optimized and controllable process design and the ways to minimize the cost that incurs in setting the synthesis plant up and running it effectively. The development of a process model is vital for identifying the most efficient process for methanol synthesis. Various mathematical models have been developed by different scientists and can be found easily by a minimalistic effort. Some of them are discussed as follows:

Experimental campaigns have been run to analyze the developed processes and to determine the model parameters for mathematical modelling of the system. Studies have been conducted to observe the reaction kinetics of the methanol synthesis process. This is usually done by running the experiment over and over again by changing the conditions in which reaction has been performed, feedstock to the reactor and the catalyst used. There are two model variations found in the literature based on the feedstock:

- One based on the synthesis using CO.
- Other using CO_2 as the feedstock.

First ever commercial plant for methanol synthesis went operational in 1923 in Germany. Operating temperature and pressures were in the range of 320–450 °C and 250–350 bar, correspondingly. A chromate-based catalyst was used in this high-pressure synthesis process. Later in the 1960s, when it was made possible due to the development of technology to produce Sulphur free synthesis gas, copper-based catalysts were started to be used. By using copper-based catalysts, the operating pressure was lowered considerably to 60–80 bar with the corresponding temperatures being 250–280 °C.

Methanol synthesis with a high-pressure process was studied with catalyst ZnO/Cr_2O_3 and its reaction kinetics were observed. The equation proposed by Natta [28] is the very first example in this domain. Hydrogenation of CO alone is observed by Natta and based on the conducted study, values of different parameters were determined. Initial assumptions were that only CO could produce methanol by going through a process of hydrogenation, but further analysis revealed that involvement of CO_2 in the reaction kinetics could not be simply neglected. The very first model for the kinetics of synthesis of methanol was presented by Leonov et al. [29] in 1973. The catalyst that was used in the study was $Cu/ZnO/Al_2O_3$. After this study had been published, various other researchers started to shift their interests towards the modelling of synthesis of methanol, and numerous other studies have been published since then. It was noticed in another study conducted by Villa et al. [30] that water-gas shift reactor must be considered in the synthesis process for methanol.

Kinetic modelling had been done for the synthesis process by considering hydrogenation of both CO and CO_2, performed by Graaf [31–34]. Water-gas shift reactor was also taken into account in these studies. In addition to these, values of the model parameters were recalculated, and the mass transport limitations were modelled as well. Equilibrium constant equations were identified by the authors by assuming ideal gas behaviours. In order to make up for the non-ideal behaviour, the Soave–Redlich–Kwong equation [35] of state was utilized.

Skrzypek et al. [36] employed the equations developed by Graaf et al. (1986) for analyzing the effect of changing the values of initial feed stock, temperature, and pressure on the concentrations of the components at the equilibrium state. Considerable influence of the above-mentioned parameters was observed on the concentrations of the components. It was observed that increasing the temperature had a negative effect on the reaction and actually slows down the production of methanol. Whereas the effect that was observed was opposite to this when pressure was increased, i.e. reaction moved in the forward direction.

Skrzypek et al. [37] also developed models for mass transportation limitations and documented them. Initially, these models were dependent on the dusty gas diffusion model which is quite a complex model to work on. Later in 1991 [38], kinetic equations for the low-pressure synthesis of methanol were developed which were of the type of Langmuir–Hinshelwood kinetic equation. These equations were developed by using the catalyst CuO (60% wt)–ZnO (30%)–Al_2O_3 (7.5%) with polished surfaces. It was found in this research that methanol production was favoured when CO_2 is used as a carbon source instead of CO. A couple of experiments were run and it was found that the methanol was produced without any difficulty when feed contained H_2 and CO_2.

Methanol, as low-cost clean fuel [22, 39], has received much attention from the research community and policymakers during the last few years [26, 40]. Various existing technologies such as phase fluidized bed converters, membrane reactors, adiabatic reactors, and isothermal reactors have been employed for the synthesis of methanol [41]. As waste to energy conversion has also been pivot to energy researchers [42–44], methanol production is also blended with waste conversion [45–47]. Methanol production from biogas has depicted substantial significance in this matter. Hernandez [48] reported biogas reformer for methanol production. The system was modelled by thermodynamic modelling (mass and energy balance); chemical and phase equilibrium. The author achieved the production cost of $1.75/gal. Santos conducted a comparative study to evaluate methanol production potential from various biomass sources such as palm oil, oil effluent, landfill, and corn cobs. The optimization studies revealed the fact that palm oil has the highest capability of methanol production while landfill lowest [49]. This study also contributes to methanol production from biogas by providing thermodynamic analysis of reverse water-gas shift reaction (rWGSR).

In this study, reverse water-gas shift reactor is selected for the conversion of CO and steam to H_2 and CO_2. There are many methods prescribed in the literature for this purpose but rWGSR is the most efficient and feasible among all. rWGSR was introduced in 1914 in the quest of hydrogen production by Bosch and Wild [50] since then rWGSR is considered as a vital process for methanol synthesis [51]. The comparative analysis of other CO_2 to liquid fuel conversion technologies reveals that rWGSR dominates the methodologies with 4.19 STF (sun to fuel) efficiency. Others including direct photosynthesis, self-contained algae, and regulated biomass have a STF efficiency of 0.70, 0.75, and 0.96. Furthermore, the anticipated STF efficiency of rWGSR, photosynthesis, self-contained algae, and regulated biomass is estimated to be 7.92, 5.46, 2.39, and 3.10, respectively [52]. In addition to STF efficiency, the technical feasibility of the rWGSR method is also an incentive for its selection in the present study [53].

2 System Description

In this section, all the individual components of the system are going to be explained. There are seven main components of the system, as depicted in Fig. 4, each one of them has its own purpose and operates at a different set of conditions.

2.1 CH_4–CO_2 Separator

Biogas is one the most widely used biofuel as a fuel for transportation, cooking, and other purposes that involve combustion of a fuel. It is mainly composed of methane (50–75%) and carbon dioxide (25–50%). CH_4–CO_2 separator is the very important component where the whole process begins. It separates the carbon dioxide from

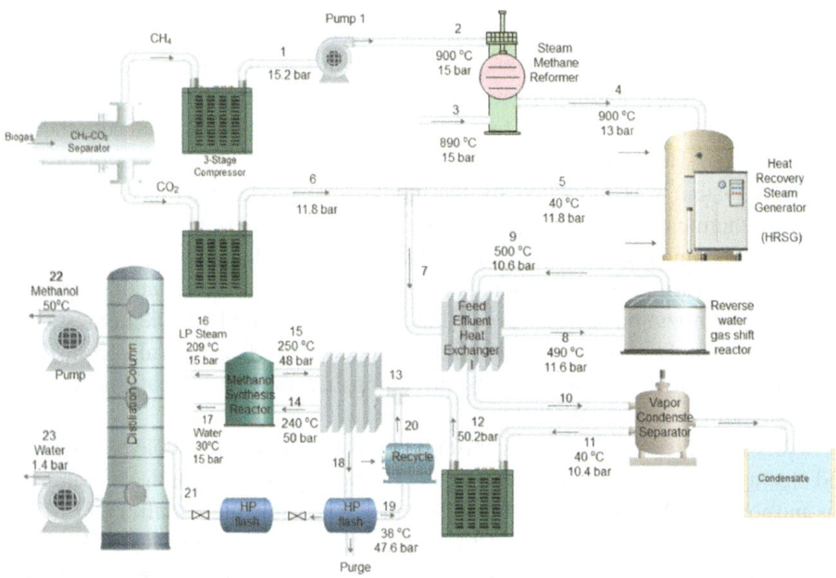

Fig. 4 Methanol production system [2]

methane and passes the methane on to steam-methane reformer. The technology that is used for the separation of methane from carbon dioxide is known as Pressure Swing Adsorption (PSA). In it, carbon dioxide is made to adsorb on a catalyst such as a zeolite or activated carbon, and the purified methane gas is extracted from the separator. Usually, there are several adsorption vessels with adsorption materials installed in them. Biogas is made to enter the vessel from the bottom end and as the gas ascends, the impurities get adsorbed on the catalyst surface. Before the catalyst surface is saturated, the adsorption phase in the first vessel is terminated and the gas mixture is transported to the other vessel to continue the adsorption process. Methane gas of almost 99.4% vol can be recovered from the gas consisting of methane and carbon dioxide in the amount of 60% vol and 40% vol, respectively.

2.2 *Steam-Methane Reformer*

The steam-methane reformer is used to convert methane and steam into a mixture of carbon dioxide and hydrogen gas. It is a catalytic process in which methane is first converted into carbon monoxide by reacting with steam endothermically. This process takes place in the presence of a nickel-based catalyst at temperatures of about 730–845 °C. In the next step, CO is further hydrolyzed to produce carbon dioxide in a shift converter in the presence of iron oxide. Methane and steam are fed into the reformer at a common reformer pressure of 15 bar whereas the temperatures of both the streams were 900 and 890 °C, respectively. Stepwise chemical equations are as follows:

$$CH_4 + H_2O \leftrightarrow CO + 3H_2 \Delta H^{\circ}_{298K} = 206.1 kJ/mol$$
$$CO + H_2O \leftrightarrow CO_2 + H_2 - \Delta H^{\circ}_{298K} = -41.15 kJ/mol$$

The equation for overall reaction can be written as

$$CH_4 + 2H_2O \leftrightarrow CO_2 + 4H_2 \Delta H^{\circ}_{298K} = 165.0 kJ/mol$$

2.3 *Heat Recovery Steam Generator (HRSG)*

Waste heat from different streams of gases can be extracted for further utilization using heat recovery steam generators. HRSGs are extensively used in power generation systems to take the waste heat out of the flowing streams of flue gases. In the system that is being studied, HRSG is used primarily for two motives: one is to cool the syngas down and secondly to generate low-pressure and high-pressure steam to be used in other parts of the system. Various types of HRSGs are available depending on the design of the system but the basic components that are present in each one of them are the same. Several tube bundles from which water flows are suspended in the path of the flue gases. As the two fluids flow, heat gets transferred from the one at a higher temperature to the colder one. The temperature and pressure of the syngas get dropped from 900 to 40 °C and 15 bar to 11.8 bar, respectively.

2.4 *Reverse Water-Gas Shift Reactor (r-WGSR)*

Water-gas shift reaction is a process of converting carbon monoxide and water into carbon dioxide and hydrogen gas. When this reaction is proceeded in the reverse direction, carbon monoxide can be formed from CO_2 and H_2. The input stream to the reactor consists of a mixture of CO_2 and H_2 (syngas) and CO_2 from the separator. In order to preheat the reactant stream, a feed effluent heat exchanger is used in which heat is exchanged between the hot products coming out of the reactor and the cold gases going in. A vapour condensate separator is used to separate out any condensate that may be generated while the gases are being cooled from 500 to 40 °C.

2.5 *Methanol Synthesis Reactor*

It is a shell and tube type reactor in which the gases coming out of the rWGSR are passed through the tubes whereas water that is to be boiled is passed through the shell side. In addition to the effluent of gases from rWGSR, some amount of unreacted gases is also fed into the reactor from the recycling system. Since the generation of methanol is an exothermic reaction, the heat generated by the reaction is taken away by the boiling process of the water. Steam generated in the reactor is at a temperature and pressure of 209 °C and 15 bar. Reactant gases are heated up to the required reactor temperature by passing it through a feed effluent heat exchanger where it exchanges heat with the product coming out of the reactor. In the direct CO_2 hydrogenation process, inflows of the reactors are the syngas that is generated in the HRSG and CO_2 that comes directly from the CH_4–CO_2 separator.

2.6 *High-Pressure (HP) Flash*

In order to have the products in the liquid phase, effluents coming out of the methanol reactor are cooled down to condense the methanol and water by passing it through a feed effluent heat exchanger. Once the outflow stream is liquified, it is made to flow through a flash system at a pressure of around 47.6 bars. The purpose of flashing the stream is to extract out any unreacted gases present in the system. These unreacted gases are fed to the reactor along with the fresh charge coming from the rWGSR.

2.7 Low-Pressure (LP) Flash

Recycled stream from the HP flash is then fed to a low-pressure flash in order to remove any remaining impurities or unreacted gases. After passing the charge from low-pressure flash at 2 bar, it is sent to a distillation tower operating at 1 bar. The condensing unit operates at 50 °C and is capable of handling both liquid and vapour phases. The liquid stream consists of two constituents: liquid distillate with 0.1 mol% water and bottoms with 0.01 mol% methanol. Vapour streams that are purged out of the condenser and flash systems could be utilized to fulfil any heating requirement.

The final product requires purification prior to its utilization. This crud product theoretically doesn't contain even a minute concentration of CO but has some traces of water and a negligibly small of carbon dioxide. In order to purify this crud product, a single-column distillation unit is used. It is also considered as a good point of reverse water-gas shift reactor that distillation process is relatively simpler as compared to other processes where a series of components such as reboilers and condensers are required [54, 55].

3 Thermodynamic Analysis

In order to discuss the sustainability and feasibility of the system, the thermodynamic analysis provides deep insight into this matter. Thermodynamic analysis of a system includes a comprehensive study of mass balance, energy balance, and exergy balance. This study focuses on mass balance and energy balance. Energy and mass share the same characteristic that they can neither be created nor be destroyed. Energy has a surplus quality of conversion from one form to another. These characteristics of mass and energy are referred to as mass and energy balance, as given in Eqs. 1 and 2 (for a steady-state system), respectively.

$$\sum m_{in} = \sum m_{out}, \tag{1}$$

$$\sum E_{in} = \sum E_{out}. \tag{2}$$

For the purpose of simplification, the system is divided into four components named as

1. Methane reformer
2. Reverse water-gas shift reactor (rWGSR)
3. Methanol synthesis
4. Purification.

3.1 Methane Reformer

Methane reforming is 'a process in which methane from natural gas is heated, with steam, usually with a catalyst, to produce a mixture of carbon monoxide and hydrogen used in organic synthesis and as a fuel' [56]. The following reactions take place in the reformer [57]:

$$CH_4 + H_2O \rightleftharpoons CO + 3H_2 \quad \Delta H_{295} = -206\,\text{kJ mol}^{-1}$$
$$CO + H_2O \rightleftharpoons \text{CO}_2 + H_2 \quad \Delta H_{298} = -41\,\text{kJ mol}^{-1}$$
$$CH_4 + 2H_2O \rightleftharpoons \text{CO}_2 + 4H_2 \quad \Delta H_{298} = 165\,\text{kJ mol}^{-1}$$

The combination of methane reformer and heat recovery steam generator (HRSG) generates the products carbon dioxide and hydrogen, hence, considered for analysis. Applying Eq. 2,

$$m_2h_2 + m_3h_3 = m_4h_4. \tag{3}$$

Moreover, $m_4 = m_5$. In order to comply with the reactions, gravimetric analysis is preferred for this study. The mass m_2 and m_3 are the mass of CH_4 and steam, respectively, while m_4 represents the combined mass of products, i.e. CO_2 and H_2.

3.2 Reverse Water-Gas Shift Reactor (rWGSR)

The carbon dioxide is introduced to the output of HRSG for efficient methanol synthesis [58]. After passing through the heat exchanger, it is transmitted to rWGSR. Energy balance can be given by Eq. 2:

$$m_7h_7 + m_9h_9 = m_8h_8 + m_{10}h_{10}, \tag{4}$$

where

$$m_7 = m_6 + m_5,$$

$$m_7 = m_6 + m_4,$$

$$m_7 = m_8 = m_9 = m_{10} = m_{11} = m_{12}.$$

Hence, Eq. 4 can be reformed as

$$h_7 + h_9 = h_8 + h_{10}. \tag{5}$$

3.3 Methanol Synthesis

Applying conservation of energy from Eq. 2 to feed effluent heat exchanger 2,

$$m_{13}h_{13} + m_{15}h_{15} = m_{14}h_{14} + m_{18}h_{18}, \tag{6}$$

where

$$m_{13} = m_{12} + m_{20},$$

$$m_{20} = m_{19} = m_{18} - m_{21}.$$

3.4 Purification

For the purpose of purification, synthesized methanol is passed through a series of flashing chamber and ultimately through a distillation chamber to get purified methanol. The energy balance for the distillation chamber is given by Eq. 7.

$$m_{21}h_{21} = m_{22}h_{22} + m_{23}h_{23}. \tag{7}$$

4 Results and Discussion

Thermodynamic analysis of methanol synthesis was performed by a number of researchers. Vita [59] performed Gibs-free energy-based analysis of the direct conversion of biogas to methane. It was inferred from the analysis that steam reforming with a high quantity of methane at the inlet is marvellous for sustainable methanol production. Borja [48] also proposed a mathematical model for methanol synthesis. The main aim of the study was to propose an economically and environmentally friendly synthesis system. Methanol at a production rate of $1.75/gal was estimated that lies in line with the economic analysis performed in the reference [48].

This book chapter also strives to provide fundamental energy analysis of the methane production system. The base of the analysis is reference [2]. The state of art of thermodynamic analysis-based results is also presented to provide the base for further studies.

The analysis shows that the concentration of the various constituents keeps on altering along the cycle at various temperatures as depicted in Fig. 5 [2]. The level of hydrogen keeps on increasing with the temperature and is highest at the highest

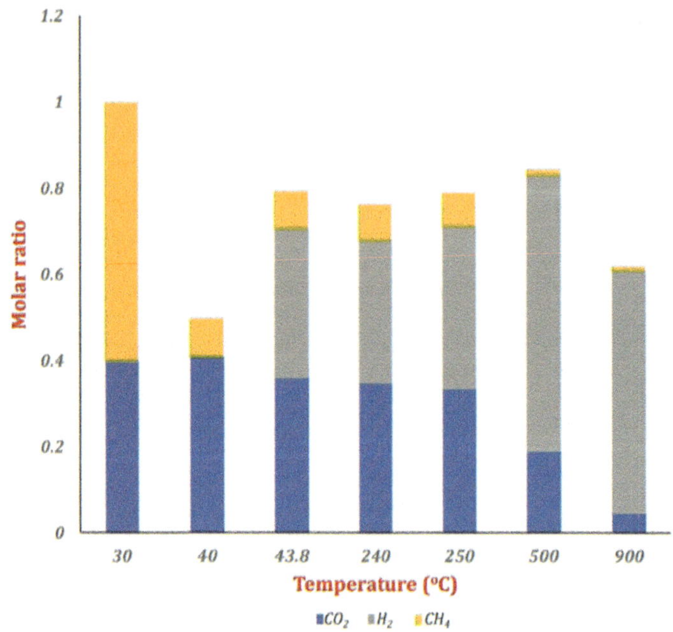

Fig. 5 The effect of temperature on the molar ratios of constituents

temperature. Contrarily, the level of carbon dioxide keeps on decreasing with increment in the temperature. Methane also shows the same trend as carbon dioxide. The aforementioned results lie in line with previous studies. The analysis performed by Gosh [2] also depicted similar results.

Another important consideration in thermodynamic analysis of methanol synthesis is concentrations of various constituents at various states. Monitoring the concentrations at different states provides substantial information for performance evaluation of the process. Figure 6 depicts the concentrations of different constituents over the cycle.

It is inferred from Figs. 5 and 6 that methanol yield is highest at the lowest level of CO [2, 60]. Moreover, the presence of CO_2 doesn't significantly decrease yield. This fact lies in line with already available literature [2, 59, 61, 62], as depicted in Fig. 7.

5 Conclusion

The thermodynamic analysis of the methanol synthesis system by reverse water-gas shift reactor is performed. The analysis shows that

1. The composition of feed varies substantially over the whole process.

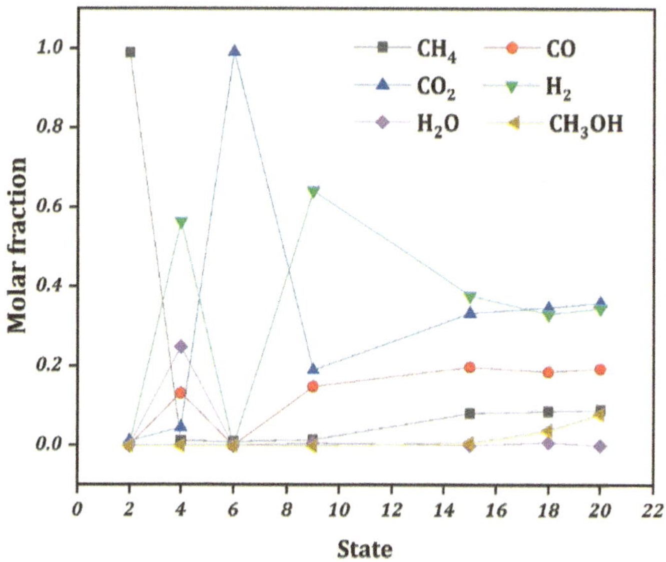

Fig. 6 Molar fractions of constituents along the process

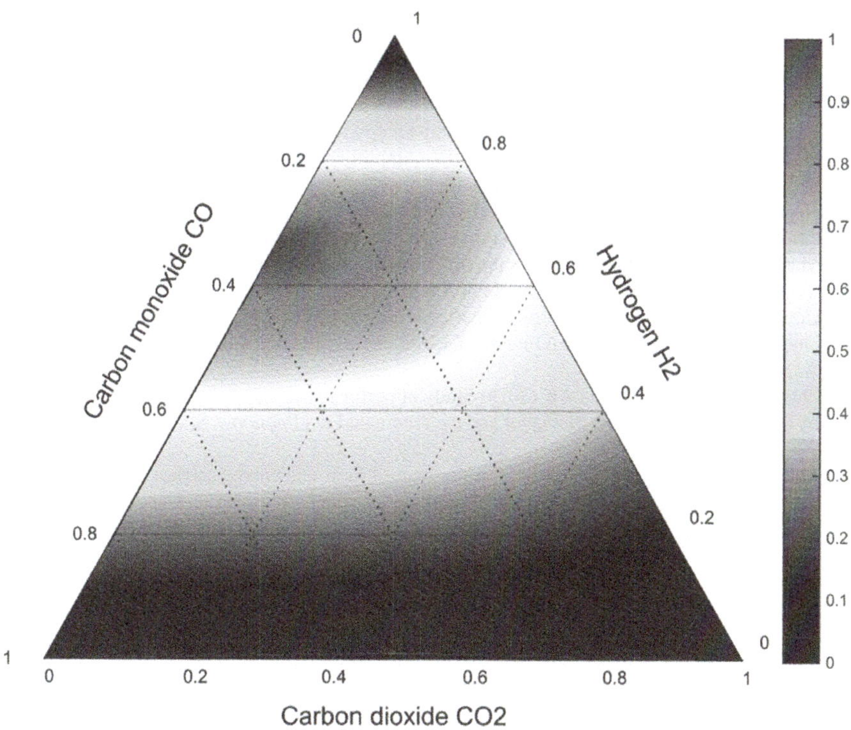

Fig. 7 Dependence of methanol yield on feed composition [59]

2. The composition quantification may provide the base for evaluation and quality management of the plant in industries.
3. Over the range of temperature in the synthesis plant, the concentration of Hydrogen increases with temperature while that of carbon dioxide and methane decreases.
4. The optimum methanol synthesis occurs at the point where the concentration of CO is minimum.
5. The concentration of CO_2 doesn't affect the yield of mesthanol substantially.

References

1. Tursunov O, Kustov L, Kustov A (2017) A brief review of carbon dioxide hydrogenation to methanol over copper and iron based catalysts. Oil & Gas Sciences and Technology-Revue d'IFP Energies nouvelles 72(5):30
2. Ghosh S et al (2019) Biogas to methanol: A comparison of conversion processes involving direct carbon dioxide hydrogenation and via reverse water gas shift reaction. J Clean Prod 217:615–626
3. Shahbaz M, Gozgor G, Hammoudeh S (2019) Human capital and export diversification as new determinants of energy demand in the United States. Energy Econ 78:335–349
4. Kahouli B (2019) Does static and dynamic relationship between economic growth and energy consumption exist in OECD countries? Energy Rep 5:104–116
5. Wordometer (2020) Current world population. https://www.worldometers.info/world-population/. Accessed 24 July 2020
6. Avtar R et al (2019) Population–urbanization–energy nexus: a review. Resources 8(3):136
7. van Ruijven BJ, De Cian E, Wing IS (2019) Amplification of future energy demand growth due to climate change. Nature Commun 10(1):1–12
8. Nepal R, Paija N (2019) Energy security, electricity, population and economic growth: The case of a developing South Asian resource-rich economy. Energy Policy 132:771–781
9. Mastrucci A et al (2019) Improving the SDG energy poverty targets: residential cooling needs in the Global South. Energy Build 186:405–415
10. Council WE (2019) World Energy Scenarios 2019. Word Energy Council, London, UK, p 1
11. Dudley B (2020) BP statistical review of world energy. BP Statistical Review, London, UK, August 2020, p 6
12. Dwivedi G, Pillai S, Shukla AK (2019) Study of performance and emissions of engines fueled by biofuels and its blends. Methanol and the Alternate Fuel Economy. Springer, New York, pp 77–106
13. Muradov NZ, Veziroğlu TN (2011) Carbon-neutral fuels and energy carriers. CRC Press
14. Pachauri RK, Reisinger A (2007) IPCC fourth assessment report. IPCC, Geneva, p 2007
15. Pachauri RK et al (2014) Climate change 2014: synthesis report. Contribution of Working Groups I, II and III to the fifth assessment report of the Intergovernmental Panel on Climate Change, Ipcc
16. Energy G (2019) CO2 Status Report—the latest trends in energy and emissions in 2018. International Energy Agency
17. IEA (2020) Global CO2 emissions in 2019. IEA, Paris
18. Ritchie H, Roser M (2020) CO_2 and greenhouse gas emissions 2019. Our world in data
19. Bose BK (2010) Global warming: energy, environmental pollution, and the impact of power electronics. IEEE Ind Electron Mag 4(1):6–17
20. Sun H, Wang W, Koo K-P (2019) The practical implementation of methanol as a clean and efficient alternative fuel for automotive vehicles. Int J Engine Res 20(3):350–358

21. Zaidi AA et al (2018) Comparative and numerical evaluation of methanol blends in CI diesel engine. In: 2018 IEEE 5th International conference on engineering technologies and applied sciences (ICETAS). IEEE
22. Nestler F et al (2020) Kinetic modelling of methanol synthesis over commercial catalysts: a critical assessment. Chem Eng J:124881
23. Stiles AB (1977) Methanol, past, present, and speculation on the future. AIChE J 23(3):362–375
24. Encyclopedia NW (2020) Methanol. 2020. https://www.newworldencyclopedia.org/p/index.php?title=Methanol&oldid=98494+3.#Credits. Accessed 28 May 2020
25. MMSA (2020) Methanol price and supply/demand. https://www.methanol.org/methanol-price-supply-demand/. Accessed 28 May 2020
26. Agarwal AK et al (2019) Introduction of methanol and alternate fuel economy. Methanol and the alternate fuel economy. Springer, New York, pp 3–6
27. Valera H, Agarwal AK (2019) Methanol as an alternative fuel for diesel engines. Methanol and the alternate fuel economy. Springer, New York, pp 9–33
28. Bussche KV, Froment G (1996) A steady-state kinetic model for methanol synthesis and the water gas shift reaction on a commercial Cu/ZnO/Al2O3Catalyst. J Catal 161(1):1–10
29. Leonov V et al (1973) Kinetics of methanol synthesis on a low-temperature catalyst. Kinet Katal 14:970–975
30. Villa P et al (1985) Synthesis of alcohols from carbon oxides and hydrogen. 1. Kinetics of the low-pressure methanol synthesis. Ind Eng Chem Process Des Dev 24(1): 12–19
31. Graaf G et al (1990) Intra-particle diffusion limitations in low-pressure methanol synthesis. Chem Eng Sci 45(4):773–783
32. Graaf G et al (1988) Kinetics of the three phase methanol synthesis. In: Tenth international symposium on chemical reaction engineering. Elsevier, New York
33. Graaf G, Stamhuis E, Beenackers A (1988) Kinetics of low-pressure methanol synthesis. Chem Eng Sci 43(12):3185–3195
34. Graaf G et al (1986) Chemical equilibria in methanol synthesis. Chem Eng Sci 41(11):2883–2890
35. Pedersen KS, Thomassen P, Fredenslund A (1984) Thermodynamics of petroleum mixtures containing heavy hydrocarbons. 1. Phase envelope calculations by use of the Soave-Redlich-Kwong equation of state. Ind Eng Chem Process Des Dev 23(1):163–170
36. Skrzypek J, Lachowska M, Serafin D (1990) Methanol synthesis from CO2 and H2: dependence of equilibrium conversions and exit equilibrium concentrations of components on the main process variables. Chem Eng Sci 45(1):89–96
37. Skrzypek J, Grzesik M, Szopa R (1985) Analysis of the low-temperature methanol synthesis in a single commercial isothermal Cu-Zn-Al catalyst pellet using the dusty-gas diffusion model. Chem Eng Sci 40(4):671–673
38. Skrzypek J, Lachowska M, Moroz H (1991) Kinetics of methanol synthesis over commercial copper/zinc oxide/alumina catalysts. Chem Eng Sci 46(11):2809–2813
39. Hain Y et al (2012) Methanol fuel as low cost alternative for emission reduction in gas turbines and utility boilers
40. Li C, Negnevitsky M, Wang X (2019) Review of methanol vehicle policies in China: current status and future implications. Energy Procedia 160:324–331
41. Bozzano G, Manenti F (2016) Efficient methanol synthesis: perspectives, technologies and optimization strategies. Prog Energy Combust Sci 56:71–105
42. Siddiqi MM et al (2019) Evaluation of municipal solid wastes based energy potential in urban Pakistan. Processes 7(11):848
43. Zaidi AA et al (2019) Combining microwave pretreatment with iron oxide nanoparticles enhanced biogas and hydrogen yield from green algae. Processes 7(1):24
44. Mushtaq K, Zaidi AA, Askari SJ (2016) Design and performance analysis of floating dome type portable biogas plant for domestic use in Pakistan. Sustain Energy Technol Assess 14:21–25
45. Iaquaniello G et al (2017) Waste-to-methanol: process and economics assessment. Biores Technol 243:611–619

46. Lam MK, Lee KT (2011) Mixed methanol–ethanol technology to produce greener biodiesel from waste cooking oil: a breakthrough for SO42 –/SnO2–SiO2 catalyst. Fuel Process Technol 92(8):1639–1645
47. Tan K, Lee K, Mohamed A (2011) Potential of waste palm cooking oil for catalyst-free biodiesel production. Energy 36(4):2085–2088
48. Hernández B, Martín M (2016) Optimal process operation for biogas reforming to methanol: effects of dry reforming and biogas composition. Ind Eng Chem Res 55(23):6677–6685
49. dos Santos RO, de Sousa Santos L, Prata DM (2018) Simulation and optimization of a methanol synthesis process from different biogas sources. J Clean Prod 186:821–830
50. Bosch C, Wild W (1914) Producing hydrogen. Google Patents
51. Xiaoding X, Moulijn J (1996) Mitigation of CO_2 by chemical conversion: plausible chemical reactions and promising products. Energy Fuels 10(2):305–325
52. Mallapragada DS et al (2013) Sun-to-fuel assessment of routes for fixing CO_2 as liquid fuel. Ind Eng Chem Res 52(14):5136–5144
53. Daza YA, Kuhn JN (2016) CO 2 conversion by reverse water gas shift catalysis: comparison of catalysts, mechanisms and their consequences for CO_2 conversion to liquid fuels. RSC Adv 6(55):49675–49691
54. Marlin DS, Sarron E, Sigurbjörnsson Ó (2018) Process advantages of direct CO_2 to methanol synthesis. Front Chem 6:446
55. Dalena F et al (2018) Advances in methanol production and utilization, with particular emphasis toward hydrogen generation via membrane reactor technology. Membranes 8(4):98
56. Dictionary CE (1995) Harper Collins Publishers. THE University
57. Chen I-T. Steam reforming of methane
58. Manenti F, Cieri S, Restelli M (2011) Considerations on the steady-state modeling of methanol synthesis fixed-bed reactor. Chem Eng Sci 66(2):152–162
59. Vita A et al (2018) Methanol synthesis from biogas: a thermodynamic analysis. Renew Energy 118:673–684
60. Ahouari H et al (2013) Methanol synthesis from CO_2 hydrogenation over copper based catalysts. React Kinet Mech Catal 110(1):131–145
61. Daza YA et al (2014) Carbon dioxide conversion by reverse water–gas shift chemical looping on perovskite-type oxides. Ind Eng Chem Res 53(14):5828–5837
62. De Falco M et al (2013) Methanol production from CO_2 via reverse-water–gas-shift reaction. CO2: a valuable source of carbon. Springer, New York, pp 171–186

Methanol as a Fuel

Methanol—A Sustainable Fuel for SI Engine

Datta Bharadwaz Yellapragada and Govinda Rao Budda

Abbreviations

A20	15% methanol + 5% bio ethanol + 80% gasoline
BMEP	Brake mean effective pressure
BOV	Blending octane values
BSFC	Brake specific fuel consumption
BTE	Brake thermal efficiency
CO	Carbon monoxide
CO_2	Carbon di-oxide
DI	Direct injection
DME	Dimethyl ether
ECU	Engine control unit
EGR	Exhaust gas recirculation
FCA	Fiat Chrysler automobiles
FTIR	Fourier transform infra-red
GC	Gas chromatography
GEM	Gasoline-ethanol-methanol
GoI	Government of India
HC	Hydro carbon
HPLC	High performance liquid chromatography
IMEP	Indicated mean effective pressure
KLCR	Knock limited compression ratio
LPG	Liquified petroleum gas
M10	Methanol 10% + 90% gasoline
M100	100% methanol

D. B. Yellapragada (✉) · G. R. Budda
Department of Mechanical Engineering, Gayatri Vidya Parishad College of Engineering (A), Visakhapatnam, India
e-mail: ydattabharadwaz@yahoo.com

A. K. Agarwal et al. (eds.), *Methanol*, Energy, Environment, and Sustainability,
https://doi.org/10.1007/978-981-16-1224-4_5

M15	Methanol 15% + 85% gasoline
M20	Methanol 20% + 80% gasoline
M30	Methanol 30% + 70% gasoline
M5	Methanol 5% + 95% gasoline
M70	Methanol 70% + gasoline 30%
M85	Methanol 85% + gasoline 15%
MAP	Manifold absolute pressure
MBT	Maximum brake torque
MPFI	Multi -point fuel injection
MTBE	Methyl tertiary butyl ether
NOx	Nitrogen oxides
OEM	Original equipment manufacturer
PFI	Port fuel injection
RON	Research octane number
SCR	Selective catalytic reduction
SI engines	Spark-ignition engines
TDI	Turbocharged direct injection
TWC	Three-way catalytic convertor
WOT	Wide open throttle
θ	Crank angle
ϕ	Equivalence ratio
λ	Excess air ratio

1 Introduction

The search for the best alternative fuels for IC engines is as old as IC engine technology. Many attempts were made throughout history to make them viable for the existing vehicle fleet. The crude oil price shocks and the need for good air quality also intensified the research in alternative fuels. The present chapter deals with one such alternative, methanol. Methanol and its derivates are one of the most used chemicals in the world which are used for many purposes such as paints, electronics, appliances, automotive, construction, electronics, fuel, and many more.[1] One more exciting feature of methanol is "polygeneration", i.e. it can be produced from any carbon-based feedstock which can be converted into synthesis gas. Some of the feedstock for methanol production are coal, natural gas, wood, biomass, CO_2, and municipal solid waste. The ability of methanol to be produced at any scale cost-effectively, the availability of methanol production technology and feedstock in many parts of the world made it a strong competitor to fossil fuels. At present, there are 90 world-scale methanol plants that can produce around 36.6 billion gallons of

[1]Methanol Institute https://www.methanol.org/wp-content/uploads/2019/02/4.-Greg-Dolan-Overview-of-Global-Methanol-Fuel-Blending.pdf.

methanol per annum. Methanol is one of the most shipped and top five traded chemicals in the world. Methanol is readily biodegradable when compared with gasoline and diesel. The 100 plus years industry experience in safe production and handling make methanol a very reliable alternative.[2] Apart from ease of production, availability, low cost and utilization of methanol has many other advantages. Some of these advantages are clean combustion, knock resistance, high compression ratios, and downsizing of the engine. Methanol is estimated to reduce greenhouse emissions by 95% if it is produced from renewable sources. This reduction is very much higher when compared with its counterparts such as gasoline, diesel, and coal.[3] By using methanol even in small blends with gasoline results in a clean-burning fuel reducing ground-level particulate matter and ozone. Utilizing methanol atleast as blends would results in huge savings at the user level and can effectively decrease the capital investment in addition to satisfying environmental constraints for oil refineries in producing premium-grade gasoline.[4] Using methanol and its blends in automobiles have forty plus years of history and many countries such as Germany, New Zealand, China, and the US successfully implemented methanol and its blends in their vehicle fleets.

Despite many advantages of methanol, it has been kept out of the picture for almost two decades due to its low energy density, toxic nature, fossil origins, corrosive nature, and formaldehyde emissions. During this time many alternatives such as hydrogen, ethanol, electric vehicles, bio-diesels have been bought centerstage emphasizing their green origins. Because of its renewable origin from crops, bio-ethanol is assumed as a viable alternative to methanol. However, serious feedstock limitations in type and quantity and slower biological processes for converting biomass to ethanol puts a limitation on its ease of scalability when compared to methanol. Methanol is also found to be a viable alternative to gasoline in many critical aspects as proposed by research agencies such as the Society of Automotive Engineers, Lloyds, the Environmental Protection Agency, and others.[5] The present chapter reviews methanol and its blends for their application in SI engines as well as their effects on performance, combustion, and emissions. The present chapter also discusses the strategies of using methanol in SI engines, the role of methanol in downsizing of engines, effect of high compression in methanol-fueled engines, knock and effect of methanol on efficiency of catalytic converters.

---

[2]Methanol Institute https://www.methanol.org/wp-content/uploads/2020/03/Future-Fuel-Strategies-Methanol-Automotive-Fuel-Primer.pdf.

[3]Methanol emissions https://www.methanol.org/wp-content/uploads/2019/01/MethanolReport.pdf.

[4]Methanol use in gasoline http://www.methanol.org/wp-content/uploads/2016/06/Blending-Handling-Bulletin-Final.pdf.

[5]Methanol significance https://platformduurzamebiobrandstoffen.nl/wpcontent/uploads/2017/09/2017_F3_Methanol-as-a-renewable-fuel.pdf.

2 History of Methanol Usage as SI Engine Fuel

Methanol has been a very widely used fuel in history, starting from racing cars to aircraft. The need for high-knock-resistant fuels also leads to the use of methanol in automobiles in the late 1940s. High latent heat of methanol ensures cooler engine operation when compared to gasoline, especially during high-performance conditions like racing. Because of this low-temperature operation, methanol was widely used in high-performance vehicles such as champ car, sprint cars, monster trucks, and many other motor championship series. The susceptibility of gasoline-fueled racing cars to get crashed because of overheating made the Indy car series to use pure methanol from 1964 as a replacement for gasoline. During World War II, methanol was used as fuel for German Military rockets and also as coolant anti-freezer in automobiles in the 1900s.[6] The crude oil crisis during the 1970s made many investigators search for the potential application of methanol in on-road vehicles. During the late 1970–1980s methanol program was seriously pursued by many countries, industries, and researchers. This increase in methanol use during the 1970s is mainly because of its cost of production, which is as close to gasoline production from crude oil (see footnote 5).

The oil crisis of the 1970s propelled many countries to pursue a methanol program for energy security and pollution control. In 1981, 40 methanol-fueled Escorts were delivered by Ford to Los Angeles County. The vehicles were driven around 200 miles and improvement in fuel efficiency and performance was found. In 1983, a fleet of around 500 methanol-fueled Escort vehicles were delivered to California and successfully ran up to around 35 million miles without any difficulty and some of them are still in working condition. Thus, the methanol fuel program in California was a huge success. But the non-availability of sufficient infrastructure for methanol refuelling led to the development of flexible fuel vehicles (FFV) which run on both methanol and gasoline. During 1985–1992 approximately ~ 700 FFV were delivered to California and Canada. Comprehensive plans were prepared in the Unites States of America for the deployment of methanol-fueled vehicles. In 1989, the United States of America made plans for introducing 500,000 methanol vehicles by 1996, 750,000 by 1997, and one million vehicles for subsequent years [1]. In 1979, the Swedish Commission for Oil Substitution understood the possible uncertainty of crude oil supply in the future and recommended forced implementation of methanol [2]. To satisfy its increasing energy needs and security, China started alternative fuel research in its 8th five-year plans from 1991to1995. Now China is one of the leading beneficiaries of methanol technology in automobiles.[7] Many other countries such as Germany, Sweden, Norway, and New Zealand also actively pursued the methanol program during the 1980s and see it as a potential substitute for gasoline.

However, the drop in crude oil prices in the years after the crisis, the introduction of MTBE (methyl tertiary butyl ether) to satisfy the octane requirements of gasoline,

---

[6]Methanol history in racing cars https://thechemco.com/chemical/methanol/.

[7]China Methanol program http://franke.uchicago.edu/bigproblems/BPRO29000-2015/Team01-TheFutureofMethanolFuel.pdf.

lack of methanol infrastructure, and lukewarm encouragement from governments resulted in fall of active methanol research. The advent of ethanol, its renewable origins, and strong lobbying made automobile manufacturers pursue ethanol as a viable alternative. Improvement in gasoline engine research in enhancing the performance and reduction in emissions in SI engines also contributed to the falling off of methanol vehicles. Despite the positive results from tests on pure methanol and its gasoline blends in many countries for millions of kilometers in their vehicle fleets, methanol has been relegated to the backburner citing the above reasons.

3 Current Status of Methanol in SI Engines

3.1 Status of Methanol as Fuel in SI Engines-World Scenario

Methanol market is steadily growing through the years, the market for gasoline blending with methanol and dimethyl ether production has gone up to 20 million tons in 10 years. This rise in methanol blending with gasoline is mostly attributed to China. The steady increase in methanol production and demand in China from the year 2000 to 2010 is shown in Fig. 1. Geely auto company of china has a production

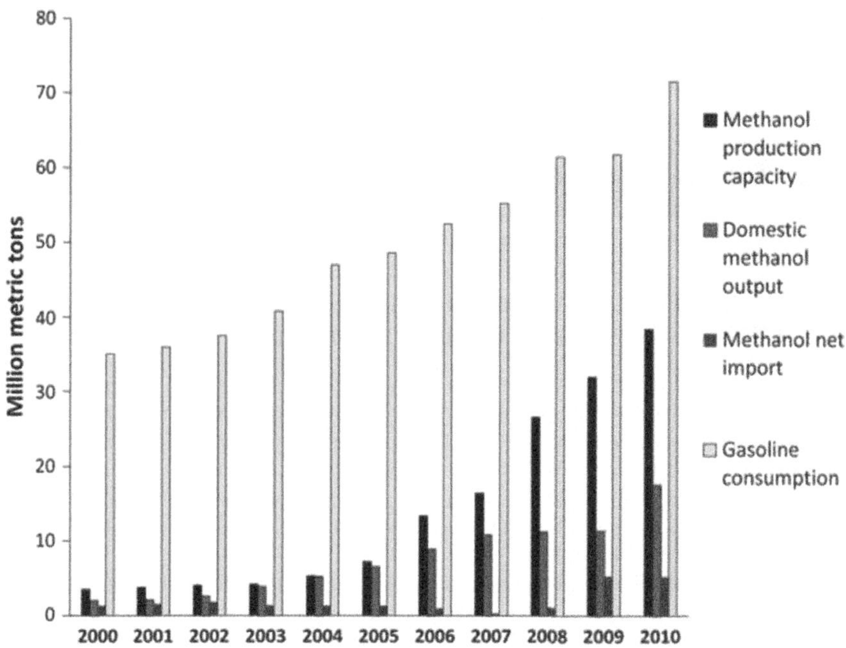

Fig. 1 Rise in methanol production and demand in China (*Source* Energy policy 41, p. 879)

capacity of 300,00-500,000 methanol vehicles per year. At present, the Geely group has two dedicated methanol engines and also supplying its M100 cars to Iceland's Carbon Recycle International (CRI). In 2013, it was estimated that about 7 million tons of methanol were blended with gasoline to achieve M15 specification in China. The infrastructure for methanol is increasing at a rapid phase, and many methanol gas stations which can dispense pure methanol and its gasoline blends are in China (See footnote 5). Apart from Geely, many other companies such as Chery Automobile, Maple Automobile, Chang'an auto group successfully demonstrated their methanol-compatible vehicles. The rise in gasoline methanol blending in China is mostly because of the taxies and passenger cars. More than 2000 taxies are fueled with M100 in the Shanxi province of China, their popularity is because of its low cost.[8]

In Europe, at present only 3% of methanol by volume is allowed for blending with gasoline under EN 228 regulations. As per the Renewable Transportation Fuel Obligation statistics report in February 2018, the current bio-methanol consumption as fuel in the UK is 16 million liters, which is almost 3% of the UK's total renewable fuel consumption.[9] In Italy, Fiat Chrysler Automobiles (FCA) and Eni agreed to develop low-carbon-emitting fuel technologies. Under this agreement, Eni developed new petrol (A20) comprising 15% methanol and 5% bioethanol content and was tested on Fiat 500 vehicle.[10] In 2019, Eni and FCA announced A20 fuel, the tests conducted on Eni Enjoy's fleet of Fiat 500's in Milan fueled with A20 fuel demonstrated a reduction in CO_2 emissions and improvement in performance. To reduce the well-to-wheel CO_2 emissions, Eni and FCA reported that they are working towards improving this A20 formula by increasing hydrocarbon content from renewable sources.[11] In the Middle East region, Israel found massive off-shore natural gas deposits from which methanol could be produced at any scale. So, it chalked out plans such as Israeli Fuel Choice Initiative (IFCI) to adopt methanol in their present vehicle fleet. Under this program, Israel conducted many trials on M15 fueled vehicles and developed a national standard for M15, SI 90-4 in 2016.[12] FCA, Dor Chemicals and IFCI jointly presented Fiat 500 as a retail-ready vehicle to run on M15 complying with Euro 6 regulations. The official Israel importer of the FCA group, MCA took up marketing of these vehicles in Israel.[13]

In 2013, FIA (Federation International de l' automobile) approved the utilization of GEM fuels (Gasoline–Ethanol–Methanol) in racing cars. In this race, about 20,000

[8]Methanol use in china https://afdc.energy.gov/files/pdfs/mit_methanol_white_paper.pdf.

[9]UK methanol consumption https://assets.publishing.service.gov.uk/government/uploads/system/uploads/attachment_data/file/678043/rtfo-year-10-report-2.pdf.

[10]Status of methanol blend in Italyhttps://www.eni.com/en-IT/media/press-release/2017/11/eni-and-fca-sign-research-agreement-for-joint-projects-to-significantly-reduce-co2-emissions-produced-by-road-transport-vehicles.html.

[11]Reformulated A20 https://www.eni.com/en-IT/media/press-release/2019/04/eni-and-fca-have-developed-a20-a-new-fuel-that-pairs-emissions-reduction-with-energy-efficiency.html.

[12]Israel M15 standardhttps://methanolfuels.wpengine.com/wp-content/uploads/2013/05/Israel-M15-Press-Release-8_2016.pdf.

[13]M15 Vehicles in Israel http://www.methanol.org/wp-content/uploads/2016/11/Dor-Press-Release.pdf.

litres of fuel is consumed, of which 42% is methanol. According to the Methanol Institute, the use of methanol has reduced CO_2 emissions by 24,445 kgs with improved performance.[14] In 2013, it was reported that in Australia, GEM fuels are actively pursued by Coogee Chemicals in collaboration with industry and academia. Coogee Chemicals is planning to introduce GEM-8, GEM-15, and GEM-56 into Australian markets commercially. GEM fuels are successfully tested in their present vehicle fleet and found that GEM fuels provide high performance at low cost.[15] During the trials conducted using on-road vehicles with GEM fuels, it is found that very few modifications are required for the engine and the cost would not exceed more than 100US$/vehicle.[16] Iceland, being one of the leading producers of renewable methanol from CO_2, as per EU/EEA quality regulations allows up to 3% of methanol blends in gasoline. By 2020, Iceland is planning to increase the methanol blending up to 10%.[17] Apart from these countries, many central Asian countries such as Uzbekistan, Azerbaijan, Tajikistan are strongly considering methanol blends as alternatives to gasoline. Recently Uzbekistan conducted tests on low methanol fuel blends M3 and M5 on their vehicles. GCC (Gulf Cooperation Council) started to consider fuel blending as an option and likely to introduce blends in their markets in the future.[18] The information about methanol status internationally collected from various sources is summarized and presented in Table 1.

3.2 Methanol as Fuel in SI Engines—India Scenario

India consumed 593 thousand barrels of gasoline per day in 2017, which is way more than the world average of 138.43 thousand barrels when averaged over 190 countries.[19] Apart from being the largest consumer of gasoline, India is facing quite a few challenges such as dependence on imports for its energy needs, everyday jump in the number of automobiles, energy security, and instability of oil prices. India ranks fourth among the largest carbon emitters in the world, raising severe concerns about its environmental health. In the Paris Summit held in 2015, India committed to reduce its dependency on fossil resources and reduce carbon emission by 2030 to 3 billion tons.[20] To reach this goal, India has devised "Methanol Economy", a comprehensive plan to reduce the load on petroleum products as fuel. Under this

[14]Methanol rally racing http://www.mefco2.eu/news/methanol-rally-racing.php.

[15]GEM fuels in Australia http://methanolfuels.org/wp-content/uploads/2013/05/Grant-Lukey-Coogee-Energy-Australia.pdf.

[16]Engine Modification cost http://www.methanol.org/wp-content/uploads/2016/07/6-Combined-Side-Deck-GFBC-1.pdf.

[17]Iceland Methanol program https://www.chemicals-technology.com/projects/george-olah-renewable-methanol-plant-iceland/.

[18]Methanol in Uzbekistan https://methanolfuels.org/public-policy/asia-pacific-middle-east/.

[19]India's gasoline consumption https://www.theglobaleconomy.com/India/gasoline_consumption/.

[20]Carbon reduction goals of India https://gbci.org/india-after-cop21-goals-and-accomplishments.

Table 1 Status of methanol and its blends at the international level

Country	Methanol blends	Status
China	M5, M15, M30, M85, M100	Using in on-fleet vehicles and many district-level standards are introduced for methanol
USA	M5	Tested flex-fuel vehicles with a range of methanol blends starting from M3 to M85
Israel	M15	The national-level standard for M15(SI 90-4)
Uzbekistan	M3 and M5	–
Iceland	Up to 3%	Allowed as per EU/EEA regulation
Italy	15% Methanol with 5% ethanol	Completed vehicle trials successfully
Europe	Up to 3%	Allowed as per EN 228
Australia	GEM fuels (Gasoline-ethanol-methanol) GEM-8, GEM-15, GEM-56	Vehicle testing completed and ready for markets

plan, India is going to introduce M15 blends in the market, reducing petrol price by 10% instantly. NITI Aayog of the Government of India (GoI) prepared a plan to substitute 10% of crude oil imports by 2030 with methanol. India is in talks with Israel for technology transfer related to methanol fuels and blending.[21] The Indian government has collaborated with Indian Oil Corporation Limited, Automotive Research Association of India (ARAI), and the Society of Indian Automobile Manufacturer to prepare roadmaps for testing and successful incorporation of methanol blends into markets.[22] Under this program, on-road vehicles are tested with M15 fuel blend over 65000 km successfully without any negative implications.[23] At present, the Bureau of Indian Standards (BIS) is working on establishing standards for M15 and M100 fuels. ARAI is currently pursuing the testing of present on-fleet vehicles fueled with methanol and their emission compliance with BS-VI norms.[24]

The production capacity of India's five methanol plants stands at 4.7 lakh tons/annum. The production capacity of methanol remained static for many years despite the rise in demand for methanol as a chemical compound. This stagnation is because of the cost of methanol production from natural gas. Government of India is planning to produce methanol from its high ash coal to meet the demand for methanol as fuel. As India is rich in coal reserves with 315 billion tons, producing

[21] India's Methanol program https://pib.gov.in/newsite/printrelease.aspx?relid=174919.

[22] Methanol Economy https://niti.gov.in/methanol-economy.

[23] India's M15 tests https://economictimes.indiatimes.com/industry/energy/oil-gas/methanol-blend-on-cards-may-trim-oil-import-bill-by-rs-5000-crore/articleshow/72929961.cms.

[24] ARAI research on Methanol https://araiindia.com/cpanel/Files/NEW_1024201820958PMARAI-Update_Q11819.pdf.

methanol from coal is an economical option. GAIL India announced that it is planning to construct a coal syngas plant for methanol production.[25] As per estimates, 1600 ton/day capacity plant can produce methanol at a competitive rate of INR 17-19/liter as against the present cost of methanol per liter of around INR 25-27. The well- to -wheel carbon emissions of methanol in IC engines is higher if methanol is produced from coal. To reduce these emissions, plans are being prepared to set up colossal coal complexes where power, methanol, and fertilizers are produced in an integrated manner to reduce the cost and emissions. Apart from coal, there are other feedstock options such as municipal solid waste, biomass, natural gas, and CO_2 from power plants. But the availability of these feedstocks is not continuous, and producing methanol from coal is the only viable option for India.[26] Government of India also recently announced to establish five methanol plants with high ash coal as raw materials and one plant based on natural gas with an estimated production capacity of 20 MMT/annum to meet the methanol demand as fuel (See footnote 22).

4 Methanol in SI Engines

This section highlights the properties of methanol as cleaner SI engine fuel, challenges in using methanol in SI engines, and the strategies that need to be applied to avail methanol as fuel.

4.1 Attributes of Methanol as Cleaner SI Engine Fuel

Methanol has a sweet alcoholic odour, clean-burning fuel with 50 wt% of oxygen. It is completely miscible in hydrocarbons such as gasoline with rich blending properties. In addition to this, methanol has many properties which makes it a proper SI engine fuel. These properties of methanol are very influential in deciding the performance, emission, and combustion in a gasoline engines. The following are some of the attributes of methanol that makes it a cleaner SI engine fuel.[27]

- Methanol is a pure substance
- High oxygen content
- High octane number
- Lower boiling temperature
- Low carbon intensity with high H/C ratio
- Zero sulphur contamination

[25] Methanol capacity of India http://content.icicidirect.com/mailimages/MethanolFuel.htm.

[26] India's methanol plans https://niti.gov.in/sites/default/files/energy/Indias-Leapfrog-to-Methanol-Economy.pdf.

[27] Clean burning properties of methanol in SI engines http://www.methanol.org/wpcontent/uploads/2016/06/Blenders-Product-Bulletin-Final.pdf.

- High latent heat of vapourization
- High RON blending values
- High volatility
- No existing gums
- Biodegradable

The high latent heat of methanol reduces the heat losses to the walls and improves the efficiency. Methanol use decreases the knocking tendency facilitating higher compression ratios and helps to maintain optimal spark timing throughout the engine operating range. Intake air density increases for methanol and its blends increasing the volumetric efficiency of the engine. Methanol burns with a high laminar flame speed which would give no time for the unburned mixture to auto-ignite avoiding knock [3]. With high blending octane value (BOV), methanol provides a cost-effective way for oil refineries to improve the octane quality of gasoline without increasing its aromatic and olefin content (see footnote 27). The range of flammability of methanol is 7.3–36 v/v % whereas gasoline has a range of 1.4–7.6 v/v %.[28] The flammability limits and EGR dilution of methanol are higher than gasoline which enable load control strategies such as WOT EGR and WOT lean-burn enhancing indicated efficiency by 5% compared to throttled stoichiometric operation [4]. Sulphur is naturally present in gasoline and it damages the conversion capabilities of catalytic converters. The risk of catalytic deactivation is decreased by using methanol. Methanol is entirely biodegradable, and the environmental half-lives of methanol in soil, surface water, and groundwater are in the range of 1–7 days (see footnote 27). So, it can be summarized that methanol has many useful properties which makes it a cleaner SI engine fuel with improved performance. The properties of methanol and gasoline are presented in Table 2.

4.2 *Issues in Utilizing Methanol as Fuel for SI Engines*

From the properties of methanol, it is clear that methanol is a clean-burning SI engine fuel. But methanol and its gasoline blends have a few shortcomings which need to be addressed for effective utilization. The following are some of the challenges with methanol as fuel.

- Poor cold starting,
- Degraded Drivability,
- Phase separation and water tolerance,
- Material compatibility,
- Poor lubrication,
- Formaldehyde emissions,
- Toxicity—Health Impact.

[28]Flammability of methanol http://www.methanol.org/wpcontent/uploads/2018/03/physicalandchemicalpropertiespart2.pdf.

Table 2 Properties of methanol and gasoline

Property	Methanol	Gasoline
Specific gravity (15 °C)	0.792	0.739
Reid vapour pressure @ 38 °C, kPa	32	53.7
Flashpoint, °C	12	−42.8
Autoignition temperature, °C	470	246–280
Boiling point, °C	64.6	38–204
Heat of combustion, Net kJ/kg	19930	46400
Latent heat of vapourization, kJ/kg @ 25 °C	1160	350–400
Stoichiometric air–fuel ratio	6.47	14.6
Higher flammability limit (Air), volume %	36	7.6
Lower flammability limit	7.3	1.4
Kinematic viscosity, cSt @15 °C	0.74	0.88

Methanol properties. (a) http://www.methanol.org/wpcontent/uploads/2016/06/Blenders-Product-Bulletin-Final.pdf, (b) http://www.methanol.org/wpcontent/uploads/2018/03/physicalandchemicalpropertiespart2.pdf
Gasoline Properties. https://www.engineeringtoolbox.com/

Poor cold starting. Due to low energy densities, high heat of vaporization, and low air–fuel ratios, it is challenging for the methanol to create a combustible mixture near the spark-plug which is very important for ignition [3]. Methanol requires 3.5 times more energy to vaporize when compared with gasoline and this adversely affects the cold start ability of methanol-fueled engines. In SI engines, rich mixtures are used at the start to create a flammable mixture. By using methanol blends, the amount of fuel added at the start is higher than the gasoline, which hampers the cold start ability. However, the cold start ability of SI engines is not significantly affected by utilizing lower blends of methanol-gasoline.[29] It is identified that adding 15% of gasoline (M85) to methanol improves the cold starting performance [5]. From the cold start tests conducted on methanol, many alternatives are proposed such as adding 10% butane or 20% iso- pentane to methanol, providing a small gasoline tank near the cold start valve to spray fuel to intake air (fuel injection system) and LPG gas starting is proposed for carbureted engines. It is also found that cold starting is not a problem for methanol- gasoline blends containing methanol less than 40% [6]. The Cold start tests conducted on gasoline-methanol blends of 10%, 30%, and 85% reported that all the test blends are cold—starting at −25 °C except M85. Tests on methanol blends also reported that the temperature should not be less than −15 °C to successfully start higher methanol blends such as M85 [7]. Over the years, much research is done to enhance the cold starting ability of methanol and many solutions have been provided. The list of solutions for cold start of methanol are summarized as follows:

[29]Methanol cold start problems http://www.methanol.org/wp-content/uploads/2016/06/Methanol-Use-in-Transportation.pdf.

Mixture enrichment. The cold-starting capability of the engine depends on the availability of combustible mixture near the spark-plug. Gardiner et al. [8] found that mixture enrichment reduced stalling of the engine at cold temperatures. They reported cold starts at subzero temperatures with equivalence ratios as high as 13 to 16. However, Nakajima et al. [9] noticed misfires in the engine due to over-rich fuel during cold start.

Engine cranking speed. The studies conducted on the cold-starting capabilities of engines suggested higher cranking speeds ensure high air velocity and easy start ability [9].

Heating fuel rails and injectors. Electrically heating the fuel rails aids in fuel vaporization and formation of combustible mixture improving the cold starting ability of methanol [3].

Ignition timing. Closer the ignition timing to TDC higher will be the chances of start ability of the engine especially during cold starts. Utilizing multiple ignitions also found to be beneficial to ensure cold start [3, 9].

Direct injection. The studies on cold starting problems with alcoholic fuels showed high pressure stratified starts and late injections are very helpful in cold starting the engine at temperatures as low as −20 °C. Two-stage direct injection is also found to be effective in improving the cold start ability of the SI engines. From the literature, it is also identified that late direct injection in compression stroke ensures better fuel evaporation [3, 10–12].

With the advancements in SI technologies and high compression ratio of methanol-fueled engines, cold starting is no longer a major problem for methanol and its blends. Also, this problem is of interest only in cold countries during non-summer months.

Drivability. The lack of stalling, surging, and vapour locks during starting, idling, and accelerating conditions of the engine is called the drivability of the vehicle (see footnote 29). Adding methanol to gasoline increase the Reid vapour pressure of the fuel and this results in more evaporation of the fuel in the supply line, eventually resulting in vapour locks and loss of drivability. To avoid this, the addition of cosolvents and special formulation of gasoline which mitigate these effects is required [10]. Changes in distillation curves of methanol gasoline blends also affect drivability. The addition of methanol to gasoline produces a flattening or "knee" effect just below the boiling temperature of the alcohols, which affects the hot engine drivability for carbureted systems (see Footnote 27). From the tests conducted on cars used in the US, it is found that these drivability issues are annoying for some drivers and completely unacceptable in some scenarios. By adding 10% of methanol to gasoline increased existing demerits. This increase in drivability issues for methanol and its blends is mostly attributed to the stoichiometry of fuel operation [11]. The recent test on vehicles with feedback emission control systems can successfully mitigate the drivability issues of low-level methanol blends (see Footnote 29). The sensitivity of drivability performance with distillation properties of high blends of methanol is reduced with high -pressure modern fuel injection systems (see footnote 27).

Phase separation and water tolerance. Like other alcohols methanol also suffers from the problem of phase separation or stability. When the water content in the methanol blend reaches its water tolerance value, separation occurs, especially during cold operations. Phase separation results in corrosion of metal components, poor vehicle performance, and fuel system damage (see footnote 27). The critical phase separation temperatures of gasoline–methanol blends increase with the increase of water content in the blend. Ethanol is added as a co-solvent to gasoline–methanol blends to preserve phase stability and it is found that ethanol is successful in reducing the critical phase separation temperatures [13]. To enhance the water tolerance of methanol blends with gasoline, ethyl acetate is added. The effectiveness of ethyl acetate was tested at three different temperatures and it was found that ethyl acetate acts as a perfect stabilizer for gasoline–methanol blends without affecting the volatility of blends [14]. In addition to the above-said co-solvents, isopropanol and tertiary butanol are also added as co-solvents to preserve phase stability.[30]

Material Compatibility. Methanol is more corrosive to metals and elastomers when compared with other alcohols. Materials should be properly selected to avoid the drastic effects of methanol corrosion. From California experience, it is established that there is no technical barrier in incorporating methanol-resistant materials in vehicles. Methanol is dangerous to aluminum and copper but most of the other commercially available metals are resistant to methanol (see footnote 8). Many components in SI engines have to be checked for material compatibility and need to be replaced with methanol-compatible materials. Components such as seals, plastics, fuel tanks, fuel rails, fuel pumps, valves, valve seats, and liners are to be checked and replaced for methanol operations. From the tests conducted on various elastomers, fluorosilicones are chosen as the best compatible materials for methanol/gas applications [15]. There are several metals and their alloys that are suitable for 100% methanol operation. For valves and metal gasket materials such as stainless steel alloys, cast iron and bronze are preferable. Aluminum and copper alloys are also being preferred in-car tanks and storage tanks. Nickel-based alloys are preferred for sensor materials in methanol applications.[31] In addition to metals, many elastomers are used in SI engines for O-rings, seals, and packings. In SI engines, Buna-N, Flurel, Fluorosilicone, Polysulfide rubber elastomers are generally used (see footnote 27). Improper selection of these materials leads to swelling, explosive decompression, hardening, and shrinkage.[32]

Poor lubrication. Methanol has poor lubrication properties resulting from low viscosity. Burning of carbon atoms in methanol produces almost 2 times water than gasoline. This water content results in wear of engine parts. The tests conducted on engines fueled with M85 have shown increased engine oil temperatures almost by 10 degrees than gasoline-fueled engines. The increase in engine oil temperatures reduces the viscosity of engine oil and reduces the life of engine components. In addition to this, high vaporization of methanol results in contamination of lube oils. When this dissolved methanol reaches action zones wearing and tearing of components takes

[30]Co-solvents for phase stability https://www.iea-amf.org/content/fuel_information/methanol.

[31]Metals compatibility with methanol http://www.methanol.org/wp-content/uploads/2016/06/Compatibility-of-Metals-Alloys-in-Neat-Methanol-Service.pdf.

[32]Elastomers compatibility http://www.methanol.org/wp-content/uploads/2016/06/Compatibility-of-Elastomers-in-Neat-Methanol-Service.pdf.

place.[33] Tribological tests conducted on a gasoline engine with methanol fuels show increased wear in valve trains, cylinder bore, and bearings. To reduce the wear of engine components, synthetic-based lube oils are formulated. These reformulated lube oils showed resistance to methanol contamination and reduced the wear rate [16]. Experiments conducted to test the lubrication of polypropylene glycol (PPG) on a 2-stroke methanol-fueled scooter engine have shown improved miscibility with methanol fuels and increased performance to piston tightening and deposit formations [17]. Four-ball wear tests conducted on methanol-fueled engines have shown high wear metal accumulation in engine oils than in conventional fuels. The high wear rate of methanol-fueled engines is attributed to the contamination of lube oil with methanol and water. The tests on methanol engines have shown improved tribological behaviour with reformulated lube oils [18]. From the above tests, it can be summarized that the problems associated with the lubrication of methanol-fueled engines can be effectively handled by methanol-resistant or reformulated lube oils.

Formaldehyde emissions. The combustion of methanol in SI engines reduces CO, NO_x, and HC emissions. But there is a possibility of high formaldehyde emissions [19]. Formaldehyde is dangerous because it is a colourless and carcinogenic gas. Exposure to formaldehyde of 0.1 ppm results in watery eyes, burning sensation in eyes, wheezing, nausea, and skin irritation. Long time exposure to formaldehyde can cause cancer in humans.[34] The effects of formaldehyde emission in the SI engine are discussed in subsequent sections (see Sect. 5.2).

Toxicity-Health Impact. Methanol is toxic when exposed or consumed in higher quantities. It occurs naturally in some fruits, vegetables, and beverages in small quantities. Proper care must be taken when handling methanol as it can be absorbed through the skin. Ingestion of methanol can result in many complications such as metabolic acidosis, permanent vision defects, Parkinson-like disease, coma, respiratory failure, circulatory failure, a complication with dialysis, and even death.[35] Some studies conducted on rats proved that ingestion of methanol could result in hypertension [20]. However, high exposure to methanol is experienced with racing fuels, camp stove fuels, shellac solvents but no negative effects were noticed. Methanol and its blends were handled safely without any side effects on vehicles from the California experience.[36] The advantages and disadvantages of methanol properties in engines is shown in the form of a flowchart in Fig. 2.

---

[33] Methanol contamination in lube oils https://patents.google.com/patent/CN103509634A/en.

[34] Formaldehyde effects https://www.cancer.gov/about-cancer/causesprevention/risk/substances/formaldehyde/formaldehyde-fact-sheet.

[35] Methanol complications https://www.ncbi.nlm.nih.gov/books/NBK482121/.

[36] Toxicity of methanol http://www.methanol.org/wp content/uploads/2016/06/Blenders-Product-Bulletin-Final.pdf.

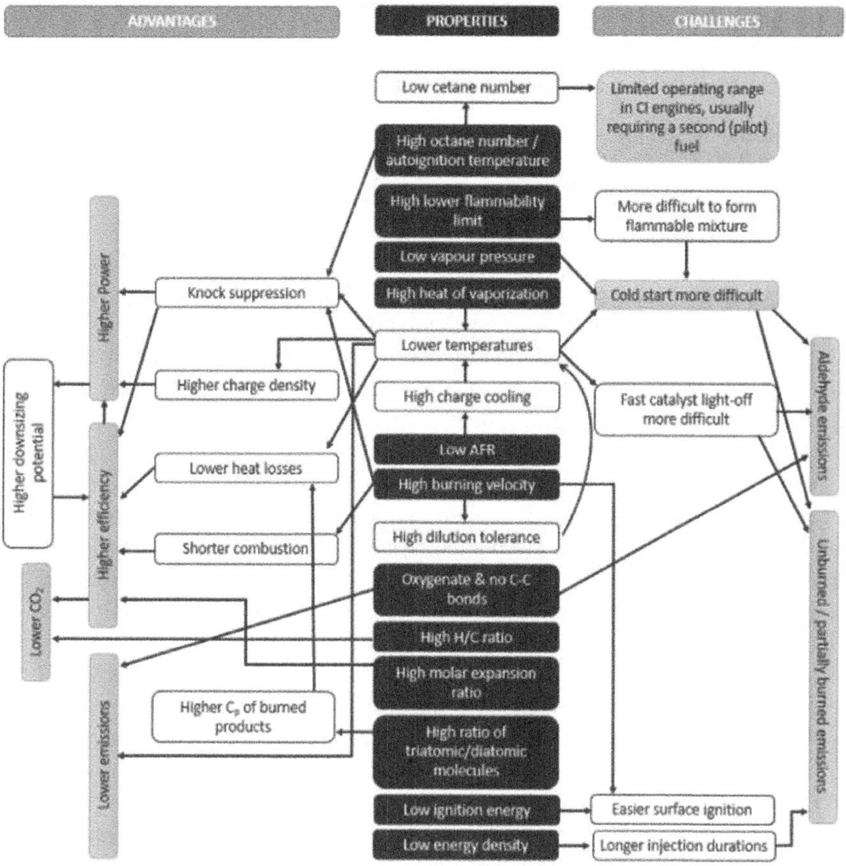

Fig. 2 Advantages and disadvantages of methanol properties (*Source* Progress in Energy and Combustion science 70, p. 83)

4.3 Strategies for Using Methanol and Its Blends in SI Engines

The various issues involved in utilizing methanol and its gasoline blends in SI engines are discussed in the previous section. These issues are mostly technical in nature.. Engineering solutions to most of the problems associated with methanol blends are already available in the literature and some of the solutions have been successfully implemented with on-road vehicles. The present section discusses the methods of adding gasoline methanol blends to SI engines and reviews the need for vehicle retrofitting for higher blends of methanol operation. Methanol and its gasoline blends can be introduced in SI engines in three ways:

- Blending,
- Dual fueling or fumigation, and
- Vehicle Modifications for methanol operation.

Blending. Methanol is being blended in various proportions with gasoline, as discussed in the previous sections. Usually, blends are prepared by using the following methods: (i) Splash blending, (ii) Sequential blending, (iii) Ratio blending, and (iv) Sidestream blending. Out of all these methods, side stream blending is advantageous and results in better mixing characteristics. In sidestream blending, the fuel blend to be added is directly metered to the main fuel lines and blending occurs upstream of the delivery meter. This process helps in the mixing of fuels thoroughly as the fuels pass through the meter.[37] Proper care must be exercised during the preparation of blends as improper blending results phase separation and in other adverse effects.

Dual fueling or fumigation. The problem of phase separation does not occur if methanol and gasoline blends are added separately. In a dual fueling system, distinct fuel systems are adopted for methanol and gasoline to meter these fuels as per the blending recipe [21]. This system requires both mechanical carburetors and fuel injection systems. In the mechanical system, two distinct carburetors are used for methanol and gasoline whose throttle opening is meticulously calibrated as per the blending ratio. In electronic injection systems, the engine control unit (ECU) is employed for appropriate metering of fuels as per the blending ratio. Despite the advantages, blends are mostly preferred over fumigation technology because of the economic aspects and complications in the design of dual fueling systems [22].

Vehicle Modifications for methanol operation. No modifications are required for SI engines with lower methanol-gasoline blends. However, retrofitting of the vehicle with dedicated methanol systems is suggested for improving the performance and drivability of higher methanol blends. SI engines modified for methanol operation are found to have increased power, lower specific fuel consumption, and low NO_x emissions. M85 blends are primed with n-heptane to reduce the cranking time and improve the cold starting capabilities [8]. Water sensitivity, methanol leaning effects, and cold start problems require high-speed electrical starters, dedicated fuel distribution systems, and modified carburetors for methanol-fueled vehicles [6]. As methanol fuel has lower stoichiometric ratios and high heat of vaporization, it cannot be used in SI engines designed for gasoline operation without extensive modifications. The emissions from gasoline–methanol blends vary based on the settings of the carburetor, rich and lean [11]. A direct injection glow plug with electronic control is developed to use neat methanol without losing drivability and cold starting. These modifications resulted in soot-free operation, compliance with stringent emission norms, and trouble-free start-up [23]. The increase in compression ratio from 8 to 10 in the SI engine resulted in improved combustion efficiency, decrease in CO and HC emissions for M50 blend [24]. Copper-coated piston crowns and cylinder head SI engine is used for M20 blends and found improved performance than conventional engines

---

[37] Blending of fuels http://info.smithmeter.com/literature/docs/tp0a015.pdf.

[25]. By employing ignition timing and EGR as load control strategy than conventional throttling, high performance and lower emissions are obtained at higher loads for pure methanol operations [26]. The high dilution tolerance of methanol facilitates higher EGR rates and by adopting variable valve timings, higher EGR rates can be effectively utilized. Besides, electrical heaters can be employed in catalytic convertors of methanol-fueled engines to deal with low light-off temperature [3]. To avail the complete advantages of methanol and its higher blends necessary modification in the current vehicle fleet is required. Along with lower methanol–gasoline blends which require no modifications in the engine, fumigation, and dedicated methanol engines should be considered as a long-term strategy for successful incorporation of methanol as an alternative fuel.

5 Methanol and Its Blends (Liquid and Gas) in SI Engines

Unlike gasoline, methanol, and its gasoline blends have superior qualities which result in enhanced performance and lower emission. The present section reviews the performance, emission, and combustion characteristics of neat methanol, methanol–gasoline blends, methanol–gaseous fuel blends, and methanol–water blends.

5.1 Effect on Performance

Methanol–liquid fuel blends. Methanol has the potential to improve the power, torque, and efficiency of SI engines. Ghent university conducted many tests on different engines and found that using methanol in SI engines resulted in enhanced power outputs by 10% compared to gasoline in the Audi engine. The use of methanol was found to have increased the loading capabilities and efficiency of an engine tuned to run on gasoline, by 10%. Methanol allows a high degree of load control with EGR and excess air (lean operation) in high compression ratio engines when compared to gasoline [27]. Using methanol blends increases the volumetric efficiency of the engine by cooling the incoming air and thereby increasing brake mean effective pressure. Tests conducted on various methanol–gasoline blends from M3 to M15 have shown M15 as the best blending combination with lower specific fuel consumption, high power, and efficiency [28, 29]. Experiments conducted on PFI engine with M10, M20, and M85 blends noticed that engine torque decreased slightly for M10 and M20 blends. The high burning heat capacity of methanol than gasoline results in low combustion temperature. As an effect, the losses of heat to the walls are reduced and efficiency is increased for methanol blends. The efficiency is increased by a percent of 10.85 for M85 in a PFI engine when compared to gasoline [30]. Alcoholic blends such as E5, E10, M5, and M10 are tested in a multi-point fuel injection (MPFI) engine on a chassis dynamometer. The results indicated a drop in brake-specific fuel consumption and a rise in brake thermal efficiency for M5 and M10 blends when compared with E5 and E10. This fine behaviour of methanol blends is attributed to the higher oxygen content of methanol than ethanol [31]. The increase of methanol content in the blend improves the volatile characteristics of blends and ensures better mixture preparation. This results in decreased flame developing period and fast-burning rates. This results in an increase of indicated mean effective pressure (IMEP) for the first 50 cycles of operation in a PFI engine during cold start [32]. The works of Elfasakhany on methanol–gasoline, ethanol–gasoline, and gasoline–ethanol–methanol blends showed that methanol–gasoline blends increased engine torque and volumetric efficiency when compared with other test fuels. The author concluded that to improve engine performance methanol should be adopted among other test fuels [33]. n-butanol is added to methanol–gasoline blends to match the vapour pressure as that of gasoline fuels. The results indicated improved volumetric efficiency and low compression work for butanol blends because of decreased exhaust gas temperature [34]. To maintain phase-stability of methanol–gasoline blends (M10 and M25) ethanol is added as a co-solvent. This study resulted in a similar fuel consumption characteristic of gasoline for the M10 blend, which contains 1.5 vol.% of ethanol. A drop in power and engine torque is observed with the increase of methanol content in the blend [13]. The tests conducted on unmodified three-cylinder PFI engine with methanol–gasoline blends resulted in a decrease in engine torque with the rise of methanol content in the blend. This study also showed that the performance of the methanol–gasoline blends could be improved by advancing the spark timings [35]. By using alternative load control strategies for methanol like EGR, the part-load efficiency is increased by 5%. With the application of the same strategy on the TDI high compression engine, the efficiencies rose to 42% without compromising on emissions [36]. Water is added to methanol in the proportions of 2.5% v/v-10% v/v to test the performance of the engine. The results have shown a significant increase in brake thermal efficiency for methanol-water blends than gasoline [37].

Methanol–gaseous fuel blends. Gaseous fuels such as hydrogen and LPG are added to methanol to improve the flame propagation and lean-burn efficiencies of SI engines. Hydrogen is added as port injection and methanol is added as a direct injection to the engine. The addition of 3-6% of hydrogen to methanol has increased the lean-burn limit of the engine. The tests conducted on hydrogen–methanol–gasoline mixtures have shown a decrease in brake-specific fuel consumption by 4% and an increase in thermal efficiency by 2% compared to gasoline. The addition of hydrogen to methanol SI engine at low speeds decreased indicated mean effective pressure (IMEP). In contrast, the addition of hydrogen at higher engine speeds resulted in improved IMEP [38–40]. To improve the cold start ability of methanol-fueled SI engine LPG is added to methanol. By adding LPG to methanol the in-cylinder temperatures and transient engine speeds are improved [41]. Various studies are conducted on the performance of methanol–gasoline, methanol–LPG, and methanol–hydrogen blends are presented in Table 3. In summary, the use of methanol and its blends with gasoline, LPG, and hydrogen opens up possibilities for achieving high efficiency, torque, and power. A setbacks of methanol fuels such as the increase in fuel consumption, drivability, and phase separation can be easily mitigated by adopting different strategies in fuel injection, ignition, and load control.

Table 3 Research summary on the performance of methanol and its blends in SI engines

Authors	Testing conditions	Injection type	Fuels used	Conclusions
Turner et al. [27]	Engine operated at MBT	PFI	M15, M100	Torque (↑), BTE (↑) for M100, Torque (↑), BTE (↑), BSFC (↑) for M15 when compared with gasoline
Abu-Zaid et al. [28]	@WOT, variable speeds in the range of 1000-2500		M3, M6, M9, M12, M15	Best performance is obtained for M15, for M15 blend: BTE (↑), BSFC (↓), Power output (↑)
Mishra et al. [29]	@ const. Torque and variable speed,	Carbureted	M5, M10, M15	Increase in methanol content in blends provides high brake power at higher speeds
Yanju et al. [30]	@ const. speed 3000 rpm	PFI	M10, M20, M85	BTE (↑), Torque (↓) for M10 and M20, Torque (↑) for M85
Eyidogan et al. [31]	Speed at 80 km/h and 100 km/h, wheel powers: 5,10,15 & 20 kW	MPFI	M5, M10	BSFC (↑), BTE (↑), methanol blends performance is better than ethanol blends
Hu et al. [32]	@ cold start and warm-up	PFI	M10, M20, M30	No starting problem for blends and mixture becomes leaner with (↑) in methanol content in the blend
Elfasakhany [33]	@ variable speeds 2600-3450 rpm; load: 1.3-1.6 kW	carbureted	M3, M7, M10	Volumetric efficiency (↑), Torque (↑)
Siwale et al. [34]	Spark timing: 18°-32° BTDC; Speed: 2500 rpm; $\lambda = 1.1$; Load (BMEP): 2.4-7.8 bars	MPFI	M20, M70, M53b17	BTE (↑) for M70 than M53b17; M53b17 exhibited low exhaust gas temperatures and higher volumetric efficiency

(continued)

Table 3 (continued)

Authors	Testing conditions	Injection type	Fuels used	Conclusions
Liu et al. [35]	Variable speed: 2000-5000 rpm; WOT condition	PFI	M10, M15, M20	Torque (↓), Power (↓), BTE (↑) for the increase of methanol content in the blend
J. Vancoillie et al. [36]	WOT lean-burn and WOT stoichiometric using EGR to control load	PFI	M100	Relative efficiency benefit about 10%; low temperatures; both WOT condition resulted in improved efficiency by 5%
Sileghem et al. [37]	$\lambda = 1$; Variable speed: 1500-3500 rpm; torque: 20-80 N-m; BMEP: 1.41-5.64	PFI	M97.5W2.5, M95W5, M90, W10	BTE is higher for methanol-water blends than gasoline
Gong et al. [38]	$\lambda = 1.4$; MAP = 93, 80, 68 kPa; Engine speed: 1200, 1800, 2400 rpm	PFI	Hydrogen % in methanol-hydrogen mixture 0,3, 6%	Addition of hydrogen increased indicated mean effective pressure
Gong et al. [42]	Engine speed: 1200 rpm; MAP: 0.07Mpa	DI methanol; PI hydrogen	Hydrogen % in methanol-hydrogen mixture 0,3, 6%	Addition of hydrogen enhances the lean-burn limit of pure methanol from 0.625 to 0.455
Yilmaz and Taştan [40]	Engine speed: 2000 rpm; A/f ratio: stoichiometric	DI	M5, M10, M15, Hydrogen is varied in 3,6,9,12 & 15%	Addition of hydrogen resulted in BSFC (↓) & BTE (↑)
Gonget al. [41]	Cold start at 8 °C	Dual-fuel port injection	M100 with LPG injection	The cold start performance of the engine can be improved by injecting LPG at 300° CA interval between methanol and LPG

5.2 *Effect on Emissions*

In this section, the effects of methanol–liquid fuel blends and methanol–gaseous fuel blends on the emission of SI engines are discussed. Utilization of methanol blends has so far proved effective in reducing the carbon emission of the engine, but there are some rising concerns regarding formaldehyde emissions. The effect of methanol gasoline blends on the efficiencies of catalytic converters is also briefly discussed in the present section. This section is an attempt to give an overview of emissions from methanol and its blends.

Carbon emissions: Addition of methanol to gasoline is found to reduce carbon emissions and helps in de-carbonizing the transportation sector. The engine speed is found to be an important parameter in influencing CO emissions of an SI engine. At lower speeds, CO emissions are low for methanol blends and at higher speeds, there is a rise in CO emissions. This rise in CO emissions with speed is attributed to the rise in BSFC with speeds [29]. Engine tests conducted with pure methanol and gasoline showed that CO_2 emissions are lower for methanol at all speeds and torques [27]. The lower CO emission for methanol and its blends is because of the low carbon content of methanol. Methanol contains only 44% of carbon whereas its counterpart gasoline contains about 86%. The presence of more oxygen in methanol blends also aids in reducing the carbon emission in the combustion phase itself [30]. Studies on carbon emission during the cold start or engine warm-up found that the cold start CO emission for methanol–gasoline blends are low and CO emissions are decreasing with the rise of methanol content in the blend [32]. The highest drop in CO emission is observed for methanol–gasoline blends when tests are conducted on ethanol, methanol, and ethanol–methanol gasoline blends. This drop in CO emission is found to be increasing with an increase in engine speed and methanol content. Further, the experiments on ethanol, methanol, and ethanol–methanol gasoline blends concluded that methanol gasoline blends are best at reducing CO emission when compared with the rest of the blends [33]. Experiments conducted on methanol–gasoline blends M20 and M70 show that CO emissions are decreasing with the rise in load and higher when compared with gasoline. This effect is attributed to the high latent heat of methanol which stops the oxidation of CO to CO_2 resulting in incomplete combustion [34]. High oxygen content of methanol helps in the decrease of CO emissions for methanol–gasoline blends. It is found that the emissions from methanol–gasoline blends are far better than gasoline after a three-way catalytic converter (TWC). This study also showed that the use of methanol–gasoline blends will not affect the conversion efficiency of TWC [35]. Emission analysis conducted on methanol blends in SI engines resulted in higher CO emission for methanol–water blends than pure methanol [37]. Methanol is enriched with hydrogen to find the effect of enrichment on performance and emissions. Hydrogen enrichment is carried with methanol in two ways: (a) by adopting port injection for both methanol and hydrogen (b) port injection for hydrogen and direct injection for methanol. Experiments concluded that port injection strategy for both methanol and hydrogen is better at reducing CO emissions [42]. Further experiments with hydrogen enrichment of methanol resulted in improved CO and CO_2 emissions as there is an enhancement in combustion with hydrogen [40]. Tests conducted on a chassis dynamometer with a vehicle with an MPFI engine resulted in lower CO emissions for methanol blends when the engine is running at 80 km/h. But when the vehicle is running at 100 km/h, a rise in CO and CO_2 is noted [43]. The addition of methanol to gasoline blends is found to be effective in reducing CO emissions in SI engines. This drop in CO emissions is due to the high oxygen content of methanol blends and also the lean operation [44].

From the literature, it is very well established that CO emissions are decreased than gasoline by using methanol and its blends than gasoline alone. CO emissions mostly occur due to the lack of oxygen availability during combustion. The high oxygen content in methanol aids in CO oxidation and complete combustion resulting in lower CO than gasoline.

HC emissions: Hydrocarbon emission appears in the tailpipe as a result of incomplete combustion. These emissions are possible because of flame quenching phenomena near the cold walls of the combustion chamber and fuel in crevices. Hydrocarbon emissions are found to be decreasing with the increase of methanol content in the blend. The amount of drop in hydrocarbon emissions with methanol content is found to be a function of the equivalence ratio [45]. Tests conducted on carbureted engines fueled with methanol–gasoline blends yield different results based on the setting of the carburetor. If the carburetor is calibrated to "rich", hydrocarbon emissions are decreased but when the setting is "lean" hydrocarbon emission increases or decreases depending on the equivalence ratio. It is also noted from the experiments that adding methanol to gasoline increased hydrocarbon emissions as the volatile characteristic of the blend is more than base fuel gasoline. This results in a large amount of unburnt fuel leaving the combustion chamber [11]. Tests conducted on PFI engine resulted in low HC emissions when compared with pure gasoline. The low HC emissions in PFI engines are attributed to the high oxygen content of methanol and better combustion [35]. The works on methanol-gasoline blends with n-butanol addition resulted in lower hydrocarbon emission when compared with gasoline and methanol–gasoline blends. The authors reported that the low C/H ratio of methanol–gasoline blend and high oxygen content are responsible for low HC emissions. When hydrogen is injected along with methanol–gasoline blends HC emissions are decreased. Hydrogen has low extinction distance, high flame temperatures, high diffusion rates which aid in further oxidation of HC, thereby reducing hydrocarbons in the exhaust [40].

To summarize, the high oxygen content and low C/H ratio of methanol–gasoline blends results in low hydrocarbon emissions than pure gasoline. Methanol is a simple molecule and it does not have any components unlike gasoline that burns at higher temperatures resulting in hydrocarbon emissions.

NO_x emissions: Nitric oxide and nitrogen dioxide are usually referred to as NO_x emissions. NO_x emissions are mainly formed because of the high-temperature reactions that happen during lightning and in IC engines. According to the European Union emission inventory report, about 40% of NO_x emissions are emitted from road transport. These NO_x emissions are dangerous and responsible for photochemical smoke and ground-level ozone. Formation of ozone at ground level has devastating effects on plant and human life. Exposure to NO_x emissions results in the inflammation of airways and lung infections.[38] To reduce the NO_x emissions from the engines after-treatment devices such as SCR and EGR are generally employed. The

[38] NO_x emissions http://www.icopal-noxite.co.uk/nox-problem/nox-pollution.aspx.

formation of NO_x and soot for methanol combustion is shown in Fig. 3. The upper shaded portion and lower shaded portion represent soot and NO_x formations, respectively, as functions of temperature and equivalence ratio. From the figure, it can be observed that the soot formation possibility for methanol operation is extremely low even for rich mixtures. Unlike soot, NO_x emissions formation for methanol operation is quite a possibility. Figure 3 also represents the steady rise in NO_x emissions formation with increase in temperatures and decrease in equivalence ratios [3, 46]. As mentioned earlier NOx emissions are direct functions of combustion temperature. The high latent heat of methanol aids in decreasing the combustion temperature and in reducing the NOx emission for methanol–gasoline blends [13]. Tests conducted on a spark-assisted direct injection methanol engine revealed that supercharging helps exhaust gas circulation at high loads decreasing NOx emission without compromising on efficiency [47]. The effect of excess air ratio on the NOx emissions of methanol and gasoline has shown that the NOx emissions are lower for methanol for all the λ values when compared with gasoline. In this study, NOx emissions are functions of oxygen concentration and cylinder temperatures. The results indicated that for the same λ (same oxygen concentration), NOx emissions are low for methanol. Low adiabatic flame temperature, intake charge cooling, and low cylinder temperatures of methanol are responsible for lower NOx emissions [48]. The effect of ignition advance angle on methanol–gasoline blends in reducing NOx emission is studied experimentally. Results indicated that the increase in ignition advance angle resulted in shorter combustion decreasing combustion temperatures and NOx emissions. This study proved that NOx emissions can be decreased by the use of methanol–gasoline blends along with appropriate ignition advances [49]. The alternative load control strategies investigated on methanol-fueled SI engines has shown that NOx emission can be effectively decreased by using EGR and spark timing load control technique than conventional load control [26]. Some studies investigate

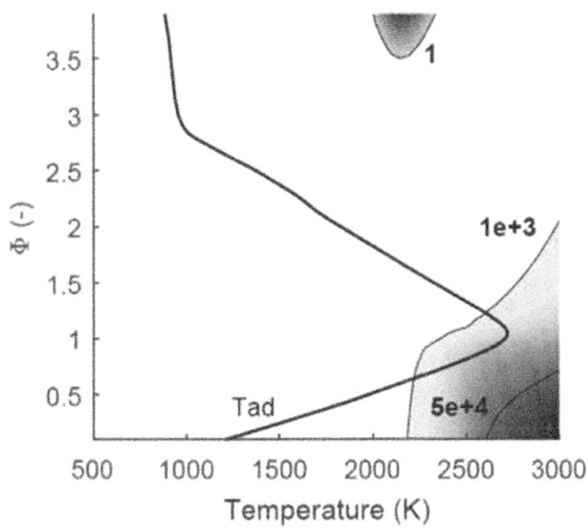

Fig. 3 Soot and NO_x formation plots as a function of temperature and equivalence ratio (*Source* Progress in Energy and Combustion science 70, p. 56)

the effects of hydrogen addition to methanol-gasoline blends. Results indicated that adding hydrogen to methanol–gasoline blends increased NOx emissions because of the higher flame temperature of hydrogen [40]. In summary, methanol and its blends reduce the NOx emission because of their inherent properties such as high latent heat, lower adiabatic flame temperatures, high oxygen content, and low in-cylinder temperatures. It is also noted from the literature that by using unconventional load control strategies such as EGR and ignition timings the NOx emissions can be further reduced. The in-cylinder handling of NOx emissions by using methanol and its blends also reduces the complexity of engine systems by avoiding expensive after-treatment devices.

Formaldehyde emissions: As discussed in Sect. 4.2, formaldehyde emissions are dangerous, carcinogenic, and have serious implications on human health if exposed for longer durations. A dual-mode combustion model is adopted to study the formation of formaldehyde emissions in methanol and its gasoline blends in SI engines. These simulation studies have shown that formaldehyde emissions increase with the methanol fraction in the blend [50]. Experimental works are carried out to find the effect of the quantity of methanol injection and ignition timing on formaldehyde emissions. The authors concluded that unburned methanol and formaldehyde emissions depend on combustion and firing. The results also have shown the opposite trend for unburned methanol and formaldehyde emissions. It is observed from the experiments that by optimizing the quantity, timing of methanol injection, and ignition timings the formaldehyde emissions can be reduced [19]. Engine tests are conducted on methanol–gasoline blends to assess the formaldehyde emissions using Fourier transform infra red (FTIR), high-performance liquid chromatography (HPLC), and gas chromatography (GC). Results indicated that formaldehyde emissions are lower for blends when loads are in the range of low to medium and increase when loads are increased from medium to high [51]. The studies conducted to find the effect of engine speed and three-way catalytic converter (TWC) on formaldehyde emission revealed that formaldehyde emissions increase with engine speed. This experiment also revealed that the functioning of TWC is different for formaldehyde emissions than regulated emissions such as CO, HC, and NO_X. TWC is found to be effective with formaldehyde with low engine speeds and negative conversion efficiencies are recorded at higher engine speeds [52]. From the studies, it can be concluded that optimizing methanol content, injection timing, ignition timing, engine speed, and TWC formaldehyde emissions can be reduced to a great extent. The research summary on emissions from SI engines fueled with methanol and its blends is shown in Table 4.

5.3 Effect on Combustion

Combustion plays a vital role in the performance and emission of fuels in internal combustion engines. The processes of combustion and quality of combustion determine the emissions from the tailpipe. In the present section, the effect of methanol and its blends on the combustion of SI engine is reviewed. Studies conducted on

Table 4 Summary of the literature on methanol blend emissions in SI engine

Authors	Injection type	Fuel types	Critical conclusions
Mishra et al. [29]	Carburetor	M5, M10, M15	NO_x (↓), CO, andCO_2 are lower at low speeds and higher at higher speeds. HC (↓)
Yanju et al. [30]	PFI	M10, M20, M85	CO (↓), NO_x (↓), THC (↓)
Hu et al. [32]	PFI	M10, M20, M30	CO (↓), HC (↓)
Liu et al. [35]	PFI	M10, M15, M20, M25, M30	CO (↓), HC (↓), NO_x (↓)
Elfasakhany [33]	Carburetor	M3, M7, M10	CO (↓), HC (↓), CO_2 (↑)
Canakci et al. [43]	MPFI	M5, M10	CO (↓), CO_2 (↓), HC (↓) & NO_x (↓) at a speed of 80 km/h
Rice et al. [44]	–	M20	CO (↓), NO_x (↓), unburned fuel (UBF) emissions for M20 (↑)
Sapre [45]	Manifold injection	Pure Methanol, M15, M30, M50, M70, M85	NO_x (↓), UBF (↑), HC (↓), Formaldehyde (↑), CO (↓) at $\phi > 1$
Geng et al. [51]	–	M15, M45	Formaldehyde emission (↑) with methanol content
Balki and Sayin [53]	carburetor	Pure Methanol	Pure methanol is tested at different compression ratios and found that methanol is effective in reducing CO, HC, and NO_x emissions when compared with ethanol and gasoline

methanol and its gasoline blends revealed that the in-cylinder pressure of methanol is higher when compared with gasoline and ethanol. The rise in cylinder pressure is due to the high volumetric efficiency, oxygen content, and high laminar flame speeds of methanol. The heat release rates of methanol have lower values and earlier combustion than gasoline because of higher flame speeds, cooling effect, and single boiling point of methanol [53]. The cylinder pressures and heat release rates are decreasing with the increase of EGR content due to the high heat capacity of exhaust gases and low temperatures in the cylinder because of recirculation. The effect of initial flame kernel radius on combustion quality of methanol is studied and found that the increase of initial flame kernel radius improves methanol combustion quality with appropriate throttle opening [54]. The studies on cold start combustion reported misfire cycles followed by firing cycles for gasoline and M20. These misfire cycles are found to be diminished for the engine when it is fueled with M30. Further, the

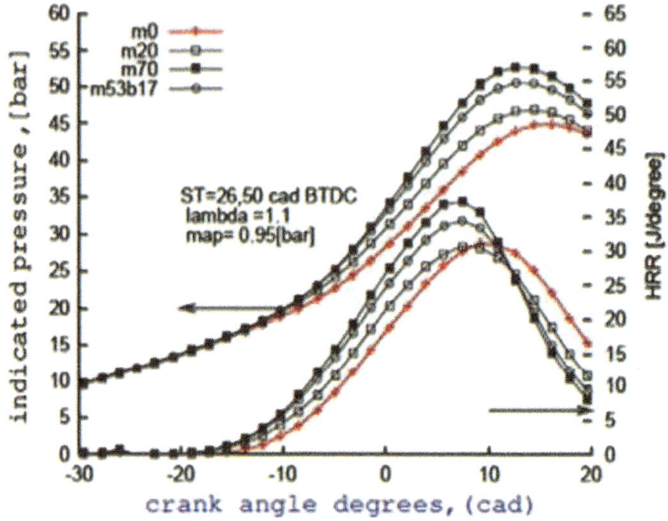

Fig. 4 Effect of methanol blending on indicated pressure and heat release rates (*Source* Fuel processing technology 118, p.322)

study on cyclic variations of combustion with pure gasoline and methanol–gasoline blends indicated improvement in cyclic variation of combustion with the increase of methanol content in the blends. Due to the fast flame speeds and burning rates, methanol–gasoline blends have shorter combustion duration and high cylinder pressures [32]. The heat release rates and indicated pressure of methanol–gasoline blends are found to be higher when compared with gasoline as shown in Fig. 4. From Fig. 4, it can also be noticed that there is a rise in indicated pressure and heat release rate with an increase of methanol content in the blend. Tests conducted on the M70 blend have shown fast-burning rates which resulted in improved combustion efficiency and reduced emissions [34].

Due to the dilution effects on the pre-mixed mixtures, misfires happen for gasoline and lower gasoline-methanol blends. Because of the lower boiling points of methanol, methanol blends can vaporize quickly and can form a better combustible mixture than gasoline. This mixture helps in reducing the dilution effects in preparing the pre-mixed mixture for combustion. The high Reid vapour pressure, lower boiling point, and high oxygen content of methanol results in a shorter flame developing period. Short flame developing period and fast burning rates aids better mixture formation and combustion [55]. Simulation studies are conducted on methanol-enriched hydrogen fuels. From the results, it is identified that turbulent flame speeds are increased with hydrogen addition to methanol during lean conditions. This increase of turbulent flame speeds improves the combustion efficiency of the methanol–air mixture, thereby reducing emissions [56]. To summarize, adding methanol to gasoline would result in higher cylinder pressures and heat release rates. This rise in in-cylinder pressures and heat release is attributed to the fast flame developing and burning characteristics of methanol. An increase of methanol fraction in the blends improves mixture formation and reduces the effects of dilution during cold start. Due to the fast-burning characteristics, methanol blends have their peak cylinder pressure close to TDC than gasoline, improving the efficiency of combustion.

5.4 Effect of High Compression and Knock

The thermal efficiency of SI engines increases with the increase in compression ratio. The only limitation in the SI engine to achieve high compressions is knock. Knock in SI engine imposes a severe limitation on other efficiency improvement techniques such as supercharging and turbocharging. The high latent heat of methanol helps in reducing the cylinder temperatures and facilitates high compression ratios, supercharging, and turbocharging. These effects help in achieving the down-sizing of the SI engine without compromising on efficiency and power. Tests conducted on a single-cylinder carbureted SI engine revealed that gasoline tends to knock when the compression ratio is increased from 6:1 to 8:1 at full throttle openings and low speeds. But no knock is observed for methanol for compression ratios ranging from 8:1 to 10:1 at full throttle and all speeds. The random formation of peaks on P - θdiagram indicate knocking. From Fig. 5, random peaks can be noticed for gasoline depicting knock and smooth curve for methanol indicating neat combustion [57].

The studies conducted to find the knock-limited compression ratio (KLCR) for various methanol–gasoline blends indicated that KLCR increased from 7.7 to 9.9 when methanol fraction increased to 50% in the blend [45]. Generally,

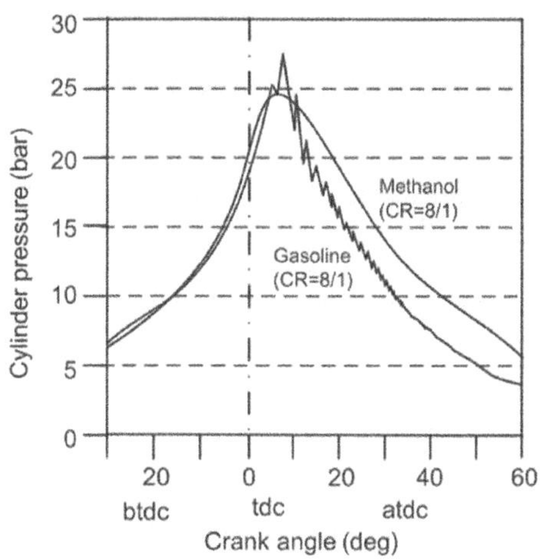

Fig. 5 P - θ diagram for gasoline and methanol indicating knock (*Source* Fuel 90, p. 1594)

methanol-fueled engines are not prone to knocking, but due to low ignition temperature, methanol–gasoline-fueled engines are prone to surface ignition. Cool flames appear before the knocking combustion in engines. However, neither cool flame not knocking combustion was observed in methanol-fueled engines [21]. Further, it is identified that laminar burning velocities, low air–fuel ratios, low adiabatic-flame temperature of methanol–gasoline blends help in improving inlet charge cooling and reducing the knocking tendency of methanol-fueled engines [27]. Methanol blends of iso-octane are studied numerically to find the effect of methanol in knocking of SI engines. From the results, it is identified that the low intensity of chemical reactions, heat release rates, and temperatures reduced the flame front and pressure wave interactions reducing the knock. The results indicated that the time for chemical reactions in the mixture increases when the methanol fraction in the blend increases from 10% to 15%. The studies further identified that, if the temperatures in the cylinder are more than 950 K the rate of chemical reaction increases creating a slight knock [58]. The multi-dimensional analysis of high compression ratio SI engine revealed that retarded spark timing, high EGR rates, rich and lean mixture effects, combustion chamber shape are essential in suppressing knock [59]. The use of high EGR rate and early spark timing has good knock resistance capabilities with less trade-off on thermal efficiency than low EGR rates and later spark timings in high compression methanol engines [60]. A compression ratio of 17.5 is used in a methanol engine simulated model to find the effect of flame kernel radius and EGR on performance and emissions. The results indicated decreased NO_x and CO emissions with an increase of initial flame kernel radius when the throttle is between 10% to 70% [54]. Knocking with detailed chemical kinetics is studied in a 17.5:1 compression ratio methanol engine using large-eddy simulations (LES). Results have shown the negligible concentration of HCO radical in knocking combustion [61]. A high compression ratio of 19.5:1 methanol engine is developed by Environmental Protective Agency's National Vehicle and Fuel Emission Laboratory. This engine demonstrated negligible knock at high loads and improved efficiencies than conventional engines [62]. The experiments conducted on Volvo T3 turbocharged DISI engine revealed that methanol is good at suppressing the knock. The results also indicated that BMEP of methanol engine can be increased to high values without knock. Further, this anti-knock capability of methanol was found to downsize the engine by ~ 10.7% by increasing the intake pressure [63]. In summary, it is found that the use of methanol can effectively reduce knocking in SI engines. The simulation studies conducted on methanol-fueled high compression SI engine revealed that high EGR rates and spark timings as efficient techniques to control knock for high compressions. A research summary on performance, emissions and combustion for various methanol blends in PFI and carbureted engines is shown in Table 5.

From Table 5, it is established that methanol and its gasoline blends can be effectively utilized in both carbureted and PFI engines. However, PFI engines have better control over the timing and quantity of fuel injection. PFI methanol engines accompanied with high compression ratios resulted in higher efficiencies over a wide operating range than any other conventional engine [62]. The high latent heat of methanol

Table 5 Research summary on performance, emission, and combustion in PFI and Carbureted Engines

Parameters	PFI engines		Carbureted Engines	
	Blends	Conclusions	Blends	Conclusions
Performance	M10, M20, M85	Brake thermal efficiency and torque increased for M10 and M20 while only torque increased for M85 [30]	M5, M10, M15	Increase in methanol content in blends provides high brake power at higher speeds [29]
	M10, M15, M20, M25, M30	Torque and power slightly decreased for methanol blends, whereas brake thermal efficiency is increased. In this study, SI engine cold start ability is improved by the addition of methanol [35]	M3, M7, M10	Utilization of methanol blends resulted in improved brake power, torque, and volumetric efficiency [33]
	M100	High compression ratio (19.5:1) methanol engine exhibited a peak brake thermal efficiency of 43% and high efficiency operating range. These values are found to be better than comparable turbocharged diesel engines [62]	M100 Gasoline E100	Pure methanol showed higher brake mean effective pressure and brake thermal efficiency when compared to gasoline and pure ethanol [53]
Emissions	M10, M20, M85	CO (↓), NO_x (↓), THC (↓). M85 blend reduced CO and NO_x emissions by 25% and 80%, respectively. [30]	M5, M10, M15	NO_x (↓), CO, and CO_2 are lower at low speeds and higher at higher speeds. HC (↓) [29]
	M100	CO, HC, and NO_x emissions decreased to very low values by using conventional after-treatment system and methanol. NO_x ~ 0.1-0.2 g/kW-hr; HC < 0.2 g/kW-hr; CO < 0.2 g/kW-hr [62]	M100	Utilizing methanol in passenger cars showed HC (↑), CO (↑), NO_x (↓), PM (↓), and formaldehyde emissions (↑) [64]

(continued)

Table 5 (continued)

Parameters	PFI engines		Carbureted Engines	
	Blends	Conclusions	Blends	Conclusions
	M10, M15, M20, M25, M30	CO (↓), HC (↓), NO_x (↓), and unregulated emissions formaldehyde and unburned methanol increase with the increase of methanol in blend. [35]	M15	THC and CO emissions decreased by 11%-34.5% and 63-84%, respectively. Whereas NO_x and formaldehyde emissions are increased by 76.9–107.7% and 16.4%– 52.5%, respectively [65]
Combustion	M10, M20, M85	The cylinder pressures are increasing with methanol content in the blend. The cylinder pressure for M85 is found to be 17.6% higher than gasoline. The burnt heat capacity of methanol is also found to be higher than gasoline [30]	M100, E100, gasoline	Pure methanol showed higher cylinder pressure and combustion efficiency compared to gasoline. The heat release rates are found to be lower for methanol [53]
	M10, M15, M20, M25, M30	Rise in cylinder pressures is observed for methanol blends. Start of combustion is advanced and the rapid burning phase is shortened with methanol blends [35]	M3, M7, M10	The combustion efficiency is improved with methanol–gasoline blends due to the high oxygen content of methanol [33]

results in poor atomization of fuel by carburetor jets. This issue effect the combustion of methanol in SI engines. Further, the stoichiometric air–fuel ratios of methanol blends are lower than gasoline. So, the use of methanol and its blends in carbureted engines tuned for gasoline operation results in incomplete combustion and drivability issues [22].

6 Catalytic Converters for Methanol-Fueled SI Engines

Methanol-fueled SI engines are free from sulphur contamination and this helps in extending the life of catalytic converters by reducing the effects of sulphur poisoning. But the use of methanol or methanol–gasoline blends significantly affects the conversion efficiency of catalytic converters due to their low-temperature exhaust. As discussed in Sect. 5.2, TWC exhibited negative conversion efficiencies for methanol blends owing to low exhaust gas temperatures. The aldehyde emissions from gasoline engines are low and high temperature in the exhaust of gasoline engine helps to reduce these aldehyde emissions. Using methanol results in low gas temperatures in the exhaust resulting in low conversion efficiencies of convertors and high emissions. So the following modifications are suggested to improve the conversion efficiency of catalytic converters: (a) the catalytic converter needs to be moved near the exhaust to make use of high exhaust temperatures, (b) dedicated catalysts designs which suits for methanol operation, (c) efficient fuel control and engine calibration, and (d) addition of excess air in the upstream of the convertor during cold start. By employing the above measures, a significant reduction in tailpipe emissions is noted [66]. Methanol-fueled vehicles need to be retrofitted with new TWC for efficient reduction of emissions. A range of tests conducted on methanol–gasoline blends proved that the vehicle retrofitted with new TWC resulted in lower emissions than OEM TWC. Moreover, the specific reactivity of retrofitted TWC is reduced for M15 blends than gasoline, stating that the emissions from methanol–gasoline blends are lower than gasoline [67].

7 Conclusions

In this chapter, a detailed review of methanol and its blends is showcased starting from their historical background to the consequences of using methanol in the present SI engines. The advantage of producing methanol from a wide variety of sources makes it a viable choice as an alternative fuel. Decades of experience in handling methanol as a chemical commodity ensures that methanol is safe like any other chemicals. The arguments on toxicity and poisonous emissions from methanol are technical and can be handled easily by using suitable technology, as discussed in this chapter. In this review, the limitation of methanol such as phase separation, cold starting, formaldehyde emissions, and poor lubrication are discussed at length. The engineering solutions to some of the drawbacks of using methanol as SI engine fuel are presented in this chapter. The use of methanol and its blends of gasoline in SI engines have shown significant improvement in brake thermal efficiency, volumetric efficiency, and exhaust gas temperatures. SI engine fueled with methanol blends has higher cylinder pressures when compared with pure gasoline which would result in higher brake thermal efficiency. The low exhaust gas temperatures improved the volumetric efficiency of methanol engines than gasoline. Methanol blends are effective in reducing emissions from petrol engines. The studies conducted on different configurations of SI engines fueled with methanol proved that methanol is effective in reducing emissions such as CO, HC, and NO_x. It is also identified that by using suitable engine parameters, unregulated emissions such as formaldehyde can be controlled effectively in methanol engines. Methanol engines require custom-made catalytic converters to handle emissions effectively. Three-way catalytic converters with a suitable catalyst, proper placement, and with excess air in the exhaust have lower emissions for methanol blends than pure gasoline. The decrease in specific reactivity of TWC for methanol blends is also a good sign indicating a reduction in emission during the combustion itself.

In summary, methanol is a highly effective SI engine fuel that can reduce emissions and improve the performance of the engine. The use of methanol at least as blends could decrease the load on gasoline and helps in reducing fuel imports. To satisfy future energy needs and to meet the goal of reduced carbon emissions, methanol is the only viable alternative for SI engines. From this review, it can be stated that methanol is a sustainable SI engine fuel to cater to the needs of the present and the future as it can be produced from the abundant coal reserves in India. The timing is also ripe since Indian economy is expected to witness exponential growth in the immediate future.

Acknowledgements The authors would like to thank Prof. D. Varadaraju and B. Sridhar Reddy for their kind suggestions and support throughout the course of this book chapter.

References

1. Nichols RJ (2003) The methanol story: a sustainable fuel for the future. J Sci Ind Res (India) 62:97–105
2. Mårald E (2010) Methanol as future fuel: efforts to develop alternative fuels in Sweden after the oil crisis. Hist Technol 26:335–357. https://doi.org/10.1080/07341512.2010.523175
3. Verhelst S, Turner JW, Sileghem L, Vancoillie J (2019) Methanol as a fuel for internal combustion engines. Prog Energy Combust Sci 70:43–88. https://doi.org/10.1016/j.pecs.2018.10.001
4. Vancoillie J, Sileghem L, Van De Ginste M, Demuynck J, Galle J, Verhelst S (2012) Experimental evaluation of lean-burn and EGR as load control strategies for methanol engines. SAE Technical Paper. https://doi.org/10.4271/2012-01-1283
5. Machiele PA (1987) Flammability and toxicity tradeoffs with methanol fuels. SAE Technical Paper. SAE International
6. Ingamells JC, Lindquist RH (1975) Methanol as a motor fuel or a gasoline blending component. SAE Technical Paper. https://doi.org/10.4271/750123
7. Chen X, Wang H, Song C, Wang W, Huang J, Liu S, Wei Y (2014) Investigation of the cold-start engine performance at a low temperature for an engine fuelled with alternative fuel. Proc Inst Mech Eng Part D J Autom Eng 228:310–318. https://doi.org/10.1177/0954407013511070
8. Gardiner DP, Caird SG, Bardon MF (1987) Spark ignition engine cold starting with methanol-based fuel blends. SAE Technical Paper. https://doi.org/10.4271/872067
9. Nakajima Y, Saito T, Takagi Y, Katoh K, Lijima T (1978) The influence of fuel characteristics on vaporization in the S.I. engine cylinder during cranking at low temperature. SAE Technical Paper, pp 2234–2250. https://doi.org/10.4271/780612
10. Adler J, Harvey AC (1987) Vehicle diveability with gasoline/alcohol blends. U.S Environmental protection agency
11. Brinkman ND, Gallopoulos NE, Jackson MW (1975) Exhaust emissions, fuel economy, and driveability of vehicles fueled with alcohol-gasoline blends. SAE Technical Paper. https://doi.org/10.4271/750120
12. Dahlander P, Gutkowski A, Denbratt I (2008) Visualization of fuel sprays for stratified cold starts in gasoline direct injection engines. In: Proceedings of the 22nd European conference on liquid atomization and spray systems - ILASS 2008. Como Lake, Italy, pp 1–7

13. Qi DH, Liu SQ, Liu JC, Zhang CH, Bian YZ (2005) Properties, performance, and emissions of methanol-gasoline blends in a spark ignition engine. Proc Inst Mech Eng Part D J Automob Eng 219:405–412. https://doi.org/10.1243/095440705X6659
14. Amine M, Zahran MAH, Awad EN, El-Zein SM, Barakat Y (2017) Influence of ethyl acetate addition on volatility, octane rating, and phase stability of methanol–gasoline blends. Pet Sci Technol 35:601–607. https://doi.org/10.1080/10916466.2016.1271809
15. Yuen PK, Villaire W, Beckett J (2010) Automotive materials engineering challenges and solutions for the use of ethanol and methanol blended fuels. SAE Technical Paper. https://doi.org/10.4271/2010-01-0729
16. Buck w. H, LLohuls JR, Murphv JA (1989) Lubrication Studies In a Methanol-Fueled spark Ignition Engine. SAE Tech Pap
17. Chadha BR, Gupta M, Gupta LR, Maji S, Singhal S (1992) Some experimental studies of lubricants in a two-stroke methanol engine. J Synth Lubr 8:295–306. https://doi.org/10.1002/jsl.3000080405
18. Shukla DS, Gondal AK, Nautiyal PC (1992) Effect of methanol fuel contaminants in crankcase oils on wear. Wear 157:371–380. https://doi.org/10.1016/0043-1648(92)90073-H
19. Li J, Gong C, Wang E, Yu X, Wang Z, Liu X (2010) Emissions of formaldehyde and unburned methanol from a spark-ignition methanol engine during cold start. Energy Fuels 24:863–870. https://doi.org/10.1021/ef9009982
20. Jahan K, Mahmood D, Fahim M (2015) Effects of methanol in blood pressure and heart rate in the rat. J Pharm Bioallied Sci 7:60–64. https://doi.org/10.4103/0975-7406.148747
21. Kowalewicz A (1993) Methanol as a fuel for spark ignition engines: a review and analysis. Proc Inst Mech Eng Part D J Automob Eng 207:43–52. https://doi.org/10.1243/PIME_PROC_1993_207_158_02
22. Valera H, Singh AP, Agarwal AK (2020) Prospects of methanol-fuelled carburetted two wheelers in developing countries. In: Singh A, Sharma N, Agarwal RAA (eds) Advanced combustion techniques and engine technologies for the automotive sector, energy. Springer, Singapore, pp 53–73
23. Hilger U, Jain G, Scheid E, Pischinger F, Bruetsch R, Bernhardt W, Heinrich H, Weidmann K, Rogers G (1990) Development of a direct injected neat methanol engine for passenger car applications. SAE Technical Paper. SAE International
24. Nuthan Prasad BS, Pandey JK, Kumar GN (2020) Impact of changing compression ratio on engine characteristics of an SI engine fueled with equi-volume blend of methanol and gasoline. Energy 191: https://doi.org/10.1016/j.energy.2019.116605
25. Murali Krishna MVS, Kishor K, Gupta AVSSKS, Murthy PVK, Narasimha Kumar S (2012) Performance of copper coated spark ignition engine with methanol-blended gasoline with catalytic converter. J Renew Sustain Energy 4:543–548. https://doi.org/10.1063/1.3664743
26. Xie FX, Li XP, Wang XC, Su Y, Hong W (2013) Research on using EGR and ignition timing to control load of a spark-ignition engine fueled with methanol. Appl Therm Eng 50:1084–1091. https://doi.org/10.1016/j.applthermaleng.2012.08.003
27. Turner JWG, Lewis AGJ, Akehurst S, Brace CJ, Verhelst S, Vancoillie J, Sileghem L, Leach F, Edwards PP (2018) Alcohol fuels for spark-ignition engines: performance, efficiency and emission effects at mid to high blend rates for binary mixtures and pure components. Proc Inst Mech Eng Part D J Autom Eng 232:36–56. https://doi.org/10.1177/0954407017752832
28. Abu-Zaid M, Badran O, Yamin J (2004) Effect of methanol addition on the performance of spark ignition engines. Energy Fuels 18:312–315. https://doi.org/10.1021/ef030103d
29. Mishra PC, Gupta A, Kumar A, Bose A (2020) Methanol and petrol blended alternate fuel for future sustainable engine: a performance and emission analysis. Meas J Int Meas Confed 155: https://doi.org/10.1016/j.measurement.2020.107519
30. Yanju W, Shenghua L, Hongsong L, Rui Y, Jie L, Ying W (2008) Effects of methanol/gasoline blends on a spark ignition engine performance and emissions. Energy Fuels 22:1254–1259. https://doi.org/10.1021/ef7003706
31. Eyidogan M, Ozsezen AN, Canakci M, Turkcan A (2010) Impact of alcohol-gasoline fuel blends on the performance and combustion characteristics of an SI engine. Fuel 89:2713–2720. https://doi.org/10.1016/j.fuel.2010.01.032

32. Hu T, Wei Y, Liu S, Zhou L (2007) Improvement of spark-ignition (SI) engine combustion and emission during cold start, fueled with methanol/gasoline blends. Energy Fuels 21:171–175. https://doi.org/10.1021/ef0603479
33. Elfasakhany A (2015) Investigations on the effects of ethanol–methanol–gasoline blends in a spark-ignition engine: Performance and emissions analysis. Eng Sci Technol an Int J 18:713–719. https://doi.org/10.1016/j.jestch.2015.05.003
34. Siwale L, Kristóf L, Bereczky A, Mbarawa M, Kolesnikov A (2014) Performance, combustion and emission characteristics of n-butanol additive in methanol-gasoline blend fired in a naturally-aspirated spark ignition engine. Fuel Process Technol 118:318–326. https://doi.org/10.1016/j.fuproc.2013.10.007
35. Liu S, Cuty Clemente ER, Hu T, Wei Y (2007) Study of spark ignition engine fueled with methanol/gasoline fuel blends. Appl Therm Eng 27:1904–1910. https://doi.org/10.1016/j.applthermaleng.2006.12.024
36. Vancoillie J, Demuynck J, Sileghem L, Van De Ginste M, Verhelst S, Brabant L, Van Hoorebeke L (2013) The potential of methanol as a fuel for flex-fuel and dedicated spark-ignition engines. Appl Energy 102:140–149. https://doi.org/10.1016/j.apenergy.2012.05.065
37. Sileghem L, Huylebroeck T, Van Den Bulcke A, Vancoillie J, Verhelst S (2013) Performance and emissions of a SI engine using methanol-water blends. SAE Tech Pap 2:xx. https://doi.org/10.4271/2013-01-1319
38. Gong C, Li Z, Chen Y, Liu J, Liu F, Han Y (2019) Influence of ignition timing on combustion and emissions of a spark-ignition methanol engine with added hydrogen under lean-burn conditions. Fuel 235:227–238. https://doi.org/10.1016/j.fuel.2018.07.097
39. Gong C, Li Z, Yi L, Liu F (2020) Experimental investigation of equivalence ratio effects on combustion and emissions characteristics of an H2/methanol dual-injection engine under different spark timings. Fuel 262: https://doi.org/10.1016/j.fuel.2019.116463
40. Yilmaz İ, Taştan M (2018) Investigation of hydrogen addition to methanol-gasoline blends in an SI engine. Int J Hydrogen Energy 43:20252–20261. https://doi.org/10.1016/j.ijhydene.2018.07.088
41. Gong C, Liu Z, Su H, Chen Y, Li J, Liu F (2019) Effect of injection strategy on cold start firing, combustion and emissions of a LPG/methanol dual-fuel spark-ignition engine. Energy 178:126–133. https://doi.org/10.1016/j.energy.2019.04.145
42. Gong C, Li Z, Yi L, Liu F (2019) Comparative study on combustion and emissions between methanol port-injection engine and methanol direct-injection engine with H2-enriched port-injection under lean-burn conditions. Energy Convers Manag 200: https://doi.org/10.1016/j.enconman.2019.112096
43. Canakci M, Ozsezen AN, Alptekin E, Eyidogan M (2013) Impact of alcohol-gasoline fuel blends on the exhaust emission of an SI engine. Renew Energy 52:111–117. https://doi.org/10.1016/j.renene.2012.09.062
44. Rice RW, Sanyal AK, Elrod AC, Bata RM (1991) Exhaust gas emissions of butanol, ethanol, and methanol-gasoline blends. J Eng Gas Turbines Power 113:377–381. https://doi.org/10.1115/1.2906241
45. Sapre AF (1988) Properties, performance and emissions of medium concentration methanol-gasoline blends in a single -cylinder spark-ignition engine. SAE Technical Paper
46. Svensson E, Li C, Shamun S, Johansson B, Tuner M, Perlman C, Lehtiniemi H, Mauss F (2016) Potential levels of soot, NOx, HC and CO for methanol combustion. SAE Technical Paper 2016-01-0887. https://doi.org/10.4271/2016-01-0887
47. Sato Y, Noda A, Sakamoto T (1997) Combustion and NOx emission characteristics in a di methanol engine using supercharging with EGR. SAE Technical Paper. https://doi.org/10.4271/971647
48. Wu B, Wang L, Shen X, Yan R, Dong P (2016) Comparison of lean burn characteristics of an SI engine fueled with methanol and gasoline under idle condition. Appl Therm Eng 95:264–270. https://doi.org/10.1016/j.applthermaleng.2015.11.029
49. Tian Z, Zhen X, Wang Y, Liu D, Li X (2020) Comparative study on combustion and emission characteristics of methanol, ethanol and butanol fuel in TISI engine. Fuel 259: https://doi.org/10.1016/j.fuel.2019.116199

50. Zhang F, Shuai S, Wang Z, Zhang X, Wang J (2011) A detailed oxidation mechanism for the prediction of formaldehyde emission from methanol-gasoline SI engines. Proc Combust Inst 33:3151–3158. https://doi.org/10.1016/j.proci.2010.07.029
51. Geng P, Zhang H, Yang S, Yao C (2015) Comparative study on measurements of formaldehyde emission of methanol/gasoline fueled SI engine. Fuel 148:9–15. https://doi.org/10.1016/j.fuel.2015.01.075
52. Wei Y, Liu S, Liu F, Liu J, Zhu Z, Li G (2009) Formaldehyde and methanol emissions from a methanol/gasoline-fueled spark-ignition (SI) engine. Energy Fuels 23:3313–3318. https://doi.org/10.1021/ef900175h
53. Balki MK, Sayin C (2014) The effect of compression ratio on the performance, emissions and combustion of an SI (spark ignition) engine fueled with pure ethanol, methanol and unleaded gasoline. Energy 71:194–201. https://doi.org/10.1016/j.energy.2014.04.074
54. Zhen X, Li X, Wang Y, Liu D, Tian Z, Wang Y (2020) Effects of the initial flame kernel radius and EGR rate on the performance, combustion and emission of high-compression spark-ignition methanol engine. Fuel 262: https://doi.org/10.1016/j.fuel.2019.116633
55. Song R, Hu T, Liu S, Liang X (2008) Combustion characteristics of SI engine fueled with methanol-gasoline blends during cold start. Front Energy Power Eng China 2:395–400. https://doi.org/10.1007/s11708-008-0081-7
56. Ji C, Yang J, Liu X, Zhang B, Wang S, Gao B (2016) A quasi-dimensional model for combustion performance prediction of an SI hydrogen-enriched methanol engine. Int J Hydrogen Energy 41:17676–17686. https://doi.org/10.1016/j.ijhydene.2016.07.146
57. Çelik MB, Özdalyan B, Alkan F (2011) The use of pure methanol as fuel at high compression ratio in a single cylinder gasoline engine. Fuel 90:1591–1598. https://doi.org/10.1016/j.fuel.2010.10.035
58. Feng H, Wei J, Zhang J (2019) Numerical analysis of knock combustion with methanol-isooctane blends in downsized SI engine. Fuel 236:394–403. https://doi.org/10.1016/j.fuel.2018.09.026
59. Zhen X, Wang Y, Xu S, Zhu Y (2013) Study of knock in a high compression ratio spark-ignition methanol engine by multi-dimensional simulation. Energy 50:150–159. https://doi.org/10.1016/j.energy.2012.09.062
60. Zhen X, Wang Y, Xu S, Zhu Y (2013) Numerical analysis on knock for a high compression ratio spark-ignition methanol engine. Fuel 103:892–898. https://doi.org/10.1016/j.fuel.2012.10.023
61. Zhen X, Wang Y, Zhu Y (2013) Study of knock in a high compression ratio SI methanol engine using les with detailed chemical kinetics. Energy Convers Manag 75:523–531. https://doi.org/10.1016/j.enconman.2013.07.001
62. Brusstar M, Stuhldreher M, Swain D, Pidgeon W (2002) High efficiency and low emissions from a port-injected engine with neat alcohol fuels. SAE Technical Paper
63. Nguyen DK, Van Craeynest T, Pillu T, Coulier J, Verhelst S (2018) Downsizing potential of methanol fueled DISI engine with variable valve timing and boost control. SAE Technical Paper, pp 1–13. https://doi.org/10.4271/2018-01-0918
64. Lawrence R. Smith CMU, TMB (1982) Unregulated exhaust emissions from methanol-fueled cars. SAE Trans 91:3171–3183
65. Li L, Ge Y, Wang M, Li J, Peng Z, Song Y, Zhang L (2015) Effect of gasoline/methanol blends on motorcycle emissions: exhaust and evaporative emissions. Atmos Environ 102:79–85. https://doi.org/10.1016/j.atmosenv.2014.11.044
66. Windawi H (1992) Controlling the exhaust emissions from alternative fuel vehicles. Platin Met Rev 36:185–195
67. Zhao H, Ge Y, Tan J, Yin H, Guo J, Zhao W, Dai P (2011) Effects of different mixing ratios on emissions from passenger cars fueled with methanol/gasoline blends. J Environ Sci 23:1831–1838. https://doi.org/10.1016/S1001-0742(10)60626-2

A Brief Review on Methanol-Fuelled Vehicles (MFV) in China and Implementation Policy

Kai Zhao, Gregory Dolan, and Chris Chatterton

1 Introduction

1.1 Current Status

Beginning in the early 1980s and into 2020, China has consistently demonstrated unwavering support for efforts to develop methanol fuel and later, methanol-fuelled vehicles. At the time of publishing this book, the largest methanol fuel production, consumption, as well as methanol vehicle componentry and manufacturing in the world can be found in China. Today, there are over 21,000 M100 (100% or "neat" methanol) methanol-fuelled vehicles in operation in China including passenger cars, buses, and trucks. These MFVs are certified for use under a targeted government-sponsored program. Two cities in China have predominantly shared popularity for the initial rollout and scale up: Gui Yang, the capital city of Gui Zhou province with 11,674 passenger vehicles and Xi An, the capital city of Xi An province with 8,124 passenger vehicles. All the passenger vehicles are commercialized in the local taxi fleets and manufactured by Geely Auto, a leading Chinese domestic automotive OEM. In the meanwhile, it is estimated another 100,000 retrofitted gasoline vehicles which are capable of high proportion methanol fuel (M85 and M100) on the road.

K. Zhao (✉)
Beijing Office, Methanol Institute, Level 26 Fortune Financial Center, no. 5 Dongsanhuan Rd, Chaoyang District, Beijing, China
e-mail: kzhao@methanol.org

G. Dolan
Washington DC Office, Methanol Institute, Washington, DC, USA

C. Chatterton
Singapore Office, Methanol Institute, Singapore, Singapore

A. K. Agarwal et al. (eds.), *Methanol*, Energy, Environment, and Sustainability,
https://doi.org/10.1007/978-981-16-1224-4_6

1.2 Rationale for China to Develop MFV

Energy Security. Methanol as a transportation fuel was first extensively researched and demonstrated as a result of the Oil Embargo of the 1970s in the US. In China, energy security (i.e. high dependence on imported crude oil) is also the first rationale for the development of methanol as a fuel. China became a net crude oil importer since 1993, and the overseas import dependence has been increasing since that time. In 2018, China consumed 126 million metric tons (MMTs) of gasoline and 156 MMTs of diesel, based on the consumption of 903 MMTs of crude oil, which was the first-time foreign dependency escalated beyond 70% [1]. From an energy resources perspective, the energy distribution can be described as "short on oil, lean on gas but rich in coal" for China. According to the BP Statistical Review of World Energy 2020 [2], by the end of 2019, the proven total world coal reserves are 1069.636 billion tons with a reserve to production ratio (R/P ratio) of 132. China has 141.595 billion tons of proven coal reserves, where the R/P ratio is 37, accounting for 13.2% in the world, the fourth largest of any country. By the end of 2019, the proven total world crude oil reserves are 1733.9 billion barrels, with an R/P ratio of 49.9. China has 26.2 billion barrels of proven crude oil reserves, with an R/P ratio of 18.7, accounting for 1.5% of the world total. By the end of 2019, the proven total world natural gas reserves are 198.8 trillion cubic meters,with the R/P ratio of 49.8. China has 8.4 trillion cubic meters of proven natural gas reserves, the R/P ratio is 47.3, accounting for 4.2% of the world.

Although the Chinese government is working to gradually decrease coal dependence within the total energy mix, coal will continue to play an important role in chemical production and energy supply. There are many reasons for this but primarily are centred on energy security, social norms as well as economics. The coal sector employs over 2.5 million people as of April 2020, down from over 5.2 million in 2013 [3], The country planned to reduce the proportion of coal in its energy mix to approximately 57.5% by 2019, according to a guideline released by the National Energy Administration. The proportion was only slightly off at 57.7% in 2019, down from 59.2% in 2018 in the power sector. Figures 1 and 2 show the world and China's

Fig. 1 World Primary Energy Consumption by Fuel (at end of 2019, exajoules) [2]

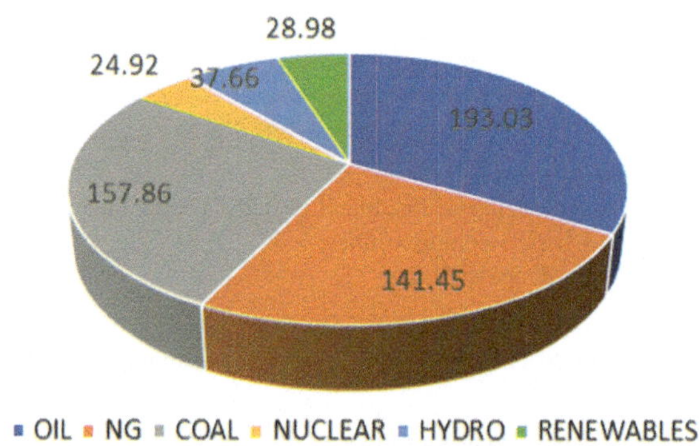

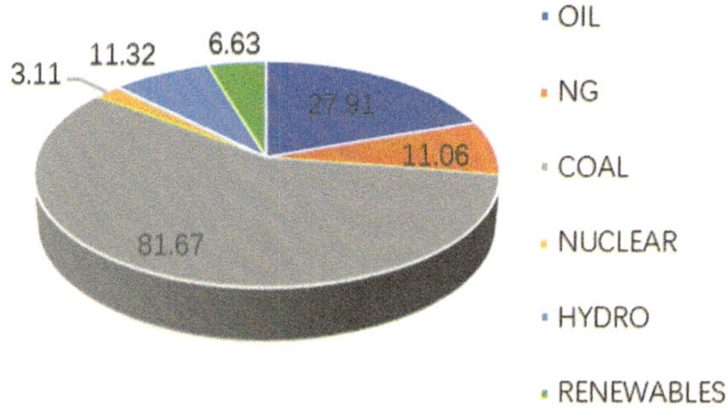

Fig. 2 Chinese Primary Energy Consumption by Fuel (at end of 2019, exajoules) [2]

primary energy consumption by fuel, it appears clearly that China still has a much higher coal dependency than the world average.

Although China is positioning itself as one of the leading countries in renewable energy production, offsetting to some degree its continued investments in coal China is still the largest importer of crude globally, increasing in 2019 to an average of 10.1 million barrels per day (b/d), which is an increase of 0.9 million b/d from the 2019 average [4]. Due to recent, increased tension between the US and China, the impact of COVID-19 and falling energy prices, China will continue to utilize coal long-term as a means to mitigate some of this volatility, which brings greater visibility to its well-developed manufacturing sector. Thus, utilizing the domestic coal supply to produce methanol as a transportation fuel, replacing crude, as an extension of that risk management process, is a practical and realistic strategy for China.

Clean Air Challenge from the Automobile Industry in China. Since 2009, China has become the world's largest automobile producer and consumer. According to the Chinese Ministry of Ecology and Environment [5], the total combined population of motors vehicles reached 327 million in 2018, with total emissions of 40.6 million tons of the 4 major pollutants (CO, hydrocarbon (HC), NOx, and particulate matters (PM)), shown in Fig. 3. Although the situation is improving as the total emissions decreased by 4% over the period 2013–2018, despite a sharp increase of motor vehicles of 32.7%. Further emissions reductions will require more expensive technologies coupled with decommissioning more vehicles with lower emission standards.

Based on the simplest alcohol molecule with a single carbon bond, Methanol Flexible Fuel Vehicles (FFV) were successfully introduced in California during the period of the early 1980s to the mid-1990s as a workaround for the 1970s Middle East Oil Embargo. China faces a similar technical challenge today, hence switching from conventional fuel to methanol can improve the situation by either introducing new technology based on methanol or retrofitting existing vehicles.

Local Economy Growth. As China is a major crude importer, most refineries are built in the coastal regions and the refined products are then distributed into the mainland. Not limited to petroleum and chemistry, coastal provinces in China have the same advantage in other sectors such as manufacturing and these provinces have developed better-performing economies than the mainland or inner regions of China. For the inner provinces in China, utilizing locally sourced feed to produce

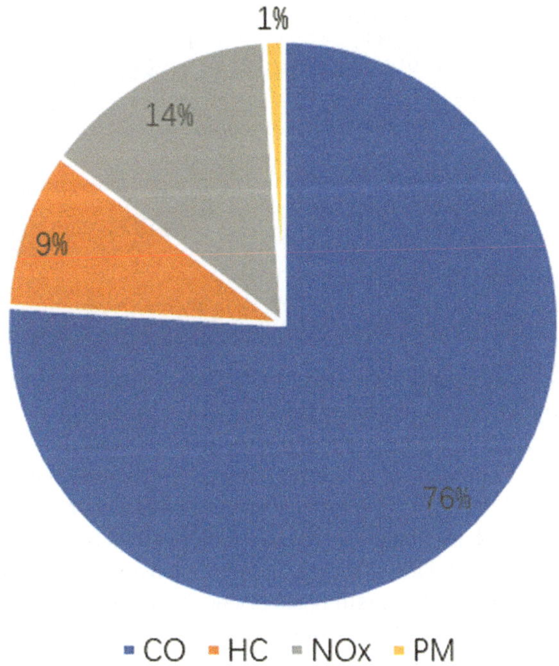

Fig. 3 The 4 Major Pollutants from Mobile Sources of China, 2018 [5]

methanol either from domestic coal or natural gas and later consumed in the local fuel pool greatly assisted in supporting a number of industries, jobs, income streams, and contributions to the local government. As a result, some provinces in China have issued various policies to further support and encourage coal chemistry and methanol-fuel-related commercial activities. Examples of such "coal-centricity" can be found in Shanxi, Gui Zhou, Shannxi, and Gansu, many of which have MFV production facilities built by Chinese automakers.

2 Development of Methanol-Fuelled Vehicles in China: A Brief History

2.1 Research and Pilots in 1980s–2000s, Mainly in Shanxi Province

Early Stage Experiments and Tests. During this period of time, methanol engines and fuel were researched, with a very limited number of vehicles for testing purposes.

Methanol-blended gasoline (M15, a blend of 15% methanol and 85% gasoline) was first researched from 1982 to 1985. A total of 475 medium-sized trucks and 14 methanol refueling stations were deployed in the experimental pilot in Shanxi province, accumulating 80 million kilometers over the course of the pilot. The blending work expanded to the whole province of Shanxi with provincial government support with peak consumption of 800,000 tons of methanol M15 in over 1,000 gasoline stations in 11 cities, in 2013. Today, M15 is used in a number of regions in China like Shanxi, Shannxi, Gui Zhou, Yun Nan, Gan Su, Jiangsu, and Zhe Jiang.

Between 1986 and 1990, a project of "Applied Technology of M100 Methanol Fuel in Spark Ignition Engines" under the National Key Sci-tech Projects was conducted on 4-cylinder engines. The selected areas of research included: (i) improving the methanol engine compression ratio, (ii) controlling maximum combustion pressure, (iii) energy recovery through preheating, (iv) selecting an appropriate spark plug, and (v) using an improved and optimized methanol carburetor. All these initiatives served to improve the thermal efficiency of methanol engines and to reduce the methanol-to-gasoline substitution ratio to below 1.8:1 (note that methanol has roughly one-half the energy content of gasoline).

Further, the active study of the ignition properties and cold-start mechanisms of methanol engines aided in developing the semiconductor ceramics for PTC (positive temperature coefficient) electric heaters and in parallel, the control systems to resolve issues related to cold starting with fuel formulation. This work included carburetion calibration to meet the requirements of burning M100 methanol fuel and experimenting with the material compatibility specifications used in the M100 methanol fuel supply system in total. Other notable findings from this study resulted from the testing of engine lubricating oil, adjusting engine working conditions, and improving lubricating oil quality to lower both the abrasion and erosion levels to that of gasoline-fuelled engines.

Between 1987 and 1994, a formal "M100 Methanol Automobile Joint Test and Research" program was conducted together with the Federal Republic of Germany. The Project was jointly initiated and supported by the then Ministry of Technology of the Federal Republic of Germany and the then China Science & Technology Committee. During the earlier stages of the Project (1987–1991), Germany's Volkswagen (hereinafter referred to as VW) provided twelve, M100 methanol engines. Under the guidance of VW experts and "Shanghai Volkswagen", a joint venture between Shanghai Automotive and VW, eight of the twelve Santana M100 methanol cars were assembled for testing. The other four were used for dynameter testing. In the later stages of the Project (1992–1994), VW produced another six methanol engines, related key components, and one Multi-Fuel Vehicle engine, which could run methanol and gasoline in different ratios.

From 1996 to 1999, China and the US completed a "3E Research of Methanol Fuel" program, also called, Life Cycle Evaluation of Transforming Coal in Shanxi Province and other Areas into Vehicle Fuels, of Economic Environment and Energy Influence. Ford and the Massachusetts Institute of Technology jointly formed a project which saw Ford donated a methanol, Flexi-fuel engine to China, which was then fully assembled and installed into a minibus for testing.

Engine Development of Vehicle Pilot. In this stage, engines and vehicles were jointly developed, followed by a small-scale commercial operation, usually referred to as a "pilot" project in China.

From 1997 to 2000, Shanxi Province launched the first methanol vehicle pilot with 55 6- to 7-meter methanol buses, with different engine developers. Further, 40 more buses were produced and put into operation, 15 of which were the "6600 Mid-Size Bus" model. This model was manufactured and used for carrying passengers and tourists from Jinzhong City to Taiyuan City and from Jinzhong City to Pingyao County. By 2000, these 55 demonstration buses had clocked 4.94 million km, with the highest mileage of a single bus being 350,000 km.

Vehicle Demonstration, more advanced engines, and additional vehicle models were demonstrated on a commercial basis. From 2003, the China National Action Plan of Clean Vehicles—Methanol Vehicle Demo Project demonstrated 120, 6-cylinder, methanol-fuelled, 10-meter buses. Out of the fleet, 20 of those methanol buses were operated across a larger, more climatically diverse territory within the province, proving again the feasibility of methanol fuel and vehicles and their advantages in technology, economy, and environmental protection.

2.2 MIIT Methanol-Fuelled Vehicle Pilot 2010–2018

Although numerous experiments, tests, and later vehicle demonstrations had been conducted, the development of MFV was not considered as the main technology solution by top decision-makers in China. The Chinese government structure had also evolved in parallel with the years of development of MFVs. As part of this evolution, the Ministry of Industry, and Information Technology (MIIT) was established in 2008 and took the role of supervising the automobile industry. From 2010, a more thorough review on a larger scale for MFVs took shape.

By 2012, similar to other new vehicle fuel types, MIIT decided to conduct a national pilot for methanol-fuelled vehicles to include the three provinces of Shanxi, Shannxi, and Shanghai. The pilot was enlarged in 2014 with two more provinces; Gansu and Guizhou. The detailed history and some of the key results of the pilot are summarized by Zhao [6], but it is also worth briefly mentioning a few breakthroughs in the pilot:

Largest in Scale. In the pilot, 1,024 vehicles included taxies, buses, and heavy trucks—all of which were commercially operated under regular service and demonstrated across 10 municipalities of China.

Definition and Normal Certification of MFV. In the pilot, MIIT certified 32 MFV models from 9 Chinese vehicle manufacturers for operation on methanol fuel. This was the first time that "MFV" was defined as any vehicle in compliance with Chinese standards for automobiles equipped with specially designed engines suitable for M85 and M100, or methanol-diesel dual fuel. When spark-ignited engines are used, the vehicle is single fuel, with methanol the main fuel for regular driving and gasoline is only used for emergency or ignition. In this case, the gasoline tank shall not be over 15L in volume. When a compression ignition engine is used, the vehicle is considered "dual fuel", whereas diesel is used for ignition and diesel-methanol dual fuel is the main fuel for regular driving conditions.

Certification acts as an ID card for the vehicle, with the model name and other important information for the entire useful life of the vehicle. Whereas prior to the pilot, all the MFVs were for the specific purpose of testing and not certified, although the local government gave the special permission to let the vehicles in operation. In the event of actual commercial application, the MFVs then were certified and licensed as "gasoline-fuelled vehicles". Such vehicles might have difficulties passing the annual vehicle quality check, which is compulsory from the public security departments. Although only within the pilot regions, the pilots served to open the door to MFVs, so that such vehicle fuel types can be certified and obtain vehicle registration plates as methanol fuelled for commercial sale, thus establishing a foundation and paving the way for future promotion.

Under the definition and regulated procedure, retrofitted gasoline-fuelled vehicles to methanol-fuelled vehicles cannot be regarded as an MFV in China as they are originally certified as gasoline model. However, retrofit kits are widely available in the Chinese market, and such a kit can be as cheap as 100 US dollars. Such a conversion not only provides fuel cost savings by sharing the same filling infrastructures but also educates the consumers which benefit the market awareness of MFVs in China.

Final Judgement on Human Health Impact. Although some tests were conducted during the different period of time when MFVs were under development over the past several decades, the issue of methanol poisoning still exists within the decision-making process. In the pilot stage, with support from the Methanol Institute, the health conditions of 1,199 people from a variety of occupations with potential methanol exposure were evaluated and no human health issues were found. Methanol vapour levels were also checked to confirm if the results were within the national standards regarding methanol concentrations of Permissible concentration-Time Weighted Average, PC-TWA 25 mg/m3, and Permissible concentration-Short Term Exposure Limit, PC-STEL 50 mg/m3.

3 Current Development and MFV Policy in 2019

3.1 Current Chinese Methanol Vehicle Development: Scaling, Technology Pathways, Model Specification Information

Current Promotion and Operation. As of the date of publishing this book, the successful MFV pilot has already begun scaling the project nationally, with the number of pilot vehicles now totaling more than 21,000. Most piloting cities have retained the current fleets of their MFVs and have plans to further scale the application locally, such as the City of Gui Yang. New cities have recently joined the campaign, including the City of Xi An and Bi Jie. The scale of the promotional campaign (i.e. current vehicle type, numbers, and application regions) is listed in Table 1.

Table 1 Methanol-Fueled Vehicles in China (by the end of July 30, 2020)

Province	City	Vehicle type	Number of vehicles
Shanxi	Jinzhong	Taxi	260
	Changzhi	HD Bus	8
Shannxi	Xi'An	Taxi	8,124
	Baoji	Taxi	585
		Mini MPV	15
	Yulin	Dump Truck	5
	Hanzhong	Taxi	20
Guizhou	Guiyang	Taxi	11,674
	Bijie	Taxi	150
Gansu	Lanzhou	Taxi	150
	Pingliang	Taxi	50
		Total	**21,041**

The promotional campaigns are mainly centred in coal-rich provinces with the M100 taxi manufactured by Geely Auto being the most popular vehicle type, which is shown in Fig. 4.

The heavy-duty bus manufactured by Yu Tong Bus Co. and dump trucks manufactured by Shannxi Auto are shown in Fig. 5 and Fig. 6, respectively.

Fig. 4 Geely Auto M100 Taxi in Xi An

Fig. 5 Zhengzhou Yu Tong M100 Bus in Chang Zhi

Fig. 6 Shannxi Auto Methanol Diesel Dump Trucks

Technology Routes. Under the definition of Chinese MFVs introduced above, there are two major technology pathways for MFVs in China.

M100 in Otto Cycle usually with gasoline as pilot fuel for cold starting in winters. In this pathway, spark ignition, methanol is the only fuel after spark-ignited either by methanol itself or gasoline. Such a design also yields extra mileage with additional gasoline which can be useful, especially as the filling infrastructure is not well established in the early stages of market development. If M85 is used in winter, gasoline is then blended in the methanol and such a modification is not required. Another technology that is currently under development with the aim of avoiding the gasoline piloting is methanol fractionation, where methanol is catalytically fractured to CO and H_2 with lower ignition energy requirement, which allows for more efficient ignition. M100 Otto cycle technology can be found in Geely's passenger vehicles and heavy-duty trucks, as well as Yu Tong Buses, with M100 heavy-duty engines produced by FAW JY Engine Co.

Diesel Methanol Compound Combustion (DMCC) in diesel cycle, invented by a team from Tianjin University of China led by Prof. Yao Chunde [7, 8]. In this pathway, diesel is the ignition fuel and methanol is injected during the combustion from the intake manifold. This technology is different from traditional Port Fuel Injection (PFI), which only allows for fuel and air mixing in the intake manifold. DMCC has two fuel injection systems; first, diesel is injected directly into the combustion chamber from ignition to normal working load and second, methanol is injected from the intake manifold mixed with air. The injection volume of diesel and methanol is tuned by the engine control unit (ECU) to maximize the methanol consumption without loss of power. DMCC also allows for a pure diesel working mode in the event of an absence of methanol fuel infrastructure. Working in diesel cycle (compression ignition) is low cost for both manufacturing and maintenance, which leads to superior competitiveness in commercial vehicles for the heavy-duty truck segment. DMCC is also available as a retrofit kit.

As vehicle electrification is rapidly developing globally, debates are ongoing regarding Internal Combustion Engine (ICE) versus Battery Electric Vehicles (BEVs) in China. Valera and Agarwal [9] comprehensively discussed an optimized strategy for India to choose between methanol for ICEs and BEVs based on a number of critical issues such as fundamental energy resources, infrastructure, cost of manufacturing, and ownership. Today China has the world's largest BEV market, which has continuously evolved through 2019, with 0.972 million BEV sales out of a total vehicle sales volume of 25.8 million, which represents a 1.2% drop from that of 2018 [10]. Due to the well-known drawbacks of BEVs, incentive policies from the Chinese government were required to sustain the rapid growth, in the form of subsidies to BEVs, preferable road policies, and special green plates which allowed for unlimited traffic limits and no need for car plate lottery for private consumers in some cities of China. The Chinese government has decided to prolong the cash subsidy to BEVs which was scheduled to end in 2020 for another two years to further support the BEV industry in China.

Unlike BEVs, methanol ICEs are another and perhaps the only, alternative in China which can be implemented at scale without subsidies. MFVs do not have cost increases compared to traditional fuel vehicles but have greater potential to meet the stringent emission regulations. The existing fuel distribution networks in China can be optimized and re-purposed with minor changes, whereas charging facilities require completely new build facilities and renovation of local electrical grids. Methanol can practically and efficiently meet the demand for greater fuel savings for private consumers and support the government in achieving the multiple targets of national energy security, environment protection, and GHG reduction by introducing renewable methanol. This is more important for diesel replacement especially in developing countries as Valera and Agarwal discussed [11].

Methanol and BEV do not instinctively repel each other. An extension of the methanol engine technology can be found in electric vehicle platforms, similar to gasoline, since methanol can also work as a range extension fuel for BEVs. Optimized methanol-fuelled engines used to power on-board generator and/or battery by producing electricity in a hybrid configuration. When storage capacity is considered, methanol has a higher energy density than many other fuels, namely compressed natural gas@200 bar, lithium batteries, and compressed hydrogen@ 700 bar [12]. It is efficient to combust methanol to produce electricity for BEVs as a range extension fuel, which also shows better emission performance than traditional fuels like gasoline. Figure 7 shows the schematic design of a methanol range extended truck developed by Olah Motors of China. An 18,000 kg gross weight electric heavy truck is being developed, which has a 700 km range with a battery capacity of 90.5kwh and a 400L methanol fuel tank. The company also designed and manufactured an integrated starter generator (ISG), which resolved and greatly enhanced cold weather ignition properties.

Vehicle Development. Since the MFV pilot was initiated by MIIT, MFVs have been continuously developed and certified. Table 2 lists all such models with manufacturer, model announcement no. and year from the pilot till the present. Considering Shanghai Maple, Zhejiang Haoqing, and Geely Commercial Vehicles are both

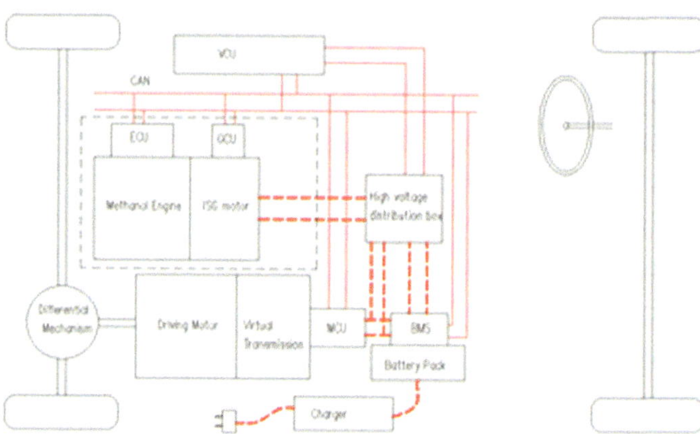

Fig. 7 Schematic design of methanol engine for EV range extension (with permission from Olah Motors)

Table 2 List of certified MFVs in China

Item	Manufacturer	Vehicle Type	Model	MIIT announcement no.	Announced year
1	SHANGHAI MAPLE	Methanol Passenger Car	SMA7151K05M	240	2012
2	SHANGHAI MAPLE	Methanol Passenger Car	SMA7181K04M	240	2012
3	SHANNXI AUTO	Methanol/Diesel Dual Fuel Self-Dumping Truck	SX3317	243	2012
4	SHANNXI TONG JIA AUTO	Lorry Truck	STJ5023X	244	2013
5	SHANNXI TONG JIA AUTO	MPV	STJ6404	244	2013
6	ZHENGZHOU YU TONG BUS	Methanol City Bus	ZK6100	246	2013
7	ZHEJIANG HAOQING AUTOMOBILE	Methanol Passenger Car	JL7152K13M	247	2013
8	ZHEJIANG HAOQING AUTOMOBILE	Methanol Passenger Car	JL7182K04M	247	2013
9	SHANXI CHENG GONG AUTO	MPV	SCH6400	251	2013
10	ZHEJIANG HAOQING AUTOMOBILE	Methanol Passenger Car	JL7152K16M	264	2014
11	ZHEJIANG HAOQING AUTOMOBILE	Methanol Passenger Car	JL7182K05M	264	2014
12	SINO TRUCK	Methanol/Diesel Dual Fuel Self-Dumping Truck	ZZ3257N 3847D 1 M	267	2014
13	SINO TRUCK	Methanol/Diesel Dual Fuel Self-Dumping Truck	ZZ3317N 4667D 1M	267	2014
14	SINO TRUCK	Methanol/Diesel Dual Fuel TrailerTruck	ZZ4257N 3247D 1BM	267	2014

(continued)

Table 2 (continued)

Item	Manufacturer	Vehicle Type	Model	MIIT announcement no.	Announced year
15	SINO TRUCK	Methanol/Diesel Dual Fuel TrailerTruck	ZZ4257N 3247D 1HM	267	2014
16	ZHEJIANG GEELY AUTO	Methanol Passenger Car	MR7152L30M	274	2015
17	ZHEJIANG GEELY AUTO	Methanol Passenger Car	MR7182L12M	274	2015
18	ZHEJIANG HAOQING AUTOMOBILE	Methanol Passenger Car	JL7152L01M	275	2015
19	ZHEJIANG HAOQING AUTOMOBILE	Methanol Passenger Car	JL7182L01M	275	2015
20	ZHEJIANG HAOQING AUTOMOBILE	Methanol Passenger Car	HQ7182L01M	276	2015
21	ZHEJIANG HAOQING AUTOMOBILE	Methanol Passenger Car	HQ7182L02M	279	2015
22	ZHEJIANG HAOQING AUTOMOBILE	Methanol Passenger Car	JL7152L02M	279	2015
23	ZHEJIANG HAOQING AUTOMOBILE	Methanol Passenger Car	JL7182L02M	282	2016
24	ZHENGZHOU YU TONG BUS	Methanol City Bus	ZK6100JG1	288	2016
25	ZHEJIANG GEELY AUTO	Methanol Passenger Car	SMA7151K05M	292	2017
26	ZHEJIANG GEELY AUTO	Methanol Passenger Car	SMA7181K04M	292	2017
27	ZHEJIANG HAOQING AUTOMOBILE	Methanol Passenger Car	JL7152K13M	292	2017
28	ZHEJIANG HAOQING AUTOMOBILE	Methanol Passenger Car	JL7152K16M	292	2017
29	ZHEJIANG HAOQING AUTOMOBILE	Methanol Passenger Car	JL7182K04M	292	2017

(continued)

Table 2 (continued)

Item	Manufacturer	Vehicle Type	Model	MIIT announcement no.	Announced year
30	ZHEJIANG HAOQING AUTOMOBILE	Methanol Passenger Car	JL7182K05M	292	2017
31	GEELY SI CHUAN COMMERCIAL VEHICLE CO	Methanol City Bus	DNC6105	299	2017
32	GEELY SI CHUAN COMMERCIAL VEHICLE CO	Methanol City Bus	JHC6105	300	2017
34	ZHEJIANG GEELY AUTO	Methanol Passenger Car	MR7182, MR7152	307	2018
36	ZHEJIANG GEELY AUTO	Methanol Passenger Car	MR7182, MR7152	320	2019

affiliated to Geely Auto, it is obvious that Geely Auto dominates the current MFV development from passenger cars to buses. Some additional models are being developed by other Chinese automakers, although they have not yet certified or announced them at the time of publication of this chapter.

3.2 Joint Ministerial Government Policy in 2019

With approved provincial/municipal pilot results reviewed by 2018, a milestone marking a new stage of the development of Chinese MFV was introduced with a policy paper issued in 2019: "Guidance for Developing Methanol Vehicle Applications in Some Parts of China" [13]. The policy was jointly supported by eight Chinese ministries and administrations, including MIIT and other important ministries such as the National Development and Reform Commission (NDRC). The policy recognizes the rationale to develop MFVs in China and encourage further development of MFVs, including:

Encourage Development of Manufacturing Capabilities of Methanol-Fuelled Vehicles, including passenger cars, commercial heavy-duty vehicles, off-road vehicles, and machinery; Critical and dedicated components for fuel injection, after treatment, filters, and lubricant oil, corrosion-resistant components; New technologies such as methanol-electric hybrid vehicles, methanol range-extended electric vehicles, and methanol fuel cell vehicles.

Promote Development of Production and Fueling Systems of Methanol Fuel, including methanol production from low calorific coal, coal bed methane, cooking gas, and exploring renewable methanol from CO2. Fuel methanol should comply with the National Standard of "Fuel Methanol for Motor Vehicles (GB/T23510-2009)" to ensure quality. Planning for the expansion of methanol fueling stations according to local conditions along with related standards and guidelines.

Accelerate Development of a Standards System, including methanol vehicles, engines, dedicated lubricant oils, and standard methanol fuel. Encourage international standards formulation and improve standards on methanol fuel and fuel delivery systems, including design and construction of methanol fueling stations, methanol fuel safe operational procedures and best practices as well as additive technology for methanol as a fuel.

Encourage Promotion of Methanol-Fuelled Vehicles, accelerate the deployment of M100 vehicles especially in the regions of Shanxi, Shannxi, Guizhou, and Gansu together with methanol production resources and leveraging onto operational experiences. Encourage the use of methanol in both passenger vehicles, service sectors, taxi fleets, and other commercial vehicle segments such as municipal duty and logistics fleets.

Execute Emission Standards Firmly, Methanol passenger vehicles should comply with China National VI emission standard, with methanol and formaldehyde emission limits of 2.5 mg/km respectively; Methanol heavy-duty vehicles should comply with China National VI emission standard of heavy-duty diesel vehicles with methanol and formaldehyde emission limits of 20 mg/kwh, respectively.

Other Promotional Measurements include normalizing methanol vehicle registration with no limitations. Additional research is required to include methanol vehicles into "Passenger car producer average fuel consumption and new energy vehicle credits", (also referred to as "Dual Credit Policy"), which is discussed later.

3.3 Future Consideration of MFV in "Dual Credit Policy"

Although significant progress has been made on MFV in China, it is hard to say MFV is the only vehicle technical solution or methanol is the only alternative fuel to answer all the challenges discussed in this chapter. China also has the largest global deployment of natural gas fuelled vehicles (NGV) and battery electric vehicles. As a result, the development of MFV is only beginning—MFV is and will, compete with other fuels and technology solutions in the land transportation vertical. The Chinese government is utilizing policy tools to include the Dual Credit Policy to encourage automakers to produce vehicles aligned with its national strategy.

The Dual Credit Policy: Passenger Cars Corporate Average Fuel Consumption (CAFC) and New Energy Vehicle (NEV) Credit Regulation was introduced in 2016 by MIIT, aiming to shape the automobile energy consumption by improving fuel economy and promotion of electrification to major Chinese automakers for their products both produced in China and imported. Like the US's CAFE, in CAFC, the Chinese government sets certain fuel economy targets for conventional energy vehicles (for ICE technology), but Chinese credit schemes allow automakers to offset their negative credits (if targeted credits cannot be met) not only by introducing fuel-saving technologies, but also by producing NEVs such like BEVs and Fuel Cell Vehicles (FCVs).

The design of the credits allocation and time span is still under discussion, for MFVs which do not enjoy or require cash subsidies from the Chinese central government, a proper credit for MFV can greatly assist automakers to consider introducing MFV models to improve the credit balance. From a longer term perspective, cash subsidies for EVs will eventually be canceled. If the battery industry cannot further reduce its production cost, producing an MFV could be a more attractive option as MFVs are not subject to notable production cost increases.

3.4 Other Policy Influences

Inspired by the MIIT's pilot and policy, some local governments in China echo the policy by providing preferential policies to MFVs including cash subsidies and other supportive rules facilitating MFVs' drivers. As a result, companies like Geely Auto have built MFV manufacturing facilities in the regions to take advantage of this. By far, Geely has established MFV manufacturing facilities in Jin Zhong, Gui Yang for passenger vehicles, and Nan Chong for commercial vehicles, with a total annual production capacity of over 300,000 units.

The province of Gui Zhou issued a provincial promotional plan in 2019, targeting 10,000 vehicles by end of 2019. The municipality of Xi An City has announced local policies to promote their own methanol fleets including vehicle subsidies, reduced driving lane limits during the peak hours.

Other regions like Shanxi, Tian Jin, and Gan Su issued policies welcoming MFVs in taxi and other logistic fleets.

4 Other Factors for Chinese MFV's Taking up in China

4.1 Chinese Methanol Industry and Methanol Supply

China is the largest methanol producer and consumer in the world. According to the China Nitrogen Fertilizer Industry Association (CNFIA) [14]. In 2019, the installed capacity of methanol in China was 89.9 MMTs, with a production of 69.9 MMTs of methanol. Methanol demand over the same period was 80.0 MMTs, with imports of approximately 10 MMTs from Iran, Saudi Arabia, Oman, and New Zealand. The growth in production capacity, volume, and imports of methanol in China for the past two decades is shown in Fig. 8. As noted, the methanol data includes methanol produced from coal and directly utilized for olefin production, also referred to as "coal to olefin" (CTO). Methanol for various energy applications such as fuel (automobiles

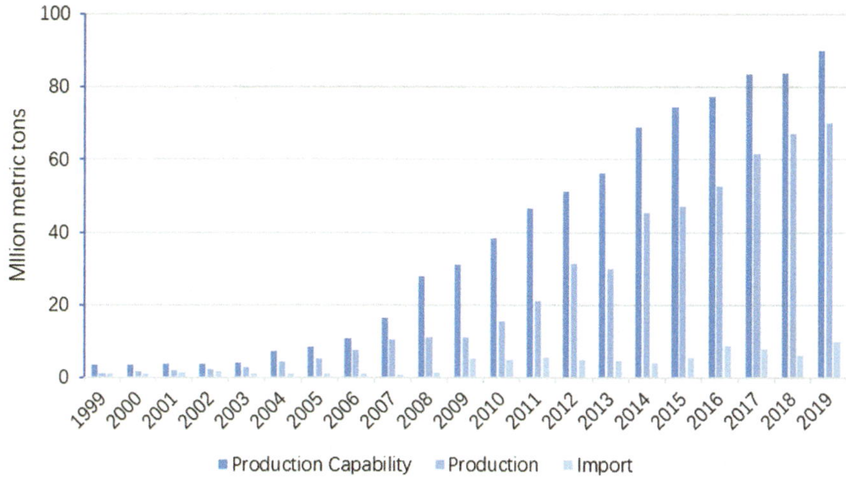

Fig. 8 Methanol production and supply in China (1999–2019) [14]

and other thermal applications like boilers and cookstoves), MTBE, biodiesel, and Dimethyl Ether (DME) comprise the second largest downstream application for methanol in China.

By 2017, coal accounts for over 80% of methanol feedstock in the production and supply chain, many times more than natural gas or coking gas, as shown in Fig. 9.

Methanol in China is largely based on available domestic resources, enjoying reliable access and long-term supply security. Imported methanol can help to reduce the environmental pressure from the supply chain of the coal chemistry and meet the demand from the coastal regions in China. Such a relatively abundant supply of methanol makes the fuel attractive from a fuel-cost perspective. As an example, an

Fig. 9 Feedstock of methanol production in China by 2017 [14]

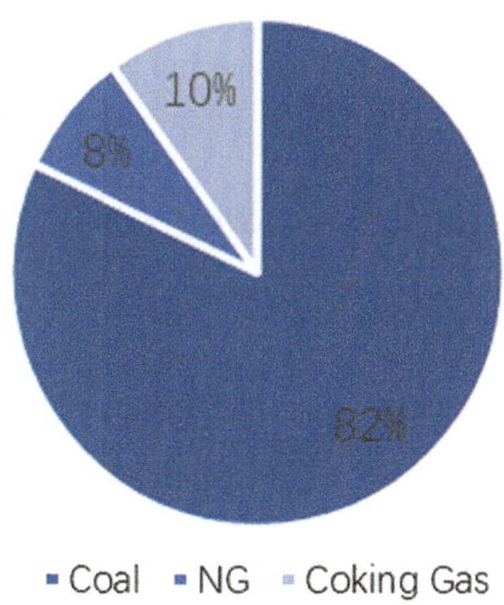

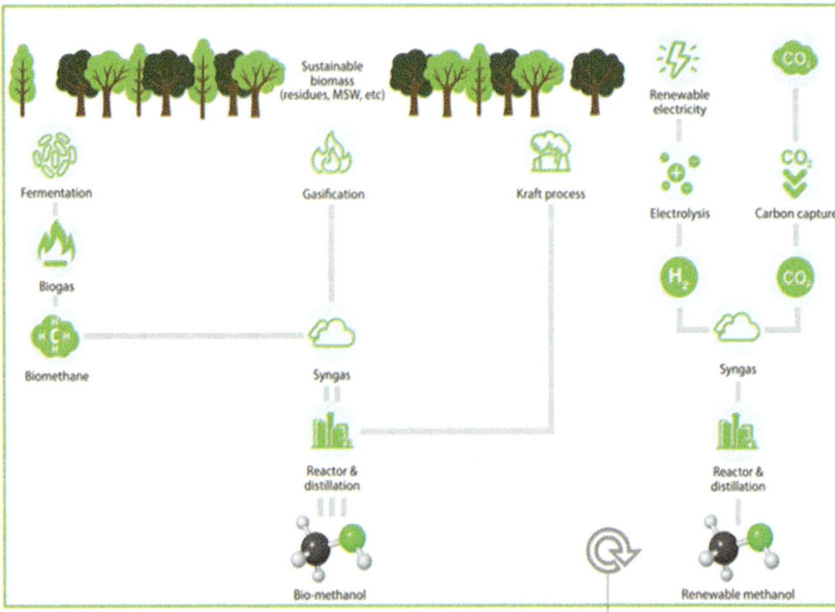

Fig. 10 Renewable methanol production processes from different feedstocks

M100 taxi can realize potential savings of over 30% of the fuel cost compared to gasoline [6].

In the longer term, as shown in Fig. 10, methanol can be produced from various renewable resources such as biomass, municipal solid waste, and renewable power, making methanol even more attractive as a sustainable solution [15]. The promotion of bio-methanol and renewable methanol production was a key component of the 2019 policy announcement from the central government. China is also studying this strategic shift and recently commissioned a solar-to-methanol project with an annual capacity of 1,400 tons in Lan Zhou, Gansu Province.

4.2 Infrastructure

The expansion of MFV requires a matching development of necessary infrastructure. Both during and after the MIIT methanol vehicle pilot, methanol fuel filling was realized by newly built, dedicated M100 filling stations and/or existing full-service gasoline stations with additional methanol fuel pumps. So far, there are over 70 M100 filling stations in China, mainly in two cities, Gui Yang and Xi An. As methanol is a liquid fuel at ambient temperature, with similar chemistry and physical properties to gasoline and diesel, new-build M100 station did not pose any significant cost increases compared to traditional fuels. Conversion from traditional fuel stations, to add a methanol fuel function is more attractive from both time and cost perspectives.

In some regions such as Shanxi province, the infrastructure was even developed earlier to support "methanol blended gasoline", usually M15. As an example, there were over 1,000 gasoline stations technically converted to venting M15 which was supported by 50 blending facilities in 15 different provinces in China. All these assets can further support MFVs by upgrading from M15 to M85 and M100.

4.3 Standardization

As the number of vehicles grows, standards and guidelines are necessary to further promote MFVs in different regions. In China, fuel standards were first developed around 2009, namely, GB/T23510-2009 Fuel Methanol for Motor Vehicles and GB/T 23799-2009 Methanol Gasoline (M85) for Motor Vehicles, prior to the pilot. During the pilot, to meet the needs of piloting cities to build new methanol filling stations, methanol filling infrastructure guidelines were also developed by MIIT, which include Guidelines for the Construction of Methanol Fuel Fueling Stations and Guidelines for Safe Operation of Methanol Fuel Fueling Stations.

With the announcement of the policy to promote MFVs, the standardization process has been accelerated, with more standards being developed and expected to be introduced in the near future as follows.

Fuel and Infrastructure related, the China Association of Alcohol and Ether Fuel and Automobiles (CAAEFA) is the main responsible body, with support from the Methanol Institute:

- National standard of M100 Fuel for Motor Vehicles
- National Standard of M100 Fuel Additives for Motor Vehicles
- National Standard of Guidelines for Safety Operation of Methanol Fuel for Motor Vehicles

MFV related, China Automotive Technology and Research Center Co., Ltd. (CATARC) is the main responsible body:

- Industry Standard of Methanol Vehicle Technical Requirement
- Industry Standard of Methanol-Fuelled Engines
- Industry Standard of Diesel/Methanol Dual Fuel Engines
- Industry Standard of Fuel System for Methanol Vehicles
- Industry Standard of Experiment Method of Fuel Consumption Test for Methanol Vehicles

National Standard of Emission Standards for Methanol Vehicles, which is being drafted by the Beijing Institute of Technology under the Ministry of Ecology and Environment of China.

5 Conclusion

For nearly 40 years, China has been developing the use of methanol as a road transport fuel. From the initial steps taken in Shanxi Province, early engagement with international automotive OEMs like Volkswagen, to large-scale pilot programs, and now the adoption of promotional policies by the central government, China leads the world in the introduction of methanol-fuelled vehicles. The 2019 adoption of a joint ministerial position paper calls for China to continue this leadership position, establishing goals to: "Maintain the international leading position of China's methanol vehicles and related industries in the fields of products, technologies and special equipment and accelerate energy diversification and Clean energy vehicle development, promote the transformation and upgrading of traditional industries, foster new economic growth points, and promote green recycling and low carbon development".

The use of methanol as a vehicle fuel meets key objectives for China in diversifying its energy supply, reducing emissions, expanding economic development, and job growth, all while providing consumers with affordable fuel and vehicles. Certainly, these goals are not unique to China, and economies around the world can benefit from the lessons gained in China on the use of methanol-fuelled vehicles. Methanol can be used in spark ignition and compression ignition engines, in passenger cars, buses, and trucks, and in new vehicles as well as retrofits. The existing liquid fueling infrastructure can be easily modified to distribute and dispense methanol, with Chinese taxi drivers finding 30% fuel-cost savings over current gasoline pricing.

As the trade association for the global methanol industry, the Methanol Institute (MI) has been an active partner with China to support the commercialization of methanol-fuelled vehicles. MI has helped share lessons from international methanol fuel activities with experts in China, who are now clearly the teachers. As China has piloted methanol-fuelled vehicles and developed safe handling standards and guidelines, MI has provided encouragement, as well as technical and financial support. Today, a key objective for the Methanol Institute is to help share the lessons from China with other global economies that can also benefit from the use of methanol as a future proof fuel.

References

1. China Petroluem and Chemical Industry Federation (CPCIF) (2019) Annual report of Chinese petroleum and chemical industry economy operation
2. BP (2020) Statistical review of World energy
3. CEIC (2020) China's No of employee: coal mining & dressing. https://www.ceicdata.com/en/china/no-of-employee-by-industry-monthly/no-of-employee-coal-mining–dressing. Accessed 2nd Aug 2020
4. U.S. Energy Information Agency (2019) China's crude oil imports surpassed 10 million barrels per day in 2019. 23 March 2020. https://www.eia.gov/todayinenergy/detail.php?id=43216. Accessed Aug 2020

5. Ministry of Ecology and Environment of People's Republic of China (2019) China mobile source environmental management annual report
6. Zhao K (2019) A brief review of China's methanol vehicle pilot and policy
7. Yao C, Cheung CS, Cheng C, Wang Y, Chan TL, Lee SC (2008) Effect of diesel/methanol compound combustion on diesel engine combustion and emissions. Energy Convers Manag 49(6):1696–1704
8. Wei H, Yao C, Pan W, Han G, Dou Z, Wu T, Liu M, Gao J, Chen C, Shi J (2017) To meet demand of Euro V emission legislation urea free for HD diesel engine with DMCC. Fuel 207:33–46
9. Valera H, Agarwal AK (2020) Future automotive powertrains for India: methanol versus electric vehicles. In: Alternative fuels and their utilization strategies in internal combustion engines. Springer, Singapore, pp 89–123
10. China Association of Automobile Manufacturers (2020) Automobile industry economy operation status. http://www.caam.org.cn/chn/4/cate_39/con_5228367.html. Accessed 2020
11. Valera H, Agarwal AK (2019) Methanol as an alternative fuel for diesel engines. In: Methanol and the alternate fuel economy. Springer, Singapore, pp 9–33
12. Verhelst S, Turner JW, Sileghem L, Vancoillie J (2019) Methanol as a fuel for internal combustion engines. Progr Energy Combust Sci 70:43–88
13. Ministry of Industry and Information Technology (MIIT) (2019) Guidance of developing methanol vehicles applications in some parts of China. http://www.miit.gov.cn/n1146295/n1652858/n1652930/n3757016/c6684042/content.html. Accessed 19 March 2019
14. China Nitrogen Fertilizer Industry Association (CNFIA). Annual Chinese methanol industry operation
15. ATA Markets Intelligence S.L. (2018) Renewable methanol report

Regulated and Unregulated Emissions from Methanol Fuelled Engines

Hardikk Valera, Jakub Čedík, Martin Pexa, and Avinash Kumar Agarwal

Abbreviations

ICEs	Internal Combustion Engines
MSW	Municipal Solid Waste
AQI	Air Quality Index
EPA	Environmental Protection Agency
OPEC	Organization of the Petroleum Exporting Countries
REmap	Renewable Energy Roadmap
MME	Ministry of Mines and Energy
MIIT	Ministry of Industry and Information Technology
UNFCCC	United Nations Framework Convention on Climate Change
GHG	Greenhouse Gas
HC	Hydrocarbon
NOx	Nitrogen oxides
CO	Carbon Monoxide
CO_2	Carbon Dioxide
PM	Particulate Matter
NO_2	Nitrogen Dioxide
NO	Nitric oxide
SO_2	Sulphur Dioxide
HCHO	Formaldehyde
CH_4	Methane

H. Valera · A. K. Agarwal (✉)
Engine Research Laboratory, Department of Mechanical Engineering, Indian Institute of Technology Kanpur, Kanpur 208016, India
e-mail: akag@iitk.ac.in

J. Čedík · M. Pexa
Department for Quality and Dependability of Machines, Czech University of Life Sciences, Prague, Czech Republic

A. K. Agarwal et al. (eds.), *Methanol*, Energy, Environment, and Sustainability,
https://doi.org/10.1007/978-981-16-1224-4_7

CH_3CHO	Acetaldehyde
C_6H_6	Benzene
HCOOH	Formic Acid
C_2H_4	Ethylene
C_4H_{10}	Butane
C_5H_{12}	Pentane
C_8H_{18}	Octane
C_2H_2	Acetylene
C_2H_5OH	Ethanol
C_2H_6	Ethane
CH_3OH	Methanol
$C_6H_5CH_3$	Toluene
$(CH_3)_2C_6H_4$	Xylene
$(CH_2{=}CH)_2$	1,3-Butadiene
IARC	International Agency for Research on Cancer
-OH	Hydroxyl
SI	Spark Ignition
CI	Compression Ignition
MoP&NG	Ministry of Petroleum and Natural Gas
WTT	Well-to-Tank
TTW	Tank-to-Wheels
DMCC	Diesel Methanol Combustion Compound
HPDI	High Pressure Direct Injection
THC	Total Hydrocarbon
BTX	Benzene, toluene and xylene
HPLC	High Performance Liquid Chromatography
PDHID	Pulsed Discharge Helium Ionization Detector
DME	Dimethyl ether
FTIR	Fourier transform infrared spectroscopy analyzer
OEMs	Original equipment manufacturers

1 Introduction

Globally, energy demand is increasing rapidly. The United States and China, being the largest energy producer, are putting adequate efforts so that their energy requirements can be fulfilled and the world can also meet its energy requirements. In 2018, their combined contribution was 54% to the energy production growth [1]. On the contrary, European Union saw a reduction in energy production because their electricity production from oil, gas, and nuclear resources was reduced. However, the utilization of energy is always at the cost of environmental degradation. This motivated researchers and policymakers of these countries alike to explore renewable fuels, which seem to be a favourable candidate to cater to increasing global energy demand. China is one of the largest energy producers, energy generator, and energy consumer. It also tops the list of installed capacity of wind power, hydropower, and solar PV. It is also the largest user of geothermal heat. In 2015, the share of renewables in its total final energy consumption was ~7%.

1.1 Air Quality Issues

European Union is a leader in spreading international awareness about the global climate challenge and is taking many strict measures to address the issue. In addition, United States Environmental Protection Agency (USEPA) had prepared an index named as "Air Quality Index" (AQI), to spread public awareness about the quality of air they are breathing and its related effects on human health. USEPA calculates "air quality index" (AQI) for five deleterious pollutants namely: (i) ground level ozone, (ii) particulate matter, (iii) carbon monoxide, (iv) sulphur dioxide, and (v) nitrogen dioxide. A higher AQI value means greater air pollution level and bigger health concern. Comprehensive details of the AQI yardstick are shown in Table 1 [2].

Table 1 AQI and its health impacts [2]

AQI values	Health concern level	Meaning
0–50	**Good**	Air quality is considered satisfactory, and air pollution poses little/no risk
51–100	**Moderate**	Air quality is acceptable; however, for some pollutants, there may be a moderate health concern for a minimal number of people, who are unusually sensitive to air pollution
101–150	**Unhealthy for Sensitive Groups**	Members of sensitive groups may experience some health effects. The general public is not likely to be affected
151–200	**Unhealthy**	Everyone may begin to experience health effects. Members of sensitive groups may experience more severe health effects
201–300	**Very Unhealthy**	Health alert: everyone may experience more severe health effects
301–500	**Hazardous**	Health warnings for emergency conditions. Entire population is likely to be affected

1.2 Energy Supply and Global Renewable Fuels Scenario

India is also following the footsteps of the European Union and is aiming to achieve an installed capacity of 175 GW renewable energy by 2022. The USA is also implementing stricter policies to introduce renewables at a strong pace. Currently, renewables share accounts for just ~8% of total energy usage in the US. In June 2018, the Organization of Petroleum Exporting Countries (OPEC) decided to increase oil production to meet the global energy demand. Therefore, the USA also reported a sharp rise of ~16.5% in crude oil production, whereas global crude oil production rose by only ~2%. Among various energy sectors, the transport sector is lagging behind in the utilization of renewable energy sources. The worldwide share of renewable energy utilization in the transport sector was only ~4% in 2015, as shown in Fig. 1 [3].

Tailpipe emissions from petroleum fuels are major contributors to global warming and climate change issues. Several countries are taking serious actions to increase the share of renewable energy utilization in the transport sector. Renewable energy options for the transport sector are mainly dominated by biofuels such as biodiesel, ethanol, and methanol. Electrification can play a major role in the transport sector if it is coupled with renewable primary energy. It is possible to cut the transport sector emissions to ~3 Gt CO_2 annually by 2050 if the renewable energy roadmap (REmap) is to be followed. Under this plan, total liquid biofuel production can exceed >900 billion liters by 2050 from ~129 billion liters in 2015. This huge increase in production is predicted because each biofuel production will be significantly upscaled. A steep increase in biofuel production requires sustainable planning of biomass feedstock supply.

Primarily, ethanol is blended with gasoline, which primarily works as an oxygenate or octane enhancer. Sometimes ethanol is used for fuelling specially designed vehicles. Ethanol is used extensively in countries like the United States and Brazil. In 2019, their ethanol production was ~15.8 billion gallons and ~8.62

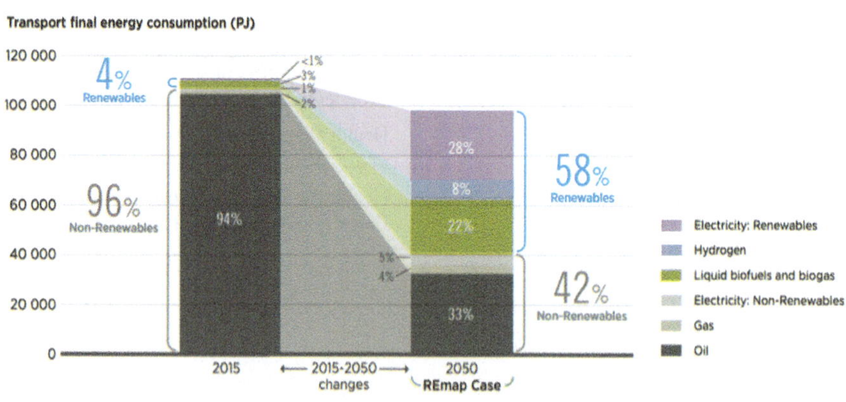

Fig. 1 Transformation of energy demand for transport sector as projected in next 35 Years [3]

billion gallons, respectively [4]. This ethanol is intended for fuelling transport engines by adding 2–5% (v/v) in conventional petroleum fuels. Ethanol is made anhydrous, i.e. the presence of water is reduced to <1% and the denatured (i.e. unfit for human consumption). Brazil's Ministry of Mines and Energy (MME) launched RenovaBio Programme in December 2016, which was later instituted as "'National Bio-Fuel Policy" [5]. In 2006, an ethanol mandate was declared, which was adding around 25% (v/v) ethanol in gasoline (E25) on a national scale. However, ethanol production is season dependent. Therefore variants of ethanol blends were introduced in Brazil e.g. E25 is made available in January-February, E20 is made available in March–October, and E23 is made available in November–December. Currently, E27 is mandated in Brazil as per Brazil Biofuels Annual Report (2019) [5]. Brazilian sugarcane ethanol production system is considered the best carbon saving production route [6–8].

Biodiesel is also an emerging alternative fuel made by transesterification of triglycerides [9]. It is a suitable alternate fuel for CI engines. Biodiesel can be produced from animal oil/fats, tallow, straight vegetable oils, and waste cooking oils. The United States, Brazil, and Indonesia are the three largest biodiesel producers in the world. In 2018, they produced 6.9 billion litres, 5.4 billion litres, and 4 billion litres of biodiesel, respectively [10]. Brazil implemented a biodiesel fuel policy for the transport sector and strictly regulated biofuel production. In Brazil, biodiesel production increased to ~5.35 billion litres for the year 2018, which was ~25% higher than in 2017. On 1st March 2020, Brazil Agency for Petroleum, Natural Gas, and Biofuels announced an increment in the blending ratio of biodiesel in mineral diesel by 1% from existing 11%. This increment was in line with the plan set out in October 2018 for 15% biodiesel blending in diesel by 2023. Indonesia is the world's largest palm oil producer and exporter, which is continuously increasing the biodiesel content of its diesel supplies. Indonesia is aiming to cut fuel imports by absorbing palm oil supplies. In 2019, Indonesia saved 43.8 trillion rupiahs from the B20 fuel policy. Thereafter, Indonesia launched '30% (v/v) palm oil biodiesel (B30) programme' in December 2019 [11].

Methanol is also considered a potential biofuel, which can be produced from renewable as well as non-renewable resources. Required materials for methanol production are available abundantly. It is a colourless, biodegradable and flammable liquid. It is a preferred cultivable renewable fuel as it offers relatively high land use efficiency. It can be used with minor adjustments to ensure material compatibility for existing engine parts hence is a favourable fuel for ICEs [12, 13]. China is the world's largest user of methanol in the transport sector. Its poor quantity coal reserves are used as a raw material for methanol production via the gasification route. Recently, eight government ministries and agencies jointly announced a promotional policy for methanol fuelled vehicles and their utilization in the transport sector. This policy enables consumers to choose methanol fuelled vehicles for their usage and Chinese original equipment manufacturers (OEMs) will ensure the support and availability. OEMs have produced 32 new models for light and heavy-duty usage, which were certified by the Ministry of Industry and Information Technology (MIIT) for commercial sales. Geely Auto is the mass producer of methanol fuelled engines. They have a facility for producing 300,000–500,000 engines annually in China. M100 fleet could reach up to 50,000 cars, trucks, and buses, which will be capable of using 500,000 metric tons of methanol in China alone [14].

1.3 Indian Strategy for Methanol Adaptation

India participated in the 21st session of the Conference of the Parties (COP21) held under the United Nations Framework Convention on Climate Change (UNFCCC), held in 2015 in Paris. In COP21, India committed to reduce greenhouse gas (GHG) emission intensity of its GDP by 33–35% below 2005 levels by 2030. Also, pledged that new policies would be adopted such that 40% of India's power will come from non-fossil fuels. It is, therefore, necessary to govern and control tailpipe emissions, which are released into the atmosphere due to fossil fuel burning. Emission standards set quantitative limits on the pollutants that are released into the atmosphere. These standards are designed to maintain the air quality in the atmosphere.

For the transport sector, India is following the trends of European emission norms and implemented these norms as "Bharat Stage Emission Standards'"as shown in Fig. 2 [15]. NITI (National Institution for Transforming India) Aayog, which was constituted to achieve sustainable development goals of the country, has already taken some precautionary steps to reduce the threats due to the excessive automotive exhaust emissions. However, the importance of road vehicles in India is increasing and it is not possible to reduce the number of the vehicle. However, NITI Aayog, a policy think tank of the Government of India thought that India's commitment towards COP21 could not be met by just introducing compliance with stricter emission norms alone. Introducing alternative fuels for use in petroleum powered vehicles is necessary, which would emit lesser GHG emissions compared to petroleum based

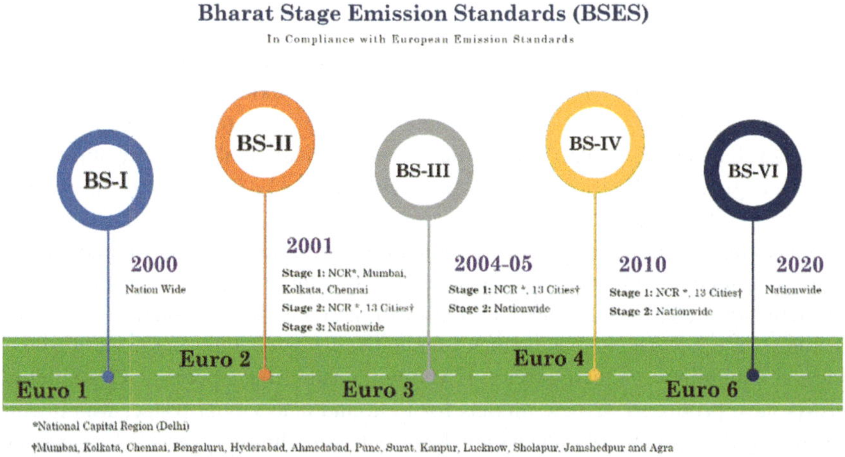

Fig. 2 Roadmap of Bharat Stage Emission Standards followed in India

fuels. Therefore, it initiated the Methanol Economy initiative to introduce methanol as fuel in the Indian transport sector. Methanol Economy program was initiated with four objectives [16]: (i) Fuel import bill reduction, (ii) GHG emissions reduction, (iii) Utilisation of Indian high ash coal and MSW by converting them into methanol, and (iv) Job creation by setting up a methanol production plants and associated economic activities. Methanol economy is projected to reduce 15% fuel import bill by 2030. It will create 5 million jobs for methanol production, distribution, and services. Five methanol plants based on high ash coal and one based on natural gas are jointly planned with Israel to produce 20 MMT/annum methanol. Standards have already been put in place for M15 by the Bureau of Indian Standards (BIS) in consultation with Indian Oil Corporation Limited (IOCL), Automotive Research Association of India (ARAI), and Society of Indian Automobile Manufacturers (SIAM). Methanol Economy is fast emerging as one of the best ways to salvage the situation arising due to fossil fuel depletion, which includes higher fuel prices, and worsening ambient air quality in Indian cities [17]. However, combustion of any fuel, whether petroleum based, or alternative, leads to the emission of pollutants. These are either regulated pollutants or unregulated pollutants. Regulated pollutants are those, which regulatory bodies have adopted as norms and these emissions are strictly regulated by those norms. Unregulated pollutants are those, which are not regulated by these regulatory bodies. In most cases, the health effects of unregulated pollutants are more severe compared to regulated pollutants however their emission dosages are relatively lower. The effect of these pollutants is attracting the serious attention of researchers and the policymakers. Some of these effects are shown in Fig. 3.

China has been running a pilot program for methanol fuelled vehicles, where 1,024 methanol fuelled vehicles were operated in the different provinces of China. It emerged that all pilot vehicles operated smoothly, their daily maintenance corresponded well with gasoline fuelled vehicles, and no technological failure was reported in this project. From Table 2, it is clear that methanol fuelled vehicles

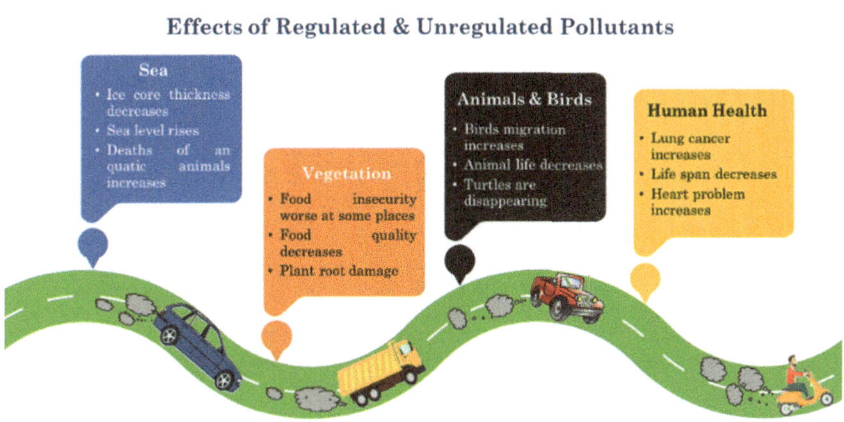

Fig. 3 Health effects of regulated and unregulated pollutants on life on the planet

Table 2 Methanol fuelled vehicle in China (By February 2019)

Province	City	Vehicle type	Vehicle no. [13]	Atmospheric conditions					
				Period (Months)		Hottest day		Cold season period	Coldest day
				Hot season	Cold season	Highest temperature (°C)	Lowest temperature (°C)	Highest temperature	Lowest temperature
Shanxi [19]	Jinzhong	Taxi	260	~4.3	~3	~32	~20	~−2	~−13
	Changzhi	HD Bus	96						
Shanghai [20]	Minhang	Taxi	18	~3.0	~3.2	~33	~27	~8	~2
Shannxi	Xi'An	Taxi	3,897	N.A.					
	Baoji	Taxi	200						
		Mini MPV	15						
	Yulin	Self-Dumping Truck	5						
	Hanzhong	Taxi	20						
Guizhou [21]	Guiyang	Taxi	5,380	~3.8	~3.1	~33	~24	~10	~0
Gansu	Lanzhou	Taxi	150	N.A.					
	Pingliang	Taxi	50						
Total			10,091						

operated successfully in China, however, the atmospheric conditions of this project undertaken in China are quite different from that in India. Indian has vast geographic diversity and topography, which makes it challenging to generalize weather conditions. The effects of pollutants emitted depend on weather conditions to a significant extent. It causes higher emission rates at lower ambient temperatures [18], dilution of pollutants at higher wind speed, and unstable thermal condition leads to vertical mixing of pollutants and influences the homogeneous and heterogeneous chemical reaction rates in the atmosphere. Therefore, it is necessary to understand the health effects of emitted regulated and unregulated pollutants from methanol fuelled vehicles before implementing the Methanol Economy initiative in India. Comprehensive details of regulated and unregulated emissions from methanol fuelled vehicles are discussed in the following sections of this chapter.

2 Regulated Emissions

Regulated emissions in India include hydrocarbons (HC), oxides of nitrogen (NOx), carbon monoxide (CO), carbon dioxide (CO_2), and particulate matter (PM). It is necessary to understand three critical parameters regarding pollutants: (i) How it forms during combustion, (ii) Its behaviour in the atmosphere, and (iii) Its negative impact on the human health. Each of these parameters related to regulated emissions is discussed at length.

Carbon Monoxide (CO): It is a poisonous, odourless, colourless, and tasteless gas. It is produced due to the incomplete combustion of fossil fuels. It has a high affinity towards the haemoglobin present in the bloodstream and forms carboxy-haemoglobin. It reduces the oxygen carrying capacity of haemoglobin in the tissues. Its exposure may damage the central nervous system due to oxygen deficiency. A high level of exposure may cause death due to a heart attack [22].

Carbon Dioxide (CO_2): CO_2 is a greenhouse gas, produced as a diesel combustion product. Inhalation of carbon dioxide in small quantity is harmless however, higher exposure causes a negative impact on sleep. An exposed human may experience increased heart rate, elevated blood pressure, convulsions, tiredness, sweating, difficulty in breathing, headaches, and dizziness [23].

Particulate matter (PM): Particulate matter is also referred to as total particulate matter (TPM) or diesel particulate matter (DPM). It is a mixture of carbonaceous soot, solid and liquid materials [24]. DPM is a pollutant that impacts human health negatively [25]. Commonly used indicators for controlling PM emissions are particulate mass and particle number (PN). Environmental protection agencies have defined PM quantitative parameters like $PM_{2.5}$ and $PM_{10.}$ They represent the concentrations of particles with a diameter of 2.5 μm and 10 μm or less, respectively. The negative health impacts of PM include respiratory and cardiovascular morbidities such as aggravation of asthma.

Oxides of Nitrogen (NOx): Oxides of nitrogen comprise nitric oxide and nitrogen dioxide, which generally form during the combustion process via oxidation of the nitrogen. NOx formation mainly depends on the peak in-cylinder temperature. Nitric oxide and nitrogen dioxide are not regulated individually but, regulated together in terms of NOx emissions.

Nitrogen Dioxide (NO_2): Nitrogen dioxide is a toxic compound and an integral part of NOx emissions. Inhalation of nitrogen dioxide cause irritation in the airways of human respiratory systems. Also, it aggravates asthma. NO_2 reacts with water, oxygen, and other chemicals. Its complex reactions lead to acid rain, which is harmful to the environment especially for lakes and forests [26]. Due to its high level of toxicity, it is likely to be regulated individually in the near future [24].

Nitric Oxide (NO): It is a colourless gas with high toxicity, also known as nitrogen monoxide. It forms via nitrogen oxidation. NO released into the atmosphere undergoes complex photochemical reactions. These complex reactions lead to the formation of photochemical oxidants, which create photochemical smog. Also, nitric oxide reacts with water vapor and forms nitric acid, which is a serious component of acid rain. The presence of high concentrations of nitric oxide depletes the ozone layer in the upper atmosphere [27].

Hydrocarbon (HC): Hydrocarbon emissions consist of carbon and hydrogen containing compounds. It forms due to incomplete combustion. Hydrocarbon vapours cause negative health impacts such as dizziness. [28].

3 Unregulated Emissions

Unregulated pollutants include several important gaseous species such as Sulphur dioxide (SO_2), Formaldehyde (HCHO), Methane (CH_4), Acetaldehyde (CH_3CHO), Benzene (C_6H_6), Formic Acid (HCOOH), Ethylene (C_2H_4), Butane (C_4H_{10}), Pentane (C_5H_{12}), Octane (C_8H_{18}), Acetylene (C_2H_2), Ethanol (C_2H_5OH), Ethane (C_2H_6), Methanol (CH_3OH), Toluene ($C_6H_5CH_3$), Xylene ($(CH_3)_2C_6H_4$), 1,3- Butadiene ($(CH_2 = CH)_2$) etc. The effect of these pollutants can sometimes be worse than that of regulated pollutants. Therefore, in order to control these emissions, it is necessary to understand the root cause of their formation. Properties of these pollutants are discussed below in detail, along with their health and environmental effects.

Sulphur Dioxide (SO_2): Sulphur dioxide (SO_2) is a colourless gas having a choking odour. It is produced during the combustion of conventional fuels (Diesel and Gasoline) due to the presence of sulphur containing compounds in the fuel. Further, the participation of lubricating oil traces in combustion also leads to its formation. However, large marine ships, locomotives, and some non-road diesel equipment emit relatively higher SO_2 into the atmosphere because of their higher fuel sulphur content compared to other applications using lower sulphur fuels. It is also responsible for other forms of sulphur compounds such as sulfurous anhydride, sulfurous oxide, sulphur oxide, and sulfurous acid anhydride [29]. Sulphur dioxide forms sulphuric acid upon reacting with moisture and is responsible for the occurrence of acid rain. Acid rain causes changes in the soil acidity, acidic lakes and streams, and decay of building materials such as marble. Exposure to sulphur dioxide can cause breathing difficulties and burning sensations in the nose and throat. Long term exposure can also result in lung function problems, heart diseases, and can be fatal upon severe exposure. It is not classified as a human carcinogen, which means, its exposure does not cause cancer.

Formaldehyde (HCHO): Formaldehyde (HCHO) is a flammable and colourless gas with a strong pungent odour. It is also known as "methanal", the simplest of aldehydes. Mainly it is formed during vapour phase oxidation of methanol. Its negative health effects include watery eyes, burning sensations in the nose and throat, coughing, nausea, wheezing, and skin irritation, when the concentration level in ambient air exceeds 0.1 ppm [30]. International Agency for Research on Cancer (IARC) has classified formaldehyde as a human carcinogen [31–33].

Methane (CH_4): Methane is a colourless and odourless gas. It is the simplest member of the paraffin series of hydrocarbons and has significant global warming potential. It is produced during the combustion of conventional fuels. Because of its long lifetime in the atmosphere, it contributes towards the greenhouse effect and climate change [34]. High level exposure to humans can cause vision problem, memory loss, facial flushing, headache, and slurred speech. However, its lower concentration does not cause any adverse health effects [35, 36].

Acetaldehyde (CH_3CHO): Acetaldehyde is an organic chemical compound and is also known as "ethanal". It is produced during the combustion of petroleum fuels in IC engines. Inhalation and exposure to acetaldehyde can cause irritation of eyes, skin, and respiratory tract. However, prolonged exposure may lead to coughing, pulmonary edema, and erythema [37, 38]. Data regarding carcinogenic effects of Acetaldehyde on humans is inadequate and only one epidemiology study is available in the open literature.

Benzene (C_6H_6): Benzene is an organic chemical compound and is present as a natural component in crude petroleum. Benzene exposure may lead to anaemia because it restricts the production of red blood cells (RBCs) into the human body. Prolonged exposure can result in white blood cells (WBC) loss and immune system damage [39, 40]. Dizziness, drowsiness, headache, confusion, tremor, unconsciousness, and irregular heartbeat are the signs and symptoms of the high level concentrations of benzene in the environment. EPA has declared benzene as a human carcinogen for all routes of exposure.

Formic Acid (HCOOH): Formic acid is the simplest of the carboxylic acids and is also known as 'methanoic acid'. It is an ocular, skin and lung tissue irritant [41]. Irritation of eyes, nose, cough, and sore throat are the signs of formic acid inhalation. High exposure can cause confusion, tightness, breathlessness, wheezing, and headache. Pain, burns, and ulcers are the signs of formic acid skin contact [42].

Ethylene (C_2H_4): Ethylene, the simplest alkene is a colourless and flammable hydrocarbon gas. Ethylene exposure causes nausea, headache, dizziness, drowsiness, unconsciousness, and weakness [43].

Butane (C_4H_{10}): Butane is strongly flammable, odourless, and colourless gas. It acts as a depressant towards the central nervous system because it reduces brain activity and human ability to respond mentally and physically [44]. Immediate side effects of exposure are aggression, confusion, blackouts, hallucinations, and slurred speech.

Pentane (C_5H_{12}): It has gasoline like odour. Its inhalation causes nose and throat irritation. Normal exposure leads to headache, light headedness, and dizziness [45]. Exposure at a higher level negatively affects the lungs and may lead to a medical emergency.

Octane (C_8H_{18}): It has a gasoline like odour. Its inhalation results in throat and lung irritation, wheezing, coughing, and shortness of breath. Exposure to certain levels leads to headache, confusion, and dizziness.

Acetylene (C_2H_2): It is a colourless, odourless, non-poisonous, and non-irritant gas and is also known as ethyne [46]. Headache, dizziness, semi-asphyxia, and brief loss of consciousness are the symptoms of its exposure [47].

Ethanol (C_2H_5OH): It is also known as ethyl alcohol and ethyl hydrate. Exposure to ethanol vapour causes fatigue, headache, sleepiness and also affects the central nervous system [48].

Ethane (C_2H_6): It is an odourless and colourless gas. Its exposure causes dizziness, nausea, headache, vomiting, and light headedness, which sometimes also leads to suffocation due to lack of oxygen.

Methanol (CH_3OH): Methanol is poisonous. Headache, severe abdominal pain, difficulty in breathing, and lack of coordination are the symptoms of its exposure. It may damage the central nervous system. However, methanol is not declared as a carcinogen [49].

Toluene ($C_6H_5CH_3$): It is a colorless gas. It is also known as methyl benzol, methylbenzene, toluol, and phenyl methane. Headache, ataxia, euphoria, dizziness, drowsiness, tremors, hallucinations, seizures, vomiting, and electrolyte imbalances are the symptoms of toluene exposure [50]. A high level of exposure may cause acid base imbalance in the body.

Xylene ($(CH_3)_2C_6H_4$): It is also known as dimethyl benzene [51]. Inhalation of xylene damages the central nervous system, depending on the level of exposure. Headache, nausea, dizziness, irritability, weakness and vomiting, clumsiness, ringing in the ears, confusion and slurred speech, loss of consciousness are the symptoms of xylene exposure.

1,3- Butadiene ($(CH_2{=}CH)_2$): It is a colourless gas with an aromatic odour and is insoluble in water. Its inhalation irritates eyes, throat, lungs, and nasal passages. Prolonged exposure may lead to cardiovascular diseases such as arterio-sclerotic heart diseases and rheumatism [52]. EPA has declared 1,3- butadiene as carcinogenic [53, 54].

4 Regulated and Unregulated Emissions from Methanol Fuelled Compression Ignition Engines

CI engines are extensively used for heavy duty transport and agricultural applications. Conventionally, diesel is considered the most suitable fuel for CI engines. However, significant efforts have been made to deploy other alternative fuels in CI engines such as biodiesel, jet fuel, compressed natural gas (CNG), and liquified petroleum gas (LPG). Methanol is a fast emerging alternate fuel, which is considered more suitable for SI engine applications. However, several studies explored the possibility of using methanol in CI engines in order to take advantage of its properties such as low carbon content, fuel-bound oxygen, and high latent heat of vaporization. Methanol can be utilized in CI engines using different techniques as reported in Table 3.

Among these techniques, HPDI technique is highly effective, since it permits higher methanol mass utilization in CI engines, as compared to other techniques. However, this technique is under development stage, where researchers are conducting laboratory level experiments in the test cells. Several studies have been conducted to assess the use of methanol for reduction of regulated and unregulated emissions and are listed in Tables 4 and 5, respectively. From Table 4, it can be observed that methanol fuelled engines have the potential to reduce regulated emissions. Port fuel injection of methanol and direct injection of pilot diesel can minimize major pollutants from CI engines, i.e. NOx and soot. NOx reduction can be attributed to higher latent heat of vaporisation and lower adiabatic flame temperature of methanol, which reduces the peak in-cylinder temperature during combustion [67]. However, blending and emulsification techniques have shown reduction of CO, HC and smoke emissions at the expense of NOx. This may be because of more complete combustion, which can be a result of superior mixing of diesel and methanol.

The studies tabulated in Table 5 demonstrate that methanol fuelled engines emit lower ethyne, 1-3 butadiene, ethene, n-butane, acetaldehyde and ethanol compared to diesel fuelled CI engines. However, higher unburned methanol and BTX emissions were observed vis-à-vis diesel fuelled engines. Moreover, the trend for formaldehyde emissions is not very clear. Increasing or decreasing trends mainly depend on formaldehyde formation and its oxidation, which were highly influenced by three factors: (i) available space in the combustion chamber, (ii) temperature in the combustion chamber, and (iii) time available for combustion.

5 Regulated and Unregulated Emissions from Methanol Fuelled Spark Ignition Engines

Gasoline fuelled vehicles are widely used in passenger car/motorcycle segments in the global market. Their production capacity increased by ~9% in 2016–17 compared to the previous year (2015–16) [70]. Older technology for fuel induction was using mechanical carburetors for homogeneous fuel–air mixture preparation. However,

Table 3 Methanol utilization techniques in CI engines

S. no	Techniques		Diesel injection	Methanol injection	Comments
1	Blending		Methanol-diesel blends can be injected into the combustion chamber	It is injected into the combustion chamber along with diesel	Methanol and diesel are immiscible in nature. Therefore, blends of methanol more than 10% (v/v) in diesel are not suitable for CI engines
2	Methanol Emulsions		It is injected into the combustion chamber with a possible concentration of methanol and emulsifying agents	It is injected into the combustion chamber in the form of a stable emulsion	Emulsifiers are used to ensure the miscibility of methanol-diesel blends
3	Port Fuel Injection of Methanol and Direct Injection of Pilot Diesel	Method 1	It is injected into the combustion chamber during each engine cycle	It is injected into the intake manifold during each engine cycle	This technique is well tested in the lab and is in nascent stage of commercialization
		Method 2	Only diesel is injected directly into the engine to start and run the engine on low load	Methanol is injected in the intake manifold along with fixed quantity of diesel direct injection at medium and high load conditions	Diesel is used to avoid cold-startability issues and methanol is used to extract the extra energy so that proportion of renewable fuel usage can be increased. This method is also known as Diesel Methanol Combustion Compound (DMCC) System

(continued)

Table 3 (continued)

S. no	Techniques		Diesel injection	Methanol injection	Comments
4	High Pressure Direct Injection (HPDI)	Method 1	It is injected into the combustion chamber using a high pressure injector	Methanol and diesel are injected into the combustion chamber using separate injectors	This technique requires modifications into the cylinder head geometry to adapt two separate injectors. However, lubrication of the high pressure methanol injector can be problematic [54]
		Method 2	It is directly injected into the combustion chamber using a co-axial injector	Diesel and methanol are injected straight into the combustion chamber using a co-axial injector	This technique is in the nascent stage for both laboratory level experiments and commercial viability
5	Methanol with Ignition Improvers		–	–	Ignition improvers are cetane improvers and can be used during adaptations of four different techniques
6	Methanol with Glow plug Concept		–	–	Glow plug works as an ignition element for easing the combustion during starting. This technique can be used during the adaptation of the above mentioned techniques

Table 4 Experimental studies on CI engines for assessment of regulated pollutants

Year	Researchers	Methanol utilization technique	Fuel used	Test result of regulated emissions
2007	Yao et al. [55]	Port Fuel Injection of Methanol and Direct Injection of Pilot Diesel	Diesel is used to start the engine. Then, methanol is injected along with a fixed quantity of diesel for medium and high engine loads	CO and HC emissions increased whereas, smoke and NOx emissions reduced using diesel methanol combustion compound system compared to conventional diesel combustion
2008	Cheng et al. [56]	Port Fuel Injection of Methanol and Direct Injection of Pilot Diesel	M0, M10, M20, M30 (Here, methanol quantity is fixed based on a percentage of power output)	NOx decreased using diesel–methanol combination compared to baseline diesel, whereas, HC and CO increased. In comparison to baseline, the highest HC increment was seven times that at low load conditions and three times that at high load conditions using M30 fuelled engine
2008	Song et al. [57]	Port Fuel Injection of Methanol and Direct Injection of Pilot Diesel	M0, M10, M20, M30, M40, M50, M60, M70 (Here, methanol quantity is varied based on a percentage of the mass fraction)	HC and CO emissions increased whereas significant reduction in smoke and NOx emissions were observed using diesel methanol combination compared to baseline diesel. An important outcome of this study was disappearance of NOx and smoke trade-off as they reduced simultaneously

(continued)

Table 4 (continued)

Year	Researchers	Methanol utilization technique	Fuel used	Test result of regulated emissions
2008	Yao et al. [58]	Port Fuel Injection of Methanol and Direct Injection of Pilot Diesel	M0, M10, M20, M30 (Here, methanol quantity represents its contribution in power output of the engine)	This study was conducted with and without oxidation catalyst. The study revealed that this technique reduced soot and NOx emissions whereas, HC and CO emissions increased compared to baseline diesel. CO, HC, NOx and soot, all reduced relatively by using oxidation catalyst
2009	Sayin et al. [59]	Blending	M0, M5, M10, M15	NOx and CO_2 emissions increased (22–69% & 14–68%) whereas, CO, UHC and smoke opacity decreased (33–52%, 26–50% and 5–22%) using diesel–methanol blend vis-à-vis baseline diesel
2010	Sayin [60]	Emulsion	M0, M5, M10	In this study, 1% (v/v) dodecanol was used during preparation of diesel–methanol blend to avoid phase separation. CO, THC and smoke opacity decreased, and NOx emissions increased using methanol-diesel blend compared to baseline diesel.

(continued)

Table 4 (continued)

Year	Researchers	Methanol utilization technique	Fuel used	Test result of regulated emissions
2011	Ciniviz et al. [61]	Blending	M0, M5, M10, M15	HC, CO decreased, and NOx emission increased with increasing methanol mass *fraction in blend compared to baseline* diesel
2016	Li [62]	Port Fuel Injection of Methanol and Direct Injection of Pilot Diesel	Diesel and Methanol quantity was varied to achieve the required load conditions	NOx and Smoke emissions reduced, whereas CO and HC emissions increased using methanol-diesel combinations, compared to baseline diesel
2017	Valera [63]	Emulsion	M0, M7, M12, M18	Mixture of oleic acid and n-butanol was used to prepare miscible methanol-diesel blends. Higher CO and THC emissions were observed, whereas lower NOx and smoke opacity were observed using methanol-diesel blends vis-à-vis mineral diesel
2018	Emiroglu et al. [64]	Blending	M0, M10	NOx emissions increased, whereas CO and smoke emissions reduced using methanol-diesel blend compared to mineral diesel
2019	Agarwal et al. [65]	Emulsion	M0, M10, M15	1% (v/v) dodecanol was used as a surfactant in diesel–methanol blends. M15 blend was the best fuel, as it reduced HC, CO and NOx emissions

(continued)

Table 4 (continued)

Year	Researchers	Methanol utilization technique	Fuel used	Test result of regulated emissions
2020	Ning et al. [66]	Port Fuel Injection of Methanol and Direct Injection of Pilot Diesel	Diesel and Methanol quantity was varied as per the requirement to achieve stable combustion	HC emissions increased and CO, NOx, soot emissions reduced using methanol-diesel mixture

M Methanol, *Ma* Methanol a% (v/v), Diesel (100-a) % (v/v)

Table 5 Experimental studies on CI engines for assessment of unregulated pollutants

Year	Researchers	Methanol utilization technique	Fuel used	Test result of unregulated emissions
2010	Zhang et al. [68]	Port Fuel Injection of Methanol and Direct Injection of Pilot Diesel	M0, M10, M20, M30 (Here, methanol percentage represents of its contribution in engine load)	Diesel-methanol mixtures reduced emissions of ethyne, 1,3-butadiene, ethene, whereas emissions of unburned methanol, formaldehyde, and Benzene, toluene and xylene (BTX) increased compared to baseline diesel
2010	Zhang et al. [69]	Port Fuel Injection of Methanol and Direct Injection of Pilot Diesel	M0, M10, M20, M30 (Here, methanol percentage represents of its contribution in engine load)	It was observed that using methanol-diesel combination, ethyne, ethene and 1,3-butadiene emissions decreased whereas, BTX, formaldehyde and unburned methanol emissions increased compared to baseline diesel
2016	Agarwal et al. [43]	Blending	M0, M5	It was observed that methane, n-pentane, acetylene emissions increased, and n-butane, formaldehyde, acetaldehyde and ethanol emissions decreased for methanol-diesel blends compared to baseline diesel

M Methanol, *Ma* Methanol a% (v/v), Diesel (100-a) % (v/v), *BTX* Benzene, Toluene and Xylene

with the introduction of an electronic fuel injection (EFI) system, fuel injectors are employed to introduce the fuel into the intake manifold/engine cylinder for homogeneous mixture preparation. Nevertheless, mechanical carburetors are still used in two-wheeler applications due to their cost-effectiveness. In future, there is a possibility that electronic carburetor may enter the market for SI engine applications. Methanol has emerged as a suitable fuel for SI engine applications in various sectors

Table 6 Different techniques for methanol utilization for SI engine applications

S. no	Techniques	Gasoline introduction	Methanol introduction	Comments
1	Blending	It is introduced in the intake manifold/engine cylinder along with methanol in the form of a blend	It is introduced into the intake manifold/engine cylinder along with gasoline in the form of a blend	This technique is widely used for methanol utilization in SI engines
2	Fumigation/Dual fueling	It is introduced separately into the intake manifold or directly into the cylinder using a separate fuel supply device	It is introduced separately into the intake manifold or directly into a cylinder using a separate fuel supply device	This technique is extensively tested on a laboratory scale and is being considered for field implementation

due to five main factors: (i) Good fuel properties, (ii) Existing fuel outlets are capable of fuel distribution without major modifications in the existing distribution system, (iii) It is more economical compared to gasoline, (iv) It reduces tailpipe emissions, and (v) Good miscibility of methanol with gasoline, which is an essential requirement for blending methanol with gasoline. Methanol can be utilized in SI engines using two different techniques, as mentioned in Table 6.

Both of these techniques have their own advantages and disadvantages. Blending is preferred over fumigation/dual fuel method because of two major reasons: (i) Miscibility of methanol and gasoline allow blending up to the highest possible concentration, and (ii) Separate fuel supply lines are required for methanol, which increases the mechanical losses and complexity of the system in case of dual fuelling. This lowers the overall mechanical efficiency of dual fuelling compared to blending [71]. On the other hand, the fumigation/dual fuel method allows making provision for varying the proportion of the two fuels according to the engine requirements. It also avoids the phase separation problem, which is generally seen in methanol-gasoline blends due to the absorption of moisture. Several studies have been attempted to investigate overall engine performance characteristics of methanol addition to gasoline in different proportions in a variety of SI engines. Many researchers have reported that the utilization of methanol-gasoline blends could reduce regulated exhaust emissions [72–77]. However, comparative analysis of unregulated emissions from methanol-fuelled engines with that of gasoline powered engines is also necessary before implementing the methanol vehicle policy. Tables 7 and 8 summarize the highlights of regulated and unregulated emissions from various studies conducted on a variety of SI engines fuelled with methanol-gasoline blends respectively.

From the experimental studies tabulated above, it can be concluded that methanol has the potential to minimize regulated emissions in the exhaust gas of SI engines. However, variations in tailpipe emissions were observed among various studies due to differences in operating conditions and design parameters of the engines

Table 7 Experimental studies on SI engines for assessment of regulated pollutants

Year	Researcher	Fuel used	Test results of regulated emissions
2007	Yanju et al. [78]	M0 M10 M20 M85	CO and NOx emissions reduced significantly with increasing methanol concentration in methanol-gasoline blends. Highest CO and NOx emission reductions obtained were 25% & 80% using M85
2014	Agarwal et al. [79]	M0 M10 M20	Methanol-gasoline blends produced lower NO and CO emissions compared to baseline gasoline. However, HC behaviour exhibited an uncertain trend (higher or lower) with the engine speed
2015	Li et al. [80]	M0 M15	M15 emitted lower HC and CO emissions in the range of 11–34.5% and 63–84% respectively, whereas NOx emissions were 76.9–107.7% higher compared to baseline gasoline in overall operating conditions
2015	Elfasakhany [81]	M0 M3 M7 M10	M3, M7, and M10 reduced CO and HC, while increased CO_2 emissions. CO emissions were 17.7, 51.5 and 55.5% lower, HC emissions were 19.6, 16 and 26% lower and CO_2 emissions were 3, 8 and 9.2% higher compared to gasoline for M3, M7 and M10 respectively
2016	Yao et al. [82]	M0 M10 M20 M30 M50 M70	Among all blends, M50 and M70 were found to be the most suitable for maximum reduction of CO (47.6 and 70.9%), NOx (29.3 and 54.5%), and non-methane hydrocarbons (NMHC) (48.5% and 79.9%) compared to baseline gasoline at all operating conditions

M Methanol, *Ma* Methanol a% (v/v), Gasoline (100-a) % (v/v)

and experimental conditions. Nevertheless, there was a unique trend for CO emission. Most studies reported a reduction in CO emission due to the addition of methanol in gasoline. CO is formed as a result of incomplete combustion of fuel. The presence of inherent fuel oxygen in methanol–gasoline blend resulted in leaner combustion, which made combustion more complete, leading to higher conversion of CO to CO_2 compared to gasoline combustion. There was no unique trend reported for HC and NOx emissions. NOx reduction was observed because of reduction in peak combustion temperature due to much higher latent heat of vaporization of methanol. However, a reduction in peak combustion temperature resulted in incomplete oxidation of hydrocarbons, thereby generating higher hydrocarbon emissions.

The above experimental studies for investigating unregulated emissions reported that formaldehyde, n-pentane, toluene, and unburnt methanol emissions increased while formic acid, ethanol, iso-butylene, benzene, acetaldehyde, and ethylene emissions decreased for methanol-gasoline blends compared to conventional gasoline fuelling for SI engines. The presence of unburnt methanol in tailpipe emissions can be understood by two possibilities: (i) formation during gasoline combustion, and (ii) due to incomplete combustion of methanol. This could be a result of experimental design parameters, where existing fuel supply devices were used for conducting

Table 8 Experimental studies on SI engines for assessment of unregulated pollutants

Year	Researcher	Fuel used	Test results of unregulated emissions
2007	Yanju et al. [78]	M0 M10 M20 M85	Methanol–gasoline blends produced relatively higher CH_3OH emissions. Also, its concentration in THC emissions was directly proportional to the methanol fraction in the fuel blends
2015	Agarwal et al. [83]	M0 M5 M15	Methanol–gasoline blends reported lower formic acid, ethanol, iso-butylene, benzene, acetaldehyde emissions, and higher n-pentane, toluene, methanol emissions compared to baseline gasoline
2015	Li et al. [80]	M0 M15	M15 emitted 16–60% lower BTEX emissions, whereas 16.4–52.5% higher CH_2O emissions vis-à-vis baseline gasoline
2016	Yao et al. [82]	M0 M10 M20 M30 M50 M70	Higher CH_3OH and CH_2O emissions were observed using methanol–gasoline blends compared to gasoline. CH_3OH emissions reached 500 ppm and CH_2O emissions were relatively four times higher for M70. Further, their increment was directly proportional to the fraction of methanol in methanol-gasoline blends. M70 released 50% lower C_2H_4 emission compared to M0. C_3H_6 emissions reduced with increasing methanol content in blends. C_4H_6 emissions were not significant, almost below 10 ppm for methanol–gasoline blends

M Methanol, *Ma* Methanol a% (v/v), Gasoline (100-a) % (v/v), CH_3OH Methanol, *BTEX*: Benzene, Toluene, Ethylbenzene, p,m,o-Xylene, CH_2O Formaldehyde, C_2H_4 Ethylene, C_3H_6 Propene, C_4H_6 Butadiene

the experiments, which were calibrated for petroleum fuels. Hence, these emissions can be minimized by calibrating existing fuel supply devices for methanol-gasoline blends such that they supply the correct air–fuel mixture according to the engine requirements. Calibration is required to be done for optimizing the engine performance along with the simultaneous reduction of emissions, in order to be able to meet strict emission norms.

Further, the accuracy of the measured quantity of any gaseous emission species heavily depends on the principle of measuring devices. For example, EPA has used high performance liquid chromatography (HPLC) method to measure formaldehyde emissions. In the past, formaldehyde was detected by the FID method [84]. Also, researchers have developed a pulsed discharge helium ionization detector (PDHID) to measure the unregulated emissions from oxygenated fuelled vehicles. PDHID is a low-cost chromatographic method [85–87]. However, frequently used methods for formaldehyde measurements are: (i) High performance liquid chromatography (HPLC), (ii) Gas chromatography (GC), and (iii) Fourier transform infrared spectroscopy analyzer (FTIR) [88–90]. Hence, it can be stated that plenty of methods are available for formaldehyde emission measurements. In this regard, Geng et al. [91] performed a comparative study on formaldehyde emission measurements using

HPLC, GC, and FTIR. They found that there were significant differences in the quantities of formaldehyde emissions obtained for similar engine operating conditions from these three instruments. The measured value of formaldehyde emission using the FTIR method was higher than that of GC and HPLC methods, whereas values by GC and HPLC methods were nearly identical. There are three major reasons for these considerable differences. First, the measurement principles of these methods are different, and temperature-pressure values have a huge effect on the accuracy of these results. Second, interferences are always present from other exhaust particles. Third, the level of calibration of instruments affects the results. Therefore, relying on any one single instrument results without defining the standard method may result in significant errors. However, Li et al. [80] revealed that although formaldehyde emissions were higher, it was far lower than the limit fixed by EPA standards; 15 mg/mile. Therefore, if formaldehyde is within the desired limit, then the vehicle can be operated using methanol as a replacement fuel.

6 Conclusions

A detailed review of methanol utilization in IC engines has been done in this chapter. Methanol utilization techniques in SI engines and CI engines are also discussed comprehensively. For SI engines, there are two methods for methanol adaptations: (i) Blending, and (ii) Fumigation. Blending is considered a more suitable method for methanol adaptation in existing SI engines. These methods allow methanol utilization, ranging from a small fraction to high fraction. Blending methods proved effective in reducing CO emission as compared to only gasoline fuelled engines. Moreover, simultaneous HC and NOx reductions can be achieved by appropriate tuning of fuel supply devices. Further, higher unburnt methanol and formaldehyde emissions were not a serious issue as they are within the limit prescribed by regulatory bodies and can be reduced further by appropriate calibration of fuel induction devices. For CI engines, there are five basic methods for methanol utilization: (i) Blending, (ii) Emulsions, (iii) Port fuel injection of methanol with direct injection of pilot diesel, (iv) High pressure direct injection of methanol using separate injector or co-axial injector, and (v) DMCC (Diesel Methanol Combustion Compound System). In addition, ignition improvers and glow plug concepts are also used to initiate methanol ignition in the combustion chamber. Among the various techniques, port fuel injection of methanol and direct injection of pilot diesel, blending, and emulsification techniques are used frequently for methanol utilization. Blending and emulsion techniques could reduce CO and HC emissions. However, this method limits the methanol utilization in CI engines due to its poor miscibility with diesel. This problem can be avoided by port injection of methanol and direct injection of pilot diesel. This technique could reduce the NOx and smoke emissions, which are serious issues for the OEMs due to stricter emission norms. Moreover, it reduced harmful unregulated pollutants such as ethyne, 1-3 butadiene, ethene, n-butane, acetaldehyde, and ethanol. Hence, it can be concluded that methanol consumption can reduce most regulated and unregulated pollutants from IC engines. Formaldehyde emission from methanol fuelled engines is not high. The standard method for its measurement is not defined by the regulatory bodies. Even then, the existing values are within the prescribed limits.

References

1. Global Energy Statistical Yearbook. https://yearbook.enerdata.net/total-energy/world-energy-production.html
2. AQI Basics. https://www.airnow.gov/aqi/aqi-basics/
3. Global energy transformation: A roadmap to 2050. https://www.irena.org/publications/2019/Apr/Global-energy-transformation-A-roadmap-to-2050–2019Edition
4. Fuel ethanol production worldwide in 2019. https://www.statista.com/statistics/281606/ethanol-production-in-selected-countries/
5. USDA Foreign Agricultural Service, Gain Report. https://apps.fas.usda.gov/newgainapi/api/report/downloadreportbyfilename?filename=Biofuels%20Annual_Sao%20Paulo%20ATO_Brazil_8-9-2019.pdf
6. Goldemberg J, Coelho ST, Guardabassi P (2008) The sustainability of ethanol production from sugarcane. Energ Pol 36:2086–2097
7. Fargione J, Hill J, Tilman D, Polasky S, Hawthorne P (2008) Land clearing and the biofuel carbon debt. Science 319:1235–1238
8. Leite RCC, Leal MRLV, Cortez LAB, Griffin WM, Scandiffio MIG (2009) Can Brazil replace 5% of the 2025 gasoline world demand with ethanol? Energy 34:655–661
9. Biodiesel. http://www.esru.strath.ac.uk/EandE/Web_sites/02-03/biofuels/what_biodiesel.htm
10. Leading biodiesel producers worldwide in 2018. https://www.statista.com/statistics/271472/biodiesel-production-in-selected-countries/
11. Indonesia launches B30 biodiesel to cut costs, boost palm oil. https://www.reuters.com/article/us-indonesia-biodiesel/indonesia-launches-b30-biodiesel-to-cut-costs-boost-palm-oil-idUSKBN1YR0D2
12. Yuen PKP, Villaire W, Beckett J (2010) Automotive materials engineering challenges and solutions for the use of ethanol and methanol blended fuels. SAE Technical Papers 2010-01-0729
13. Hu TG, Wei YJ, Liu SH, Zhou LB (2007) Improvement of spark ignition (SI) engine combustion and emission during cold start fueled with methanol/gasoline blends. Energy Fuels 21:171–175. https://doi.org/10.1021/ef0603479
14. Zhao K (2019) A Brief Review of China's Methanol Vehicle Pilot and Policy. https://www.methanol.org/wp-content/uploads/2019/03/A-Brief-Review-of-Chinas-Methanol-Vehicle-Pilot-and-Policy-20-March-2019.pdf
15. Bharat Stage Emission Standards. https://en.wikipedia.org/wiki/Bharat_stage_emission_standards
16. NITI Aayog_Methanol Economy. https://niti.gov.in/methanol-economy
17. Valera H, Agarwal AK (2020) Future automotive powertrains for India: methanol versus electric vehicles. In: Alternative fuels and their utilization strategies in internal combustion engines. Springer, Singapore, pp 89–123. https://doi.org/10.1007/978-981-15-0418-1_7
18. Chang TY, Michigan JM (1983) Vehicular CO emission factors in cold weather. J Air Pollut Control Assoc 33(12):1188–1189. https://doi.org/10.1080/00022470.1983.10465711
19. Shanxi Weather Conditions. https://weatherspark.com/y/136165/Average-Weather-in-Shanxi-China-Year-Round
20. Shanghai Weather Conditions. https://weatherspark.com/y/137446/Average-Weather-in-Shanghai-China-Year-Round
21. Guizhou Weather Conditions. https://weatherspark.com/y/122068/Average-Weather-in-Guizhou-China-Year-Round

22. Acute Exposure Guideline Levels for Selected Airborne Chemicals: Volume 8. https://www.ncbi.nlm.nih.gov/books/NBK220007/
23. Wisconsin of Health Department Services_Carbon Dioxide. https://www.dhs.wisconsin.gov/chemical/carbondioxide.htm)
24. Rounce PL (2011) Engine performance and particulate matter speciation for compression ignition engines powered by a range of fossil and biofuels.Ph.D. thesis. The University of Birmingham, UK
25. Bharj RS, Singh GN, Valera H (2020) Role of diesel particulate filter to meet bharat stage-VI emission norms in India. In: Advanced combustion techniques and engine technologies for the automotive sector. Springer, Singapore, pp 215–228. https://doi.org/10.1007/978-981-15-0368-9_10
26. Environmental Protection Agency_Nitrogen Dioxide Pollution. https://www.epa.gov/no2-pollution/basic-information-about-no2
27. Nitric Oxide_Chemical Compound. https://www.britannica.com/science/nitric-oxide
28. Reşitoğlu İA, Altinişik K, Keskin A (2015) The pollutant emissions from diesel-engine vehicles and exhaust aftertreatment systems. Clean Techn Environ Policy 17:15–27. https://doi.org/10.1007/s10098-014-0793-9
29. Wisconsin Department of Health Services. https://www.dhs.wisconsin.gov/chemical/sulfurdioxide.htm
30. National Cancer Institute. https://www.cancer.gov
31. U.S. Environmental Protection Agency, Office of Air and Radiation. Report to Congress on Indoor Air Quality, Volume II: Assessment and Control of Indoor Air Pollution. https://irp-cdn.multiscreensite.com/c4e267ab/files/uploaded/kt34RqduTlGgjxf3soeQ_EPA_Report%20to%20Congress%20on%20Indoor%20Air%20Quality_Volume%20II_Assessment
32. International Agency for Research on Cancer (2004) IARC Monographs on the Evaluation of Carcinogenic Risks to Humans Volume 88 (2006): Formaldehyde, 2-Butoxyethanol and 1-tert-Butoxypropan-2-ol. http://monographs.iarc.fr/ENG/Monographs/vol88/index.php
33. National Toxicology Program (2011) Report on Carcinogens, Twelfth Edition. Department of Health and Human Services, Public Health Service, National Toxicology Program
34. Methane Chemical Compound. https://www.britannica.com/science/methane
35. Public Health England_Protecting and improving the nation's health_Methane. https://assets.publishing.service.gov.uk/government/uploads/system/uploads/attachment_data/file/769766/Methane_PHE_general_information__070119.pdf
36. Hazardous Substance Fact Sheet https://nj.gov/health/eoh/rtkweb/documents/fs/1202.pdf
37. U.S. Environmental Protection Agency (1987) Health Assessment Document for Acetaldehyde. EPA/600/8-86- 015A. Environmental Criteria and Assessment Office, Office of Health and Environmental Assessment, Office of Research and Development, Research Triangle Park, NC. https://nepis.epa.gov/
38. U.S. Department of Health and Human Services (1993) Registry of toxic effects of chemical substances (RTECS, online database). National Toxicology Information Program, National Library of Medicine, Bethesda, MD. https://www.ncbi.nlm.nih.gov/books/NBK45471/
39. Centers of Disease Control and Preventions. https://emergency.cdc.gov/agent/benzene/basics/facts.asp
40. Benzene. https://www.epa.gov/sites/production/files/2016-09/documents/benzene.pdf
41. Final report on the safety assessment of formic acid. https://journals.sagepub.com/doi/pdf/10.1080/109158197227161
42. Public Health England_Protecting and improving the nation's health_Formic Acid. https://assets.publishing.service.gov.uk/government/uploads/system/uploads/attachment_data/file/769719/Formic_acid_PHE_GI_070119.pdf
43. Agarwal AK, Shukla PC, Patel C, Gupta JG, Sharma N, Prasad RK, Agarwal RA (2016) Unregulated emissions and health risk potential from biodiesel (KB5, KB20) and methanol blend (M5) fuelled transportation diesel engines. Renew Energy 98:283–291. https://doi.org/10.1016/j.renene.2016.03.058
44. Butane-Drug and Alcohol Information. https://www.drugs.ie/butane

45. Hazardous Substance Fact sheet_Pentane. https://nj.gov/health/eoh/rtkweb/documents/fs/1476.pdf
46. Williams NR, Whittington RM (2001) Death due to inhalation of industrial acetylene. J Toxicol Clin Toxicol 39(1):69–71. https://doi.org/10.1081/CLT-100102882
47. Documentation of the threshold limit values and biological exposure indices, 6th edn, vol 1. American Conference of Governmental Industrial Hygienists Inc, Technical Affairs Office, Kemper Meadow Drive, Cincinnati, Ohio 45240
48. Nadeau V, Lamoureux D, Beuter A, Charbonneau M, Tardif R (2003) Neuromotor effects of acute ethanol inhalation exposure in humans: a preliminary study. J Occup Health 45(4):215–222. https://doi.org/10.1539/joh.45.215
49. Medina E, Wellon GC, Evegren F (2017) Methanol safe handling manual. http://www.methanol.org/wp-content/uploads/2017/03/Safe-Handling-Manual.pdf
50. Toxic Substances Portal—Toluene. https://www.atsdr.cdc.gov/MMG/MMG.asp?id=157&tid=29
51. Toxicological Profile for Xylene (1993) U.S Department of Health and Human Services, Public Health Service, Agency for Toxic Substance and Disease Registry
52. Agency for Toxic Substances and Disease Registry (ATSDR) (1992) Toxicological Profile for 1,3-Butadiene. Public Health Service, U.S., Department of Health and Human Services, Atlanta, GA
53. U.S. Environmental Protection Agency (2009) Integrated Risk Information System (IRIS) on 1,3-Butadiene. National Center for Environmental Assessment, Office of Research and Development, Washington, DC
54. Bereczky Á, Lukács K, Khan, Z (2018) Investigation of methanol utilization in Ci engines. In: MATEC Web of Conferences.235;00036. EDP Sciences. https://doi.org/10.1051/matecconf/201823500036
55. Yao C, Cheung CS, Cheng C, Wang Y (2007) Reduction of smoke and NOx from diesel engines using a diesel/methanol compound combustion system. Energy Fuels 21(2):686–691. https://doi.org/10.1021/ef0602731
56. Cheng CH, Cheung CS, Chan TL, Lee SC, Yao CD (2008) Experimental investigation on the performance, gaseous and particulate emissions of a methanol fumigated diesel engine. Sci Total Environ 389(1):115–124. https://doi.org/10.1016/j.scitotenv.2007.08.041
57. Song R, Liu J, Wang L, Liu S (2008) Performance and emissions of a diesel engine fuelled with methanol. Energy Fuels 22(6):3883–3888. https://doi.org/10.1021/ef800492r
58. Yao C, Cheung CS, Cheng C, Wang Y, Chan TL, Lee SC (2008) Effect of diesel/methanol compound combustion on diesel engine combustion and emissions. Energy Convers Manag 49(6):1696–1704. https://doi.org/10.1016/j.enconman.2007.11.007
59. Sayin C, Ilhan M, Canakci M, Gumus M (2009) Effect of injection timing on the exhaust emissions of a diesel engine using diesel–methanol blends. Renew Energy 34(5):1261–1269. https://doi.org/10.1016/j.renene.2008.10.010
60. Sayin C (2010) Engine performance and exhaust gas emissions of methanol and ethanol–diesel blends. Fuel 89(11):3410–3415. https://doi.org/10.1016/j.fuel.2010.02.017
61. Ciniviz M, Köse H, Canli E, Solmaz O (2011) An experimental investigation on effects of methanol blended diesel fuels to engine performance and emissions of a diesel engine. Sci Res Essays 6(15):3189–3199
62. Li G, Zhang C, Li Y (2016) Effects of diesel injection parameters on the rapid combustion and emissions of an HD common-rail diesel engine fueled with diesel-methanol dual-fuel. Appl Therm Eng 108:1214–1225. https://doi.org/10.1016/j.applthermaleng.2016.08.029
63. Valera H (2017) An experimental study on the physio-chemical properties, performance parameter in a CI engine fuelled with methanol blended diesel (MBD). M. Tech Thesis
64. Emiroğlu AO, Şen M (2018) Combustion, performance and emission characteristics of various alcohol blends in a single cylinder diesel engine. Fuel 212:34–40. https://doi.org/10.1016/j.fuel.2017.10.016
65. Agarwal AK, Sharma N, Singh AP, Kumar V, Satsangi DP, Patel C (2019) Adaptation of Methanol–Dodecanol–diesel blend in diesel genset engine. J Energy Resour Technol 141(10). https://doi.org/10.1115/1.4043390

66. Ning L, Duan Q, Chen Z, Kou H, Liu B, Yang B, Zeng K (2020) A comparative study on the combustion and emissions of a non-road common rail diesel engine fueled with primary alcohol fuels (methanol, ethanol, and n-butanol)/diesel dual fuel. Fuel 266: https://doi.org/10.1016/j.fuel.2020.117034
67. Yao C, Pan W, Yao A (2017) Methanol fumigation in compression-ignition engines: a critical review of recent academic and technological developments. Fuel 209:713–732. https://doi.org/10.1016/j.fuel.2017.08.038
68. Zhang ZH, Cheung CS, Chan TL, Yao CD (2010) Experimental investigation on regulated and unregulated emissions of a diesel/methanol compound combustion engine with and without diesel oxidation catalyst. Sci Total Environ 408(4):865–872. https://doi.org/10.1016/j.scitotenv.2009.10.060
69. Zhang ZH, Cheung CS, Chan TL, Yao CD (2010) Experimental investigation of regulated and unregulated emissions from a diesel engine fueled with Euro V diesel fuel and fumigation methanol. Atmos Environ 44(8):1054–1061. https://doi.org/10.1016/j.atmosenv.2009.12.017
70. Sakthivel P, Subramanian KA, Reji M (2018) Indian scenario of ethanol fuel and its utilization in automotive transportation sector. Resour Conserv Recycl 132:102–120. https://doi.org/10.1016/j.resconrec.2018.01.012
71. Valera H, Singh AP, Agarwal AK (2020) Prospects of methanol-fuelled carburetted two wheelers in developing countries. In: Advanced combustion techniques and engine technologies for the automotive sector. Springer, Singapore, pp 53–73. doi:https://doi.org/10.1007/978-981-15-0368-9_4
72. Akutsu Y, Toyoda F, Tomita K, Yoshizawa F, Tamura M, Yoshida T (1991) Effect of exhaust from alcohol fuel on ozone formation in the atmosphere. Atmos Environ Part A Gen Top 25:1383–1389. https://doi.org/10.1016/0960-1686(91)90247-5
73. Ozsezen AN, Canakci M (2011) Performance and combustion characteristics of alcohol-gasoline blends at wide-open throttle. Energy 36:2747–2752. https://doi.org/10.1016/j.energy.2011.02.014
74. Pischinger F (1983) Alcohol fuels for automotive engines. In: 11th World petroleum congress, RTD 6 (3). London. https://www.onepetro.org/conference-paper/WPC-20331
75. Bata RM, Elord AC, Rice RW (1989) Emissions from IC engines fueled with alcohol gasoline blends: a literature review. Trans ASME 111:424e431
76. Palmer FH (1986) Vehicle performance of gasoline containing oxygenates, Book, Mech. Eng. Publication Limited
77. Gravalos I, Moshou D, Gialamas T, Xyradakis P, Kateris D, Tsiropoulos Z (2011) Performance and emission characteristics of spark ignition engine fuelled with ethanol and methanol gasoline blended fuels. Alternat Fuel:85
78. Yanju W, Shenghua L, Hongsong L, Rui Y, Jie L, Ying W (2008) Effects of methanol/gasoline blends on a spark ignition engine performance and emissions. Energy Fuels 22(2):1254–1259. https://doi.org/10.1021/ef7003706
79. Agarwal AK, Karare H, Dhar A (2014) Combustion, performance, emissions and particulate characterization of a methanol–gasoline blend (gasohol) fuelled medium duty spark ignition transportation engine. Fuel Process Technol 121:16–24. https://doi.org/10.1016/j.fuproc.2013.12.014
80. Li L, Ge Y, Wang M, Li J, Peng Z, Song Y, Zhang L (2015) Effect of gasoline/methanol blends on motorcycle emissions: exhaust and evaporative emissions. Atmos Environ 102:79–85. https://doi.org/10.1016/j.atmosenv.2014.11.044
81. Elfasakhany A (2015) Investigations on the effects of ethanol–methanol–gasoline blends in a spark-ignition engine: performance and emissions analysis. Eng Sci Technol Int J 18(4):713–719. https://doi.org/10.1016/j.jestch.2015.05.003
82. Yao D, Ling X, Wu F (2016) Experimental investigation on the emissions of a port fuel injection spark ignition engine fueled with methanol–gasoline blends. Energy Fuels 30(9):7428–7434. https://doi.org/10.1021/acs.energyfuels.6b01586
83. Agarwal AK, Shukla PC, Gupta JG, Patel C, Prasad RK, Sharma N (2015) Unregulated emissions from a gasohol (E5, E15, M5, and M15) fuelled spark ignition engine. Appl Energy 154:732–741. https://doi.org/10.1016/j.apenergy.2015.05.052

84. Methods for determination of formaldehyde in air of public places. China Standard- GB/T 18204.26-2000. https://www.chinesestandard.net/PDF/English.aspx/GBT18204.26-2000
85. Westerholm R (1996) Christensen A (1996) Regulated and unregulated exhaust emissions from two three-way catalyst equipped gasoline fuelled vehicles. Atmos Environ 20:3529–3536. https://doi.org/10.1016/1352-2310(96)00048-9
86. Yang BB, Yao MF, Cheng WK, Zheng ZQ (2014) Yue L (2014) Regulated and unregulated emissions from a compression ignition engine under low temperature combustion fuelled with gasoline and n-butanol/gasoline blends. Fuel 120:163–170. https://doi.org/10.1016/j.fuel.2013.11.058
87. Zhu L, Cheung CS, Zhang WG, Fang JH, Huang Z (2013) Effects of ethanol–biodiesel blends and diesel oxidation catalyst (DOC) on particulate and unregulated emissions. Fuel 113:690–696. https://doi.org/10.1016/j.fuel.2013.06.028
88. Xu X, Su R, Zhao X, Liu Z, Li D, Li X, Zhang H, Wang Z (2011) Determination of formaldehyde in beverages using microwave-assisted derivatization and ionic liquid-based dispersive liquid–liquid microextraction followed by high-performance liquid chromatography. Talanta 85(5):2632–2638. https://doi.org/10.1016/j.talanta.2011.08.037
89. Ahamad T (2012) Alshehri SM (2012) TG–FTIR–MS (evolved gas analysis) of bidi tobacco powder during combustion and pyrolysis. J Hazard Mater 199–200:200–208. https://doi.org/10.1016/j.jhazmat.2011.10.090
90. Hu HC, Tian YX, Chai XS, Si WF, Chen C (2013) Rapid determination of residual formaldehyde in formaldehyde related polymer latexes by headspace gas chromatography. J Ind Eng Chem 19:748–751. https://doi.org/10.1016/j.jiec.2012.11.012
91. Geng P, Yao CD (2015) Comparative study on measurements of formaldehyde emission of methanol/gasoline fueled SI engine (vol 148, pg 9, 2015). Fuel 161:384–385

Utilization Aspects

Low-Temperature Combustion Kinetics of Methanol-Blended Gasoline and Methanol Synthesized Dimethyl Ether

Sayan Biswas and Vyaas Gururajan

Abbreviations

ACI	Advanced compression ignition
AHRR	Apparent heat release rate
aTDC	After top dead center
B	Bore diameter
CA	Crank angle referenced to main TDC [°]
CA10, CA50, CA90	10, 50, 90% cumulative burn angle
CO	Carbon monoxide
CO_2	Carbon dioxide
COV	Coefficient of variation
DME	Dimethyl ether
EVC/EVO	Exhaust valve close/open
H	Atomic hydrogen
H/C	Hydrogen-to-carbon ratio
H_2O	Water
H_2O_2	Hydrogen peroxide
HC	Hydrocarbon
HO_2	Hydroperoxyl
HTHR	High-temperature heat release
IMEP	Indicated mean effective pressure
ITE	Indicated thermal efficiency
ITHR	Intermediate-temperature heat release
IVC/IVO	Intake valve close/open

S. Biswas (✉)
University of Minnesota Twin Cities, Minneapolis, MN 55455, USA
e-mail: biswas@umn.edu

V. Gururajan
Argonne National Laboratory, Lemont, IL 60439, USA

A. K. Agarwal et al. (eds.), *Methanol*, Energy, Environment, and Sustainability,
https://doi.org/10.1007/978-981-16-1224-4_8

LHV	Lower heating value
LLNL	Lawrence Livermore National Laboratory
LTC	Low-temperature combustion
LTHR	Low-temperature heat release
N_2	Nitrogen
NO	Nitric oxide
NO_x	Nitrogen oxide
NTC	Negative temperature coefficient
NVO	Negative valve overlap
O	Atomic oxygen
O_2	Molecular oxygen
O_3	Ozone
OH	Hydroxyl
PAE	Plasma assist effectiveness
PID	Proportional, Integral, Derivative
PM	Particulate matter
PMT	Photomultiplier tube
ppm	Parts per million
PVO	Positive valve overlap
QOOH	Isomer of alkyl hydoperoxide
R	Alkyl radical,
RGF	Residual gas fraction
RI	Ringing intensity
RO	Alkyloxy radical
RO_2	Alkylperoxy
ROC	Rate of consumption
RON	Research octane number
ROOH	Alkyl hydroperoxide
rpm	Revolutions per minute
RXN	Reaction, used for reaction numbering
SACI	Spark assisted compression ignition
SI	Spark ignition
ST	Spark timing
T	Temperature
TDC	Top dead center
TP	Transient plasma
UHC	Unburned hydrocarbon
E	Electric field
R	Specific gas constant
k	Reaction rate
n	Electron density
p, P	Pressure
T	Time period
α	Plasma assist
ϕ	Equivalence ratio

Subscripts

e	Electron
f	Fuel
int	Intake
max	Maximum

1 Introduction

Stringent regulations of oxides of nitrogen (NOx) and particulate matter (PM) emissions, depleting fuel reserves, and unstable fuel prices have forced the automotive manufacturers worldwide to develop alternative combustion strategies to reduce emission levels while maintaining high efficiency and engine power [1–3]. Current emission reduction technology heavily relies on exhaust after-treatment systems with little control of in-cylinder reaction kinetics associated with pollutant generation pathways [4]. However, the proposed future legislation with stricter emission standards is pushing the engine manufacturers to explore pollutant controlling strategies involving in-cylinder combustion processes [5]. In this regard, low-temperature combustion (LTC) is one of the promising advanced in-cylinder combustion strategies capable of reducing NOx and PM emissions via decreased cylinder temperature [6]. LTC also offers high efficiency along with low specific fuel consumption (SFC).

However, LTC suffers from several significant shortcomings that are incredibly challenging to overcome. Despite the low NOx and PM emissions benefits, LTC strategies produce considerably higher carbon monoxide (CO) and unburned hydrocarbon (UHC) emissions [7]. The emission of UHC takes a toll on fuel efficiency. Furthermore, loss of combustion control at high load forbids LTC strategies from attaining brake mean effective pressure (BMEP) greater than 6–7 bars. Therefore, additional strategies, along with the LTC approach, are required to increase the range of engine operating conditions and reduce UHC and CO emissions.

There are several approaches to employ LTC concepts in an engine such as homogeneous charge compression ignition (HCCI), premixed charge compression ignition (PCCI), reactivity-controlled compression ignition (RCCI), active radical combustion (ARC), multiple stage diesel combustion (MULDIC), spark-assisted compression ignition (SACI), etc. [8–12]. While every LTC concept has its advantages and challenges, the performance and emissions characteristics of any LTC strategy strongly depend on the choice of fuel. The LTC behavior is coupled with the autoignition characteristics of the fuel. A fuel that exhibits resistance to autoignition or knocking, i.e., high research octane number (RON) and rich low-temperature chemistry, is considered ideal for LTC strategies. While gasoline or gasoline-like fuels possess high RON, the reactivity of regular gasoline at low temperature lacks sufficient low-temperature chemistry. Hence, an in-depth look into conventional and sustainable alternate fuels to create tailored fuel blends is necessary.

Methanol and dimethyl ether (DME) are two alternate renewable fuels that present several attractive attributes and hold great promise in LTC [13]. Both methanol and DME can be derived from biomass such as plants, wood, agricultural waste, etc., using biochemical conversion processes. The presence of oxygen in methanol enables complete combustion and reduces the toxic emission of unburned hydrocarbon (UHC) and carbon monoxide (CO) [14, 15]. Compared to gasoline, methanol has a higher latent heat of vaporization that facilitates greater thermal efficiency via intake boosting. The hydrogen to carbon ratio of methanol is higher than that of gasoline, but the high-oxygen content reduces the energy content, i.e., heating value of methanol. The high-oxygen content also leads to a low stoichiometric air-fuel ratio. However, the other important physical properties of methanol, such as viscosity, thermal diffusivity, etc., are close to gasoline. Like methanol, DME has several fuel properties that make it attractive for use in diesel engines. DME exhibits rich low-temperature chemistry enabling easy implementation of LTC strategies. Under normal atmospheric conditions, DME is a colorless gas. DME has a high cetane rating, greater than 50. Unlike methane, DME does not require an odorant as it has a sweet smell. DME and propane have similar handling requirements due to similarities in vapor pressures. Like propane, DME must also be kept in pressurized storage tanks at atmospheric conditions. DME has a well-to-tank efficiency slightly lower than petroleum and higher than biomass [16].

A limited number of studies explored the prospect of methanol-blended gasoline in engine combustion. Wei [17] found that methanol-blended gasoline significantly improved the brake thermal efficiency (BTE) of the engine. Shenghua [18] investigated the performance and emission characteristics of a gasoline engine with 10–30% methanol-blended in gasoline. Shenghua's results supported Wei's findings that with methanol addition, the BTE improved with the increase of methanol percentage in the fuel blend, but the engine power and available torque somewhat dropped. Bilgin [19] varied the methanol percentage in the methanol-gasoline blend to find that a 5% methanol blend produced the highest brake mean effective pressure (BMEP). Altun [20] examined emissions characteristics of a gasoline engine running on a fuel blend with 10% methanol and observed a 15 and 10% drop in UHC and CO emissions, respectively, with a slight increase in carbon dioxide (CO_2). Sapre [21] found that the benefit of increased thermal efficiency soon started to diminish with a higher amount of methanol, 70% and above, in the fuel blend. Bardaie [22] reported a 5% power loss from an engine operating using methanol instead of gasoline. Abu-Zaid [23] suggested that methanol–gasoline blends were more suitable for engines with higher compression ratios due to the enhanced octane rating of the fuel blend due to methanol addition. Overall, 5–30% of methanol addition had proved to be beneficial in improving the thermal efficiency and reducing UHC and CO emissions. However, most of the studies on methanol-blended gasoline in engine combustion examine specific conditions or high-temperature spark ignition conditions. Hence, the present research focuses on the low-temperature combustion of methanol-blended gasoline. Similarly, a limited number of studies can be found on the application of DME in engine combustion. A diesel engine fueled with DME was found to meet and surpass California's ultra-low emissions vehicle regulation for medium-duty vehicles [24].

Hansen [25] has noted the possibility of using DME for "late-stage ignition" two-stroke HCCI engines. Park and Lee [26] have discussed the applicability of DME in compression ignition engines and have found that certain engine components like the fuel feed pump, engine sealing materials need to be redesigned to handle DME's low lubricity. The studies above clearly indicate the benefits of methanol-blended gasoline and DME in LTC. However, an in-depth understanding of fuel blends kinetic behavior on combustion performance, and emission characteristics are necessary before introducing them to production engines.

Non-equilibrium plasma discharges have the potential to augment and improve LTC strategies. A low-temperature plasma discharge creates non-equilibrium plasma that results in virtually zero erosion of the electrodes over its lifetime. In low-temperature plasma, the energy relaxation time is sufficiently prolonged, causing the number of inelastic collisions between electrons and surrounding molecules to far exceed the number of elastic collisions. The temperature of such electrons can reach tens of thousands of Kelvins, while the surrounding molecules experience very little gas heating (a typical increase of 10–20 K) [27]. The inelastic collisions by electrons cause molecules to reach certain excited states, which upon deexcitation result in radical species. These radicals play a vital role in initiating chain branching reactions leading to ignition. This possibility, therefore, provokes the following question: "Can the large population of radicals formed in a low-temperature plasma discharge facilitate low-temperature combustion?"

The impact of LTP on kinetic pathways of fuel oxidation and ignition at low temperatures has been studied quite extensively. Applications of diverse types of plasma discharge on lean flames, high-speed flows, gas turbine combustion, pulsed detonation engines, and HCCI combustion have thoroughly been reviewed by Starikovskiy [28]. The coupling of low-temperature plasma with low-temperature fuel chemistry has been explored by a handful of studies. Filimonova [29] showed that the negative temperature coefficient (NTC) behavior could be nearly eliminated for propane-air mixtures when subjected to low-temperature plasma discharges. However, the present chapter would find out conditions for which Filimonova's observation would break down. Sun [30] measured key species in DME in non-premixed flames that show a distinct period of plasma and ground-state kinetics coupling. However, a detailed understanding of plasma-chemistry mechanisms for fuels with practical applications is far from complete. Hence, the knowledge of the current study is of immense importance towards a better understanding of plasma-chemistry coupling.

The first part of the present study investigated the performance and emissions characteristics of an HCCI operation with 0–50% methanol-blended gasoline. The effect of engine speed, intake temperature, equivalence ratio, and methanol percentages was explored in detail on the low-temperature autoignition kinetics of the methanol-blended gasoline. The temperature sensitivity of the critical reactions was assessed for different methanol/gasoline fuel blends. The second part of the study examined the interplay between low-temperature plasma discharges and low-temperature chemical kinetics of DME. Low-temperature plasma kinetics was introduced here to "kickstart" the low-temperature kinetics of DME via hydrogen abstraction and to observe how this affected the prospects of ignition. To understand the role of such rapid hydrogen abstraction, fuels of comparable molecular weight were also subjected to plasma discharges. Fuels that exhibited low-temperature chemistry were by far the most responsive to plasma assistance, but only within a certain range, due to a thermally controlled bottleneck reaction that is explained in detail. The knowledge obtained in this chapter can be leveraged to design tailored fuel blends with controlled LTC behavior.

2 Computational Methodology

Simulations were performed in a 0-D, adiabatic, closed, internal combustion engine reactor (IC-engine) module in ChemkinPro [31] using methanol/gasoline fuel blends for five different methanol concentrations—10, 20, 30, 40, and 50% by volume. The performance and emission characteristics of each fuel blend in the well-mixed HCCI type of combustion environment were estimated. The engine specifications and operating conditions are presented in Table 1.

A surrogate fuel blend was defined for the research-grade gasoline RD587 with a research octane number (RON) of 92.1 and an octane sensitivity (OS) of 7.3 at the Sandia National Laboratories based on the actual fuel composition and was validated with the experimental findings [32]. The constituents of RD587 closely resembles regular E10 gasoline [10, 11]. The mole fraction of each class of hydrocarbon present in the RD587 was included in the surrogate fuel based on a detailed hydrocarbon analysis. Different components of RD587 and the surrogate fuel are listed in Table 2. The key physical and thermodynamic properties of RD587 are summarized in Table 3.

Chemical simulations were carried out using a detailed reaction kinetic mechanism for gasoline surrogates with methanol chemistry [33] from Lawrence Livermore

Table 1 Engine specifications and operating conditions

Displaced volume (cm^3)	551
Bore/Stroke/Connecting Rod (mm)	86/95.1/166.7
Geometric compression ratio	13:1
Intake valve open/close (CA)[a]	343/−145
Exhaust valve open/close (CA)[†]	160/−343
Intake/exhaust pressure (kPa)	100/110
Intake temperature (°C)	60–90
Engine speed (rpm)	800–1600
Fueling rates (mg/cycle)	10.2–17.9
Equivalence ratio, ϕ	0.55–0.9
Residual gas fraction (RGF) (%)	10.5–20

[a]0 CA corresponds to the top dead center (TDC) of the compression stroke

Table 2 Components of the surrogate fuel formulation for RD587

RD587		Surrogate fuel	
Hydrocarbon class	Mole percentage (%)	Species	Mole percentage (%)
Paraffins	16.4	**n-Pentane** **n-Heptane**	9 9
Iso-paraffins	23.6	**Iso-octane**	29
Aromatics	21.1	**Toluene**	20
Oxygenated	19.9	**Ethanol**	20
Cyclo-paraffins	12.2	**Cyclo-pentane**	7
Olefins	5.8	**1-Hexene**	6

Table 3 RD587 gasoline fuel properties

Density @ 288 K (g/L)	749
LHV (MJ/kg)	41.9
H/C ratio	1.972
O/C ratio	0.033
Research octane number	92.1
Octane sensitivity	7.3
10%/50%/90% boiling points (°C)	57/98/155

National Laboratory (LLNL). The mechanism consisted of 2027 species and 8619 reactions, including 215 reactions for NOx chemistry. The chemical mechanism used in this study is an updated edition of the open-source version available on the LLNL website [34], which had been validated widely against experimental data [35].

CVODES [36] was used for the time integration of the species and energy conservation equations in the 0-D isobaric batch reactor model. Cantera [37] was used for ground-state species rate calculations and ZDPLASKIN [38] for plasma kinetic rate calculations. The combustion chemistry models from Burke [39] for DME, methane, ethane, and Merchant [40] for propane were employed. The transient plasma kinetic model is similar to that used by Kosarev [41] for methane, ethane, and propane. For DME, the electron scattering cross-sections of ethane were utilized following Sun's [30] recommendation. It was assumed that reactions involving electrons and ions do not contribute to any temperature increase in the gas. This was a reasonable assumption since, for a typical non-equilibrium plasma, most of the electron collisions are inelastic in nature.

The time integration of the Boltzmann equation for electron transport (known as ZDPLASKIN via BOLSIG + [40]) was carried out for every time-step for the time-period transient plasma discharge was "ON." This implies that plasma kinetics were solved concurrently along with combustion kinetics. However, the objective of the current study was not to compare the effects of distinct types of plasma discharges, e.g., dielectric barrier, corona, streamer, etc., on combustion chemistry. Rather, a sinusoidal 50 ns pulse of transient plasma discharge with a reduced electric field, *E*

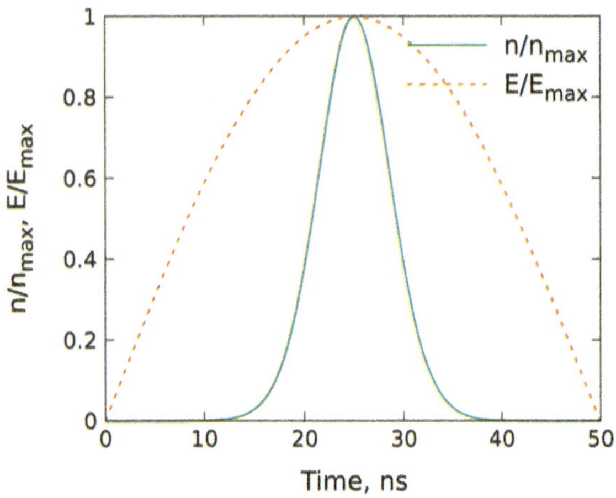

Fig. 1 Variation of electric field and electron density over a single discharge

(i.e., electric field per molecule) was implemented:

$$\frac{E}{E_{max}} = \left|sin\left(\frac{2\pi t}{\mathbb{T}}\right)\right| \tag{1}$$

where $\mathbb{T}$ denotes time period, and the half-period of the transient plasma discharge was 100 ns. E_{max} was set equal to 100 Townsends for all cases. The time-varying electron concentration, n_e (i.e., number of electrons per unit volume) can be written as

$$\frac{n}{n_{max}} = \exp\left[-10sin^2\left(\frac{2\pi t}{\mathbb{T}} - \frac{\pi}{2}\right)\right] \tag{2}$$

Figure 1 shows the time-varying electric field and electron density for a single discharge. Note that the electric field varied over a much longer time compared to the electron density.

The maximum electron density was varied from 10^{11} to 10^{13} cm^{-3} and corresponded to the "weak TP" and "strong TP" cases, respectively. To distinguish "weak TP" from "strong TP," a term called "plasma assist effectiveness"s (PAE) was defined. Figure 2 explains how PAE was calculated from ignition delay measurements. In most practical applications, it was easier to tune the plasma discharge energy to achieve desired ignition characteristics. Hence, PAE values denoted the increase in discharge energy deposited in the fuel/air mixture. A high value of PAE corresponded to a lower ignition delay due to additional energy provided by the transient plasma discharge.

All computations were carried out at a pressure of 10 atm. Argon was used as the inert to avoid complications arising from rotational and vibrational states of nitrogen. Sensitivity analysis was carried out using the code sensBVP [42]. sensBVP was used to capture both the first and second-stage ignition delay sensitivities.

Fig. 2 The plasma-assisted effectiveness at a given temperature was calculated as the ratio of the vertical distances "a" and "b"

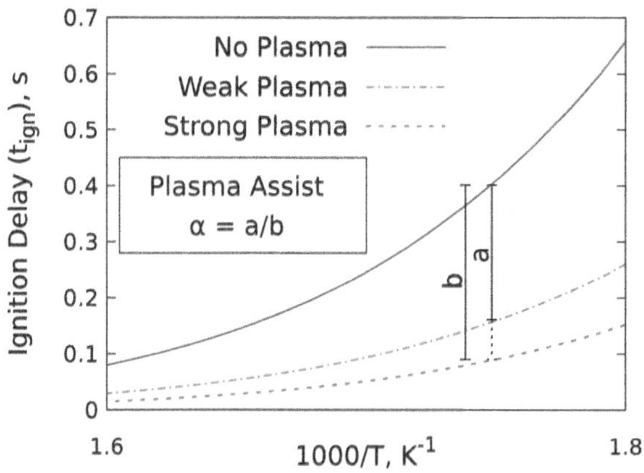

3 Results and Discussion

The results and discussion section is divided into two major subsections. Section 3.1 examined the performance and emissions characteristics of methanol-blended gasoline. First, the effect of methanol addition was discussed in detail for a baseline operating condition of 1200 rpm, intake temperature 75 °C, $\phi = 0.75$. Later, the impact of engine speed sweep, equivalence ratio sweep, and intake temperature sweep on the engine performance and emissions was examined. Finally, the temperature sensitivity of the critical reactions was assessed for different fuel blends. Section 3.2 studied the low-temperature kinetics of DME, employing a non-equilibrium plasma ignition simulation strategy. A zero-dimensional modeling approach was used, accompanied by sensitivity analysis of ignition delay from two-stage ignition. The effect of the negative temperature regime and the dynamics of the thermal bottleneck affecting the efficiency of plasma processes had been discussed.

3.1 Methanol-Blended Gasoline Operation

3.1.1 Effect of Methanol Addition

Methanol addition to gasoline altered the autoignition characteristics of the fuel blend. Figure 3 plots the effect of 10–50% methanol addition to gasoline on cylinder pressure, apparent heat release rate (AHRR), and temperature for a baseline 1200 rpm, intake temperature 75 °C, $\phi = 0.75$ operating condition. Note that the neat gasoline is denoted by M0, i.e., 0% methanol addition. The justification for not pushing further than 50% methanol addition was due to the loss in indicated mean effective pressure (IMEP) or excessive increase in the SFC and UHC that voids the purpose of the present study. Also, the addition of a high percentage of methanol decreased the heating value, and the reactivity of the fuel blend substantially. For the

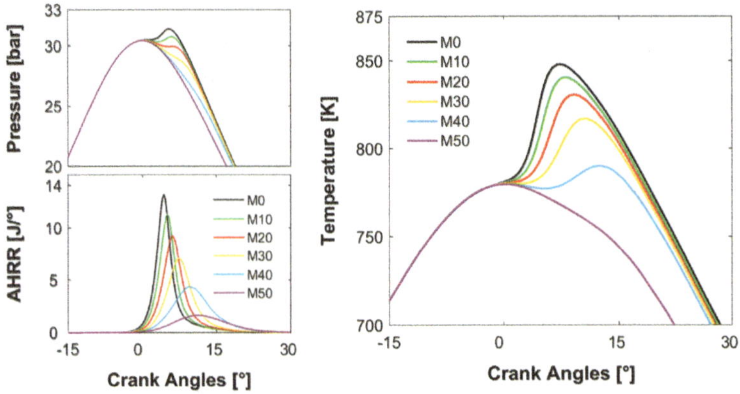

(a) (b)

Fig. 3 In-cylinder (**a**) pressure (top), AHRR profiles (bottom), and (**b**) temperature of methanol-blended gasoline with different methanol percentages for a baseline 1200 rpm, intake temperature 75 °C, $\phi = 0.75$ operating condition

baseline operating condition, ignition was not achieved with 80% methanol-blended gasoline. To ignite a fuel blend with a higher methanol percentage required elevated charge temperature via intake heating and excess RGF that caused a decrease in efficiency.

Figure 3 illustrates that cylinder pressure, temperature, and AHRR dropped with increased methanol percentages in the fuel blend. For the baseline condition, the neat gasoline had a peak temperature of 846 K compared to the peak temperature of the M50 blend, 778 K—a decrease of 68 K. Similarly, the peak AHRR of the M50 blend decreased nearly 80% from the AHRR of neat gasoline. The AHRR curve broadened with methanol addition due to loss in reactivity, which served as a limiting factor towards the further addition of methanol. Moreover, the combustion phasing started to shift away from the TDC with methanol addition. So even though the methanol addition decreased the peak combustion temperature, excessive methanol could hurt the heat release rate leading to misfire and higher cycle-to-cycle variation.

Figure 4 illustrates the effect of 0–50% methanol addition on the performance parameters—IMEP, indicated thermal efficiency (ITE), SFC, and burn angles, CA10, CA50, and CA90–10, 50, and 90% cumulative burn angle, respectively. Burn angles express the combustion phasing in relation to TDC. For example, "CA50 10°" indicates 50% of total heat release due to combustion occurs at 20° CA after TDC. The IMEP decreased linearly from M0 to M40 with a decrease of 0.15 bar per 10% methanol addition. However, the IMEP dropped sharply for M50 indicated a significant change in the chemistry of the M50 blend that led to a sharp drop in ITE and increased SFC as well. The burn angles, CA10, CA50, CA90, started to diverge with a higher concentration of methanol in gasoline, especially for M40 and M50. Methanol addition affected the emissions as well. Figure 5 reveals the effect of methanol addition on NOx, CO, and UHC emissions. Even though NOx and CO

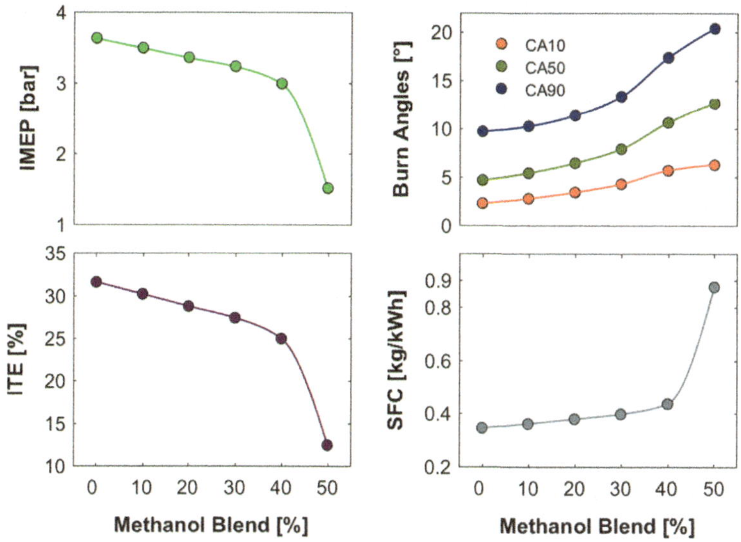

Fig. 4 The effect of 0–50% methanol addition on performance parameters—IMEP, ITE, SFC, and burn angles

Fig. 5 The effect of 0–50% methanol addition on NOx, CO, and UHC emissions

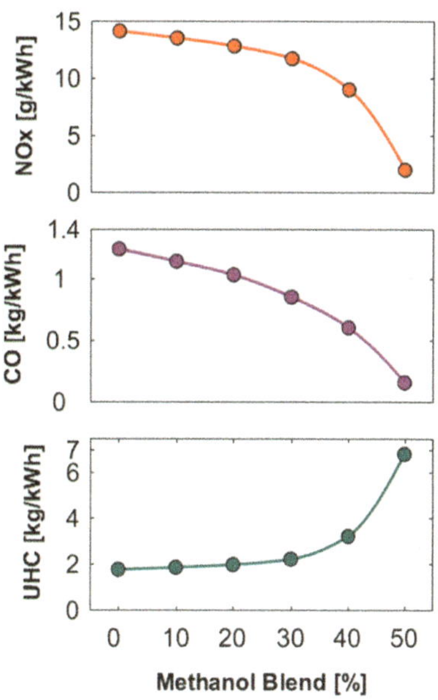

emissions decreased gradually with the addition of methanol, UHC increased significantly after M40. Compared to the neat gasoline, the UHC emissions of M40 and M50 were 1.8 and 3.9 times higher, respectively, due to incomplete burn of the fuel blend with higher methanol percentages.

Based on Figs. 4 and 5, two methanol percentages, 20 and 40%, were selected for detailed study in the following sections. M20 produces extremely UHC emissions while maintaining high ITE and combustion efficiency. On the other hand, M40 was clearly too much of methanol that brought a drastic drop in IMEP and ITE. The emissions values of M40 were not reliable since the combustion was extremely inefficient.

3.1.2 Engine Speed

Figure 6 compares the AHRR (top) and temperature (bottom) of 0, 20, and 40% methanol-blended gasoline for an engine speed sweep (800–1600 rpm) and fixed ϕ = 0.75, T_{int} = 75 °C. The engine speed affected the autoignition chemistry by varying the cycle time available for complete combustion. A greater cycle time allowed more time for complete combustion, and hence the AHRR of M20 had a tighter and higher peak at 800 rpm compared to 1600 rpm. A higher AHRR led to higher cylinder

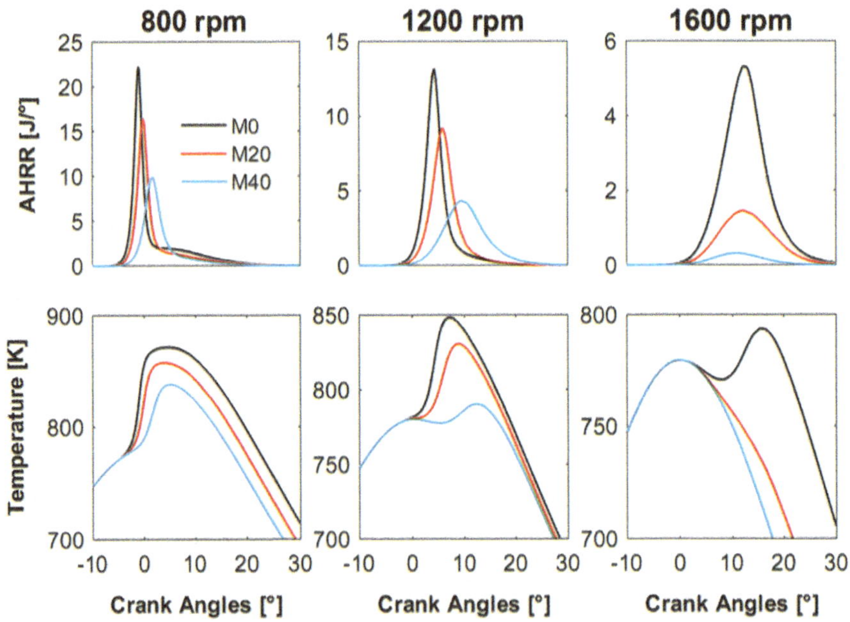

Fig. 6 Comparison of AHRR (top) and temperature (bottom) from 0, 20, and 40% methanol-blended gasoline for a sweep of engine speeds (800–1600 rpm) and fixed ϕ = 0.75, T_{int} = 75 °C

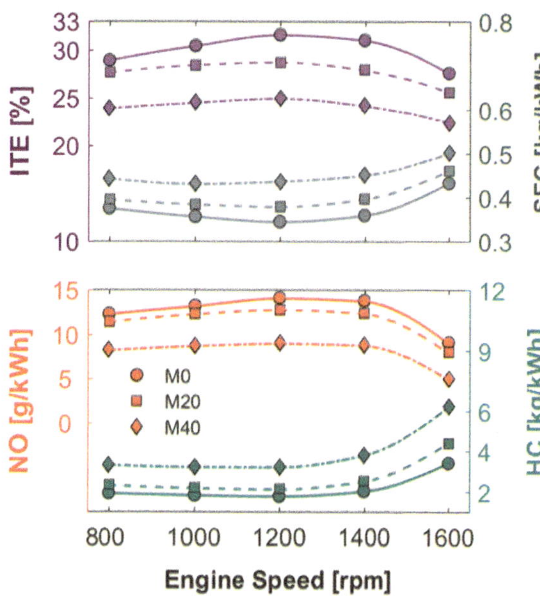

Fig. 7 Effect of engine speed on ITE, SFC (top), NO, and HC (bottom) in 0, 20, and 40% methanol-blended gasoline

temperature as well. A higher methanol addition (M40 compared to M20), showed lesser heat release, and at higher rpm, the heat release was exceedingly small. This behavior was exhibited via the temperature distribution of M40 at 1600 rpm.

Figure 7 shows the effect of engine speed on the performance and emissions characteristics of 0, 20, and 40% methanol-blended gasoline. The ITE dropped a few points with methanol addition, with a slight increase in SFC compared to neat gasoline. NOx emissions decreased significantly, but the UHC emissions grew with methanol addition. Interestingly, for the current range of engine speed, the highest ITE and lowest NOx emissions occurred at 1200 rpm. This showed that the performance and emission characteristics of methanol-blended gasoline were coupled with engine speed and worked best for a speed range of ~1200 rpm. Above 1600 rpm, the available cycle times were largely insufficient for complete combustion led to an excessive increase in UHC emission.

3.1.3 Equivalence Ratio

Figure 8 compares the AHRR (top) and temperature (bottom) of 0, 20, and 40% methanol-blended gasoline for charge equivalence ratio sweep, $\phi = 0.55 - 0.9$ and fixed 1200 rpm, $T_{int} = 75\,^{\circ}\mathrm{C}$. As expected, the AHRR of methanol-blended gasoline dropped with lean mixture; however, the AHRR at lean conditions was tighter, i.e., took the lesser crank angles to complete. Moreover, at $\phi = 0.55$, the difference in peak AHRR between M20 and M40 was less pronounced compared to $\phi = 0.9$. The AHRR of methanol-blended gasoline at the lean condition demonstrated that the true

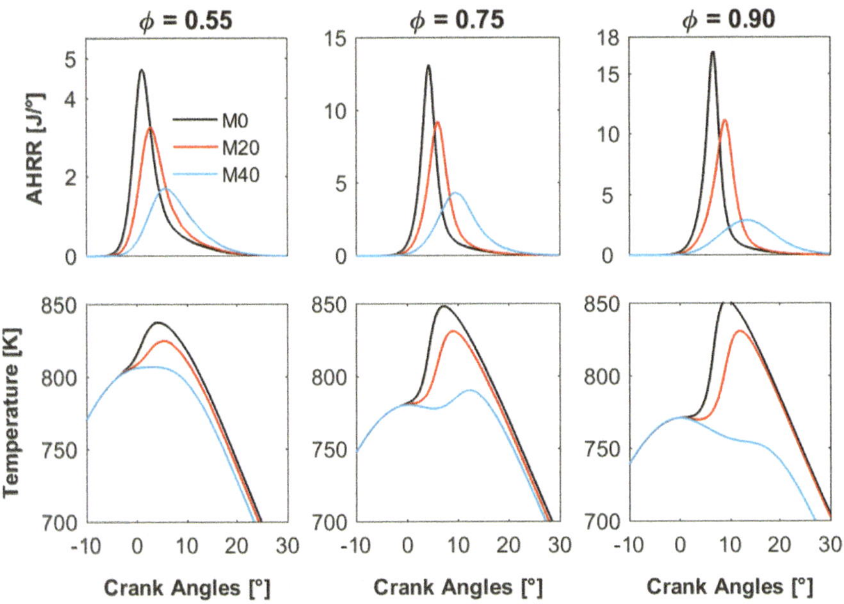

Fig. 8 Comparison of AHRR (top) and temperature (bottom) from 0, 20, and 40% methanol-blended gasoline for a sweep of equivalence ratio, $\phi = 0.55 - 0.9$ and fixed 1200 rpm, $T_{int} = 75\ °C$

potential of methanol addition could be realized with the lean operation. The peak temperature of M0, M20, and M40 at $\phi = 0.55$ were much tighter (36 K difference) compared to the wider peak temperature differences at $\phi = 0.9$. Another interesting fact was that the peak temperature of M40 decreased with the equivalence ratio due to reduced reactivity with a high percentage of methanol presented in the fuel blend that led to incomplete burn. However, M20 did not show this trend. The authors speculate that there existed a limiting methanol percentage between 20 and 40% for which the peak temperature would be constant with an equivalence ratio serving as a limiting methanol percentage in gasoline.

3.1.4 Intake Temperature

Figure 9 compares the AHRR (top) and temperature (bottom) of 0, 20, and 40% methanol-blended gasoline for a sweep of intake temperatures (60–90 °C) and fixed 1200 rpm $\phi = 0.75$. The fuel reactivity increased with higher intake temperature, and the cylinder temperature increased slightly. However, the AHRR decreased because a higher intake temperature also decreased the air density, and hence the IMEP and efficiency dropped. If the IMEP was stayed fixed, the AHRR would have increased with an increase in the intake temperature. However, Fig. 9 illustrates a key point that the variation in the peak temperatures was remarkably close to each other at a higher

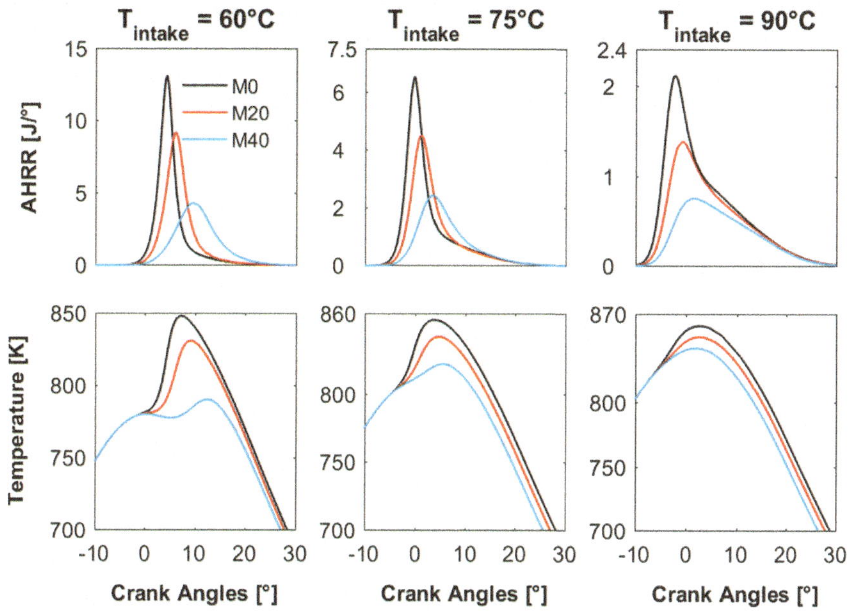

Fig. 9 Comparison of AHRR (top) and temperature (bottom) from 0, 20, and 40% methanol-blended gasoline for a sweep of intake temperatures (60–90 °C) and fixed 1200 rpm $\phi = 0.75$

intake temperature, 90 °C. This implied that the reactivity of the fuel blend reached a plateau with increased temperature. This behavior can be leveraged in designing the intake heating system of the engine.

3.1.5 Sensitivity Analysis

A sensitivity analysis was performed to identify the effect of temperature on the most significant reactions for three different fuel blends, M0, M20, and M40. The logarithmic sensitivity coefficients, S_i were estimated as

$$S_i = \frac{k_i}{T} \frac{\partial T}{\partial k_i} \tag{3}$$

where k is the reaction rate constant, T is the temperature at 0° CA, and subscript i denotes the ith species. The detailed chemical mechanism consisting of 8619 reactions was too large for the sensitivity analysis. Hence, a skeleton reaction mechanism [34] with 538 species and 784 reactions derived from the original detailed mechanism was employed for the sensitivity analysis.

Figure 10 illustrates the temperature sensitivity coefficients of major reactions associated with M0, M20, and M40 methanol blends at the baseline condition of 1200 rpm, 1 bar, 75 °C, $\phi = 0.75$. At low-temperature, methanol addition initiated

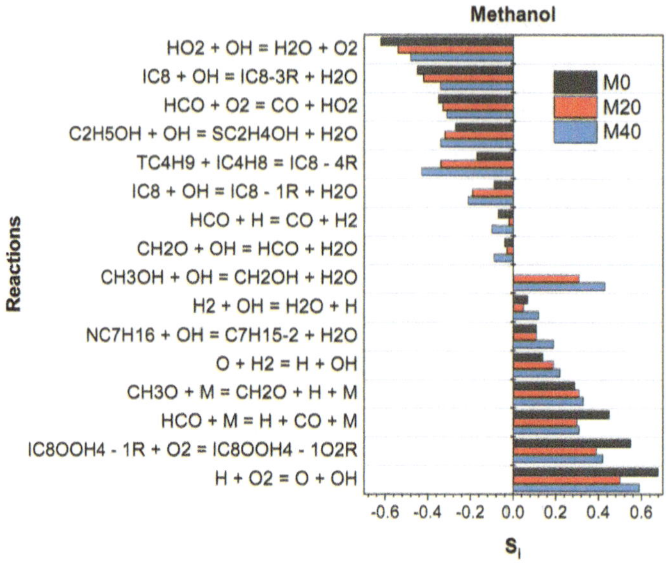

Fig. 10 Logarithmic sensitivity coefficients, S_i of temperature for M0, M20, and M40 fuels blends at the baseline 1200 rpm, 1 bar, 75 °C, $\phi = 0.75$

a pathway for OH to react with methanol that was absent for neat gasoline. Hence the sensitivity coefficients for reaction $CH_3OH + OH = CH_2OH + H_2O$ were zero for M0. Reactions involved larger (IC_8, C_4H_9, C_7H_{16}, etc.) and small (H, OH, etc.) species—both played a key role in methanol-blended gasoline combustion. Several reactions associated with iso-octane (IC8), e.g., $IC_8 + OH = IC_8 - 1R + H_2O$, $IC_8 + OH = IC_8 - 3R + H_2O$, became more important for blended methanol fuels. The temperature sensitivity of the aldehyde reaction, $HCO + M = H + CO + M$, was lessened with methanol addition. The reactions involving H, OH, and O, were similar for gasoline and methanol-added gasoline. However, their temperature sensitivity changed, and the low-temperature chemical reactions got triggered with the methanol addition.

3.2 DME Kinetics and Its Coupling with Low-Temperature Plasmas

3.2.1 The Negative Temperature Coefficient

To understand the transient plasma effects on the ignition, first, a brief overview of the low-temperature chemistry is presented [43–45]. Figure 11 outlines the main steps involved in typical low-temperature hydrocarbon combustion. One of the most important reactions influencing low-temperature chemistry was

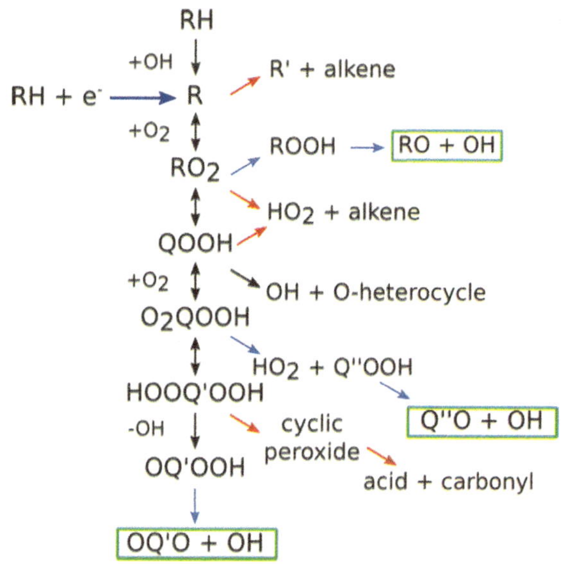

Fig. 11 Main pathways for low-temperature hydrocarbon ignition adapted from Merchant [40]). Blue arrows indicate chain branching, and red arrows indicate chain termination. Chain propagation is designated by black arrows

$$R + O_2 \leftrightarrow RO_2 \quad \text{(RXN1)}$$

where R and RO_2 denote an alkyl and Alkylperoxy radical, respectively. When the ignition temperature was sufficiently low, the forward rate of RXN1 was larger compared to the following internal isomerization reaction of RO_2.

$$RO_2 \leftrightarrow QOOH \quad \text{(RXN2)}$$

Here, QOOH denotes the isomer of alkyl hydroperoxide, where Q stands for H abstracted R. After reaching a sufficient concentration, RO_2 initiated the ROOH formation, which was ultimately responsible for chain branching.

The backward reaction rate of RXN1 increased with temperature, and at a certain temperature, the backward reaction exceeded the forward reaction rate. Only fuels that exhibited NTC behavior possessed such attributes. Figure 12 indicates that both propane and DME retained this feature. The reversal of RXN1 carried important implications for chain branching. Fuel oxidation was forced to proceed via the slower reaction RXN2 as ROOH formation progressively ceased with an increase in temperature. Autoignition could occur if OH formation was sustained via chain branching, which depended on the accumulation of QOOH. Subsequently, OH initiated ignition via H abstraction. However, the two-stage ignition occurred when instead of QOOH, OQ'OOH started forming, which disabled the sustainable OH formation pathway. The autoignition chemistry was vastly different at elevated temperatures, where nearly the entire chain branching happened through hydrogen peroxide radical, HO_2.

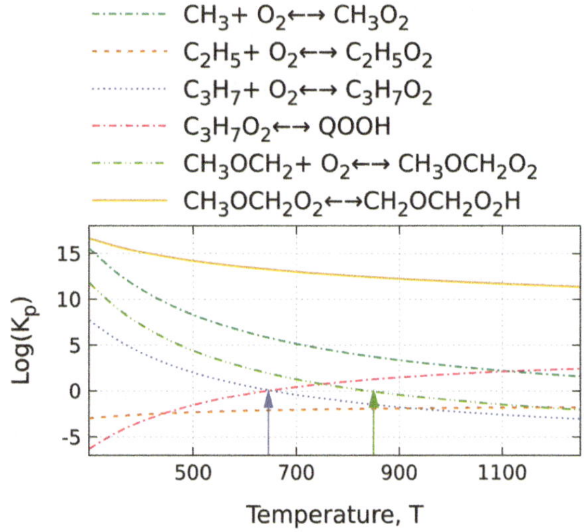

Fig. 12 Equilibrium constants at 10 atm pressure for **a** RXN1 ($R + O_2 \leftrightarrow RO_2$) for methane, ethane, propane, and DME, **b** RXN2 ($RO_2 \leftrightarrow QOOH$) for propane and DME. Notice the temperatures at the crossover junction

3.2.2 Comparison of Ignition Delays

The ignition delays of methane, ethane, propane, and DME with and without plasma assists were plotted in Fig. 13 for stoichiometric fuel/oxidizer/argon mixtures at an initial pressure of 10 atm. The ignition delays were shown as a region bounded between two curves. Purely thermal autoignition was represented by the upper curve, while autoignition in the presence of strong transient plasma was indicated by the lower curve. The continuum of cases in between these two curves was shown as a shaded region, with the ignition delay decreasing with increasing transient plasma intensity. As expected, the ignition delay followed the increasing trend, $\tau_{propane} < \tau_{ethane} < \tau_{methane}$. Interestingly, the dissociation energies of the C-H bond of propane, ethane, and methane were also varied in the same order [46]. DME was an outlier as it manifested the least ignition delay compared to the other fuels, as the presence of O atom in DME weakened the molecular bonds.

The ignition delays were most affected by transient plasma discharge when it was directly coupled with the mixture's chain branching. For instance, methane acted as an inhibitor of chain branching due to its removal of H radicals via

$$CH_4 + H \leftrightarrow CH_3 + H_2 \tag{RXN3}$$

which suppressed the following chain branching reactions responsible for O and OH production.

$$CH_3 + O_2 \leftrightarrow CH_3O + O \tag{RXN4}$$

$$CH_3 + O_2 \rightarrow CH_2O + OH \tag{RXN5}$$

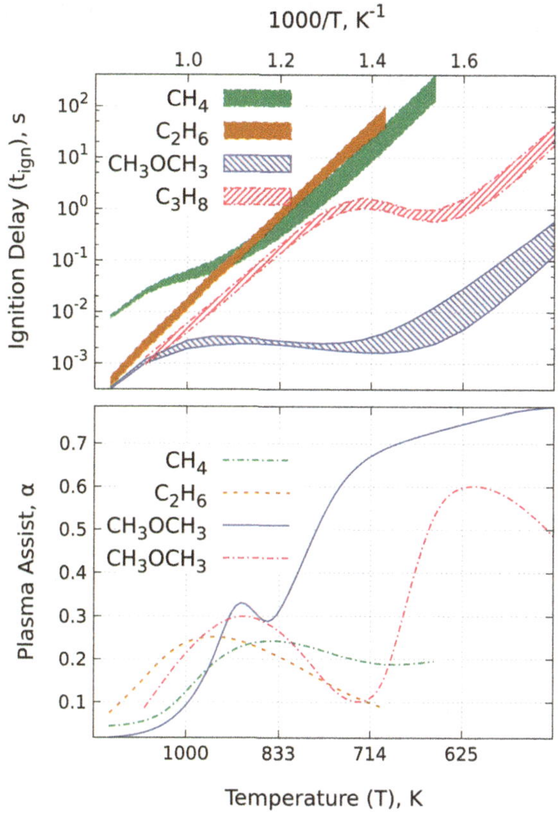

Fig. 13 The ignition delays and plasma assists, α of various fuel/oxidizer/argon mixtures, **a** methane/oxygen/argon (1/2/7.52), **b** ethane/oxygen/argon (1/3.5/13.76), **c** propane/oxygen/argon (1/5/18.8), and **d** DME/oxygen/argon (1/3/11.28). Filled regimes in the ignition delay plot denoted the infinitude of ignition delay times as a function of increasing plasma intensities

Since the primary effect of transient plasma was the formation of alkyl radicals via electron collisions similar to RXN3, the effect of inhibition was once again repeated. This was why transient plasma did not decrease the ignition delay significantly. Note that the PAE was poor for methane due to an increase in the backward rate of

$$CH_4 + HO_2 \leftrightarrow CH_3 + H_2O_2 \quad \text{(RXN6)}$$

Hydrogen peroxide (H_2O_2) was the dominant precursor to chain branching at elevated temperatures for all alkanes. H_2O_2 participated in the chain branching via

$$H_2O_2 + M \leftrightarrow OH + OH + M \quad \text{(RXN7)}$$

Decreasing temperature reduced the PAE via RXN6. Transient plasma-induced alkyl radicals increased the backward RXN6 rate, which implied a decrease in H_2O_2 formation. RXN7 was an entirely different pathway for chain branching compared to those that occurred at lower temperatures. The PAE increased and then decreased near 1000 K as the chain branching regime rapidly changed around this temperature.

Transient plasma discharges broke ethane (C_2H_6) into smaller fuel fragments and radicals such as C_2H_5, C_2H_3, and C_2H_4. The PAE diminished due to the enhancement of the backward rate of

$$C_2H_4 + (H, OH) \leftrightarrow C_2H_3 + (H_2, H_2O) \quad \text{(RXN8)}$$

Transient plasma discharges could effectively couple with the fuels that showed NTC behavior. When the temperature was low enough, RXN1 proceeded forward, forming ROOH, which eventually resulted in chain branching. The next subsection further discusses why the PAE was so high at low-temperatures ignition of DME. DME ignition coupled more effectively with transient plasma discharges compared to propane because DME could not undergo the following reaction:

$$RO_2 \leftrightarrow \text{alkene} + HO_2 \quad \text{(RXN9)}$$

Figure 14 (a) and (b) emphasizes the great distinction in chemical pathways in the first and second-stage ignition of DME by plotting the ignition delay sensitivities for different reactions from both stages. The first stage was dominated by QOOH and RO_2 pathways, while the second-stage was essentially controlled by H_2O_2 formation and dissociation kinetics.

3.2.3 Thermal Bottleneck

At a temperature for which the equilibrium constant of RXN1 went below unity, the production of alkyl radicals in the presence of transient plasma discharges reduced the PAE for both propane and DME. In the case of slightly higher temperatures, OH chain branching relied almost entirely on the accumulation of QOOH radicals. RO_2 must be available in sufficient amounts for this to happen. But if the temperature was high enough, the RXN1 equilibrium reversed, making the addition of alkyl radicals produced by transient plasma discharges inefficient. The fuel was primarily consumed by reactions like RXN6 and RNX8, where OH and HO_2 took part in H abstraction, forming alkyl radicals. Figure 15 shows that in the absence of transient plasma, the temperature profile exhibited two stages of ignition for an initial temperature of 775 K. The radical profiles, on the other hand, displayed three stages. These three stages must be distinguished from the ignition stages. In stage 1, fuel was mainly absorbed by HO_2 as compared to OH. At the starting of the second stage, ample HO_2 existed to initiate

$$HO_2 + HO_2 \leftrightarrow H_2O_2 + O_2 \quad \text{(RXN10)}$$

This was exhibited as a slower growth rate in its concentration profile. DME did not display such a two-stage HO_2 profile since it produced CH_2O much earlier compared to propane. CH_2O was responsible for consuming HO_2 instead of RXN10.

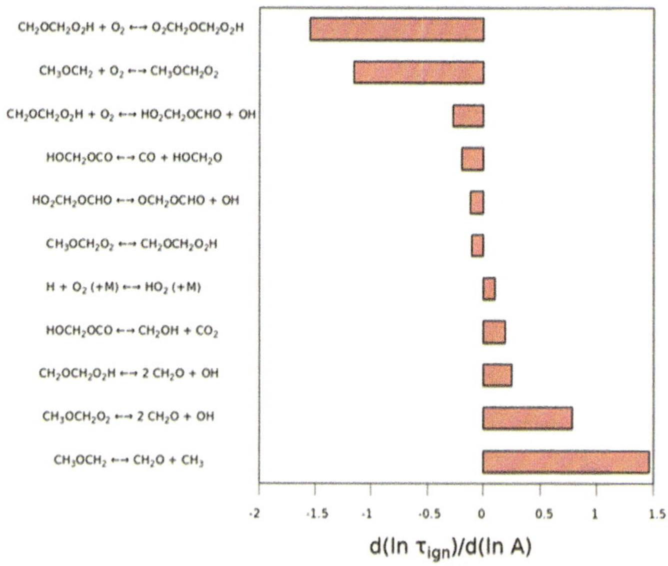

(a)

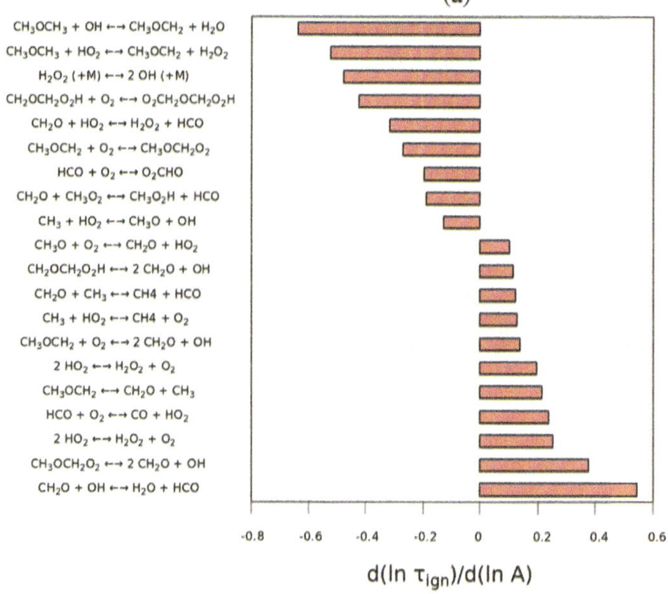

(b)

Fig. 14 **a** First-stage and **b** second-stage ignition delay sensitivities for DME

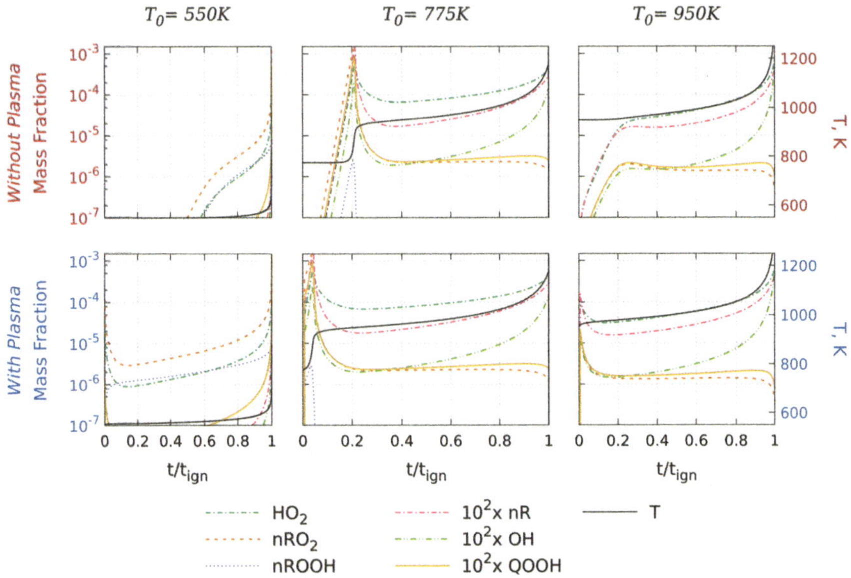

Fig. 15 Key radical concentrations for DME/oxygen/argon ignition with and without transient plasma discharge in stoichiometric mixtures at 10 atm pressure. The time on the x-axis has been scaled by the corresponding ignition delays

Transient plasma substantially reduced the ignition delay of the first-stage ignition due to the presence of alkyl radicals, which increased the formation of RO_2, QOOH, resulting in a faster heat release. Nevertheless, this did not shorten the second-stage ignition delay, as explained by Merchant [40]. As the first-stage delay shortened with the addition of transient plasma, heat release from combustion happened faster. Therefore, the following reactions were pushed backward, resulting in the accumulation of OQ'OOH.

$$nR + O_2 \leftrightarrow nRO_2 \quad \text{(RXN11)}$$

$$QOOH + O_2 \leftrightarrow O_2QOOH \quad \text{(RXN12)}$$

An increase in OQ'OOH concentration led to OH consumption via reaction pathways that promoted fuel consumption and subsequent chain branching reactions. Eventually, the OQ'OOH concentration reached a tipping point, and OH started releasing rapidly, which led to the first-stage ignition. The transient plasma discharge did not influence the first-stage ignition processes, as evident from Fig. 15, due to the excessive accumulation of OQ'OOH. At higher temperatures, the two-stage radical profiles were pronounced, but without any associated temperature rise from RXN10. Since OQ'OOH did not act as a bottleneck in this region, PAE recovered for a brief period, before succumbing to the reverse reactions of RXN6.

3.2.4 Variation with Pressure and Equivalence Ratio

Figure 16 illustrates the variation of ignition delay of DME/oxygen/argon and propane/oxygen/argon mixtures for the stoichiometric condition at two different pressures, 1 and 10 atm. At elevated pressures, the NTC region moved toward higher temperatures due to the pressure-dependent nature of RXN1. Consequently, it was expected that the range of high PAE would also expand. Figure 16 supports this observation.

Figure 17 demonstrates the effect of equivalence ratio on ignition delays of DME/oxygen/argon at 10 atm pressure. Increasing the equivalence ratio for both propane/oxygen/argon and DME/oxygen/argon mixtures decreased ignition delay. This behavior could be attributed to the increase in the concentration of active radicals at higher equivalence ratios. The effectiveness of transient plasma increased accordingly, as shown in Fig. 17.

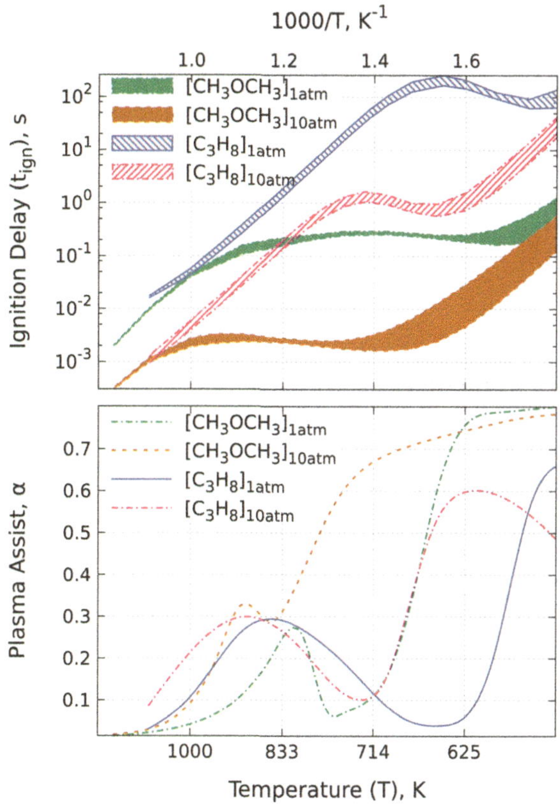

Fig. 16 The variation of ignition delay of DME/oxygen/argon and propane/oxygen/argon mixtures for the stoichiometric condition at two different pressures, 1 and 10 atm

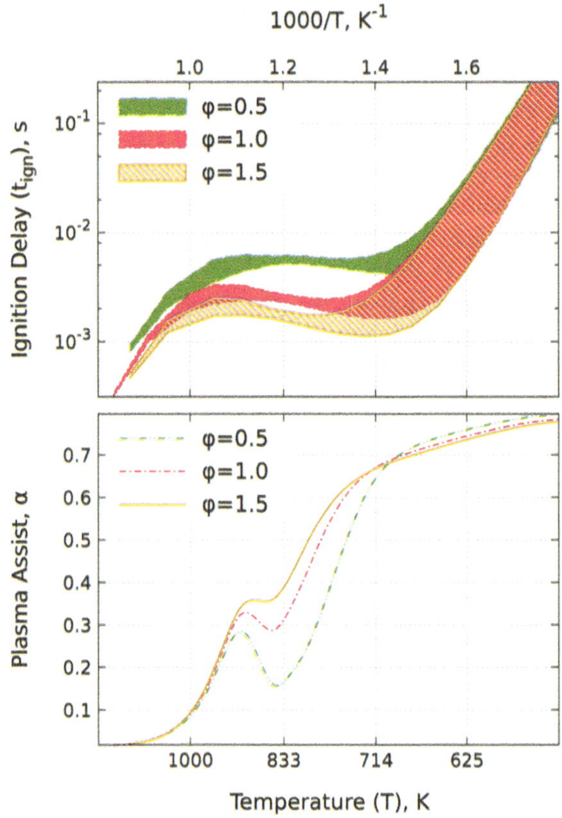

Fig. 17 The effect of equivalence ratio on ignition delays of DME/oxygen/argon at 10 atm pressure

4 Conclusions

In the first part of the present study, the performance and emissions characteristics of homogeneously mixed engine combustion of the effect of 0–50% methanol-blended gasoline was studied numerically for a baseline operating condition of 1200 rpm, intake temperature 75 °C, $\phi = 0.75$. The effect of engine speed, equivalence ratio, and intake temperature was examined in detail. The temperature sensitivity of the key reactions was assessed for different methanol/gasoline fuel blends. The significant findings are as follows:

- For an identical operating condition, with methanol addition, the peak heat release rate decreased, and the heat release distribution broadened compared to neat gasoline due to loss in reactivity. For a methanol percentage of 40% or higher, the ITE and IMEP dropped significantly, with an increase in UHC. However, the NOx and CO emissions decreased with methanol addition. The LHV of the fuel blend reduces significantly with 40% methanol addition is the key reason that resulted in a substantial loss in engine performance. Intake heating or fuel additive can solve this problem by increasing the mixture reactivity.

- For the range of engine speed studied, maximum ITE and minimum NOx was achieved for an engine speed of 1200 rpm. Above 1600 rpm, the available cycle times were largely insufficient for complete combustion led to an excessive increase in UHC emission.
- The variation in the peak combustion temperatures was marginal with intake temperature sweep, 60–90 °C. This implied that the reactivity of the fuel blend reached a plateau with increased intake temperature.
- At low temperature, methanol addition created several alternate pathways for OH production that led to low-temperature heat release and enabled autoignition kinetics.

In the second part of the present study, the interplay between plasma and DME low-temperature kinetics was examined in detail. Major findings are as follows:

- Fuels exhibiting negative temperature coefficient behavior exhibited a higher inclination to effectively couple with transient plasma discharges at low temperatures.
- A thermal bottleneck occurred for negative temperature coefficient ignitions, which could not be overcome by transient plasma discharges alone. The presence of transient plasma discharge could shorten the first-stage ignition delays but failed to do so for the second stage.
- Transient plasma discharges were ineffective at higher temperatures as the chain branching was dominated by hydrogen peroxide formation via several thermally controlled stages after transient plasma formed fuel radicals.

In this study, methanol-blended gasoline and DME demonstrated improved combustion and emissions characteristics due to low-temperature reaction kinetics. The individual components in the fuel blend not only affected the physical property of the fuel, such as heating value, H/C ratio, etc., but also altered the key intermediate reaction pathways that appear promising in designing tailored fuels. In the years to come, renewable fuels like methanol and DME could create a potential pathway for developing low carbon tailored fuel blends with higher efficiency and lesser pollutant emission for automotive applications.

References

1. Caterpillar (2003) Cat estimates $40 million expense for EPA non-conformance penalties (NCPs) for Post-October, 2002 diesel engines. Diesel Fuel News 7(5): 7–10
2. United States, E.P.A., Office of Transportation and Air Quality (2012) Non-conformance penalties for heavy-duty diesel engines subject to the 2010 NOx emission standard. U.S. Environmental Protection Agency, Office of Transportation and Air Quality, Washington, D.C.
3. Agarwal RA, Agarwal AK, Gupta T, Sharma N (eds) (2019) Pollutants from energy sources. energy, environment, and sustainability. Springer, Singapore
4. Agarwal AK, Gupta JG, Sharma N, Singh AP (eds) (2019) Advanced engine diagnostics. energy, environment, and sustainability. Springer, Singapore

5. Shurpali N, Agarwal AK, Srivastava VK (eds) (2019) Greenhouse gas emissions: challenges, technologies and solutions. energy, environment, and sustainability. Springer, Singapore
6. Glassman I, Yetter RA (eds) (2008) Combustion. Academic Press, Amsterdam, Boston, p 773
7. Krishnamoorthi M, Malayalamurthi R, He H, Kandasamya S (2019) A review on low temperature combustion engines: performance, combustion and emission characteristics. Renew Sustain Energy Rev:116
8. Stanglmaier RH, Roberts CE (1999) Homogeneous Charge Compression Ignition (HCCI): benefits, compromises, and future engine applications. J Engines: SAE Int 108(3):2138–2145
9. Dec JE (2009) Advanced compression-ignition engines—understanding the in-cylinder processes. Proc Combust Inst 32(2):2727–2742
10. Biswas S, Ekoto I (2019) Detailed investigation into the effect of ozone addition on spark assisted compression ignition engine performance and emissions characteristics. SAE Technical Paper 2019-01-0966
11. Biswas S, Ekoto I (2019) Spark assisted compression ignition engine with stratified charge combustion and ozone addition. JSAE 20199089
12. Biswas S, Ekoto I (2019) Ozone added spark assisted compression ignition. advanced combustion techniques and engine technologies for the automotive sector, ed. Singh AP, Sharma N, Agarwal RA, Agarwal AK. Springer, Singapore
13. Beatrice C, Bertoli C, Giacomo ND (1998) New findings on combustion behavior of oxygenated synthetic diesel fuels. Combust Sci Technol 137:31–50
14. Zhang C, Wu H (2016) Combustion characteristics and performance of a methanol fueled homogenous charge compression ignition (HCCI) engine. J Energy Inst 89(3):346–353
15. Rakopoulos CD, Kyritsis DC (2001) Comparative second-law analysis of internal combustion engine operation for methane, methanol, and dodecane fuels. Energy 26(7):705–722
16. Semelsberger TA, Borup RL, Greene HL (2006) Dimethyl ether (DME) as an alternative fuel. J Power Sources 156:497–511
17. Wei J, Liu H, Li S, Yang R, Liu J, Wang Y (2008) Effects of methanol/gasoline blends on a spark ignition engine performance and emissions. Energy Fuels 22(2):1254–1269
18. Shenghua L, Clemente E, Tiegang H, Yanjv W (2007) Study of spark ignition engine fueled with methanol/gasoline fuel blends. Appl Therm Eng 27:1904–1910
19. Bilgin A, Sezer I (2008) Effects of methanol addition to gasoline on the performance and fuel cost of a spark ignition engine. Energy Fuels 22:2782–2788
20. Altun S, Oztop H, Oner C, Varol Y (2013) Exhaust emissions of methanol and ethanol-unleaded gasoline blends in a spark ignition engine. Thermal Sci 17(1):291–297
21. Sapre R (1989) Properties, performance and emissions of medium concentration methanol–gasoline blends in a single-cylinder, spark-ignition engine. SAE Trans 97(3):1105–1126
22. Bardaie Z, Janius R (1984) Conversion of spark-ignition engine for alcohol usage - comparative performance. Agric Mech Asia, Africa Latin Am 15(2):31–40
23. Abu-Zaid M, Badran O, Yamin J (2004) Effect of methanol addition on the performance of spark ignition engines. Energy Fuels 18(2):312–315
24. Fleisch T, McCarthy C, Basu A, Udovich C, Charbonneau P, Slodowske W, Mikkelsen S, McCandless J (1995) A new clean diesel technology: demonstration of ULEV emissions on a navistar diesel engine fueled with dimethyl ether. SAE Technical Paper 950061
25. Hansen HR, Persen TD, Schramm J (2011) Development of HCCI engines for dimethyl ether. DTU Mechanical Engineering
26. Park SH, Lee CS (2014) Applicability of dimethyl ether (DME) in a compression ignition engine as an alternative fuel. Energy Convers Manag 86:848–863
27. Starikovskaia SM, Plasma assisted ignition and combustion. J Phys D: Appl Phys:39
28. Starikovskiy A, Aleksandrov NL (2013) Plasma-assisted ignition and combustion. Prog Energy Combust Sci 39(1):61–110
29. Filimonova EA (2015) Discharge effect on the negative temperature coefficient behaviour and multistage ignition in C3H8-air mixture. J Phys D: Appl Phys 48:015201
30. Sun W, Uddi M, Won SH, Ombrello T, Carter C, Ju Y (2012) Kinetic effects of non-equilibrium plasma-assisted methane oxidation on diffusion flame extinction limits. Combust Flame 159(1):221–230

31. Reaction Design. Reaction Workbench 15131 San Diego, 2013
32. Lopez Pintor D, Dec J, Gentz G (2019) Φ-Sensitivity for LTGC engines: understanding the fundamentals and tailoring fuel blends to maximize this property. SAE Technical Paper 2019-01-0961, p 24
33. Mehl M, Pitz W, Westbrook CK, Curran HJ (2011) Kinetic modeling of gasoline surrogate components and mixtures under engine conditions. Proc Combust Inst 33(1):193–200
34. Mehl M, Wagnon S, Tsang K, Kukkadapu G, Pitz WJ, Westbrook CK, Tsang Y, Curran HJ, Atef N, Rachidi MA, Sarathy MS, Ahmed A (2017) A comprehensive detailed kinetic mechanism for the simulation of transportation fuels. In: 10th US National combustion meeting, College Park, MD
35. Mehl M, Pitz WJ, Westbrook CK, Yasunaga K, Conroy C, Curran HJ (2011) Autoignition behavior of unsaturated hydrocarbons in the low and high temperature regions. Proc Combust Inst 33(1):201–208
36. Cohen SD, Hindmarsh AC (1996) CVODE, A Stiff/Nonstiff ODE Solver in C. Comput Phys 10:138–143
37. http://www.cantera.org/docs/sphinx/html/index.html
38. Pancheshnyi S, Eismann B, Hagelaar GJM, Pitchford LC (2008) Computer code ZDPlasKin. http://www.zdplaskin.laplace.univ-tlse.fr
39. Burke U, Somers KP, O'Toole P, Zinner CM, Marquet N, Bourque G, Petersen EL, Metcalfe WK, Serinyel Z, Curran HJ (2015) An ignition delay and kinetic modeling study of methane, dimethyl ether, and their mixtures at high pressures. Combust Flame 162:315–330
40. Merchant SS, Goldsmith CF, Vandeputte AG, Burke MP, Klippenstein SJ, Green WH (2015) Understanding low-temperature first-stage ignition delay: propane. Combust Flame 162:3658–3673
41. Kosarev IN, Aleksandrov NL, Kindysheva SV, Starikovskaia LS, Starikovskii AY (2009) Kinetics of ignition of saturated hydrocarbons by nonequilibrium plasma: C2H6- to C5H12-containing mixtures. Combust Flame 156(1):221–233
42. Gururajan V, Egolfopoulos FN (2019) Direct sensitivity analysis for ignition delay times. Combust Flame 208:478–480
43. Zádor J, Taatjes CA, Fernandes RX (2011) Kinetics of elementary reactions in low-temperature autoignition chemistry. Prog Energy Combust Sci 37(4):371–421
44. Rousso AC, Hansen N, Jasper AW, Ju Y (2018) Low-temperature oxidation of ethylene by ozone in a jet-stirred reactor. J Phys Chem 122:8674–8685
45. Ekoto I, Wolk B, Northrop W (2017) Energy analysis of low-load low-temperature gasoline combustion with auxiliary-fueled negative valve overlap. SAE Int J Engines 10(3):1238–1255
46. Gribov LA, Novakov IA, Pavlyuchko AI, Vasil'ev EV (2006) Spectroscopic calculation of CH bond dissociation energy in the series of chloro derivatives of methane, ethane, and propane. J Struct Chem 47:635–641

Ignition and Extinction of Hydrogen and Gasoline Blended Methanol Flames

Sayan Biswas

Abbreviations

Φ	Equivalence ratio
ξ	Strain rate
ρ	Density
λ	Thermal conductivity
BMEP	Brake mean effective pressure
BTE	Indicated thermal efficiency
CO	Carbon monoxide
CO_2	Carbon dioxide Subscripts
c_p	Constant pressure specific heat
FWHM	Full-with half maximum
HC	Hydrocarbon
H/C	Hydrogen-to-carbon ratio
HCNG	Hydrogen added compressed natural gas
HCO	Aldehyde
H_2O	Water
HO_2	Hydroperoxyl
H_2O_2	Hydrogen peroxide
HR	Heat release
IC8	Iso-octane
ITE	Indicated thermal efficiency
κ	Reaction rate
L_f	Laminar flame thickness
Le	Lewis number
LHV	Lower heating value

S. Biswas (✉)
University of Minnesota Twin Cities, Minneapolis, MN 55455, USA
e-mail: biswas@umn.edu

A. K. Agarwal et al. (eds.), *Methanol*, Energy, Environment, and Sustainability,
https://doi.org/10.1007/978-981-16-1224-4_9

LLNL	Lawrence Livermore National Laboratory
m	Mass
NOx	Nitrogen oxide
O	Atomic oxygen
O_2	Molecular oxygen
OH	Hydroxyl
OS	Octane sensitivity
P	Pressure
Q	Power
R	Alkyl radical
RON	Research octane number
S	Logarithmic sensitivity coefficient
S_L	Laminar flame speed
SI	Spark ignition
T	Temperature

Subscripts

UHC	Unburned hydrocarbon
ad	Adiabatic
b	Burned
ext	Extinction
f	Fuel
i	ith species
ign	Ignition
min	Minimum
mix	Mixture
u	Unburned

1 Introduction

Over a century, burning fossil fuels—petroleum, natural gas, oil, and coal–has helped to meet our ever-growing energy needs. Even today, fossil fuels contribute to 80% of the world's total energy demand in transportation, industrial, residential, and commercial sectors [1–3]. Heavy reliance on fossil fuels has taken an immense toll on society and the environment—from pollutant emissions to climate change. In the past decade, sharp depletion in the fossil fuel reserve, fluctuating oil prices, and stringent emissions regulations have compelled the search for viable renewable alternatives. Two such renewable fuels are methanol and hydrogen—a potential pathway for developing low carbon tailored fuel blends with higher efficiency and lesser pollutant emission [4–6]. Both methanol and hydrogen can be derived from biomass such as plants, wood, agricultural waste, etc., using biochemical conversion processes. The presence of oxygen in methanol enables complete combustion and reduces the toxic emission of unburned hydrocarbon (UHC) and carbon monoxide (CO) [7–9]. Compared to gasoline, methanol has a higher latent heat of vapourization that facilitates greater thermal efficiency via intake boosting but also poses difficulty in cold start to an engine running entirely on methanol. On the other hand, hydrogen is an extremely reactive fuel with high diffusivity that is easy to ignite and results in a faster-propagating flame. Hence, it would be interesting to study the effect of methanol addition to gasoline and hydrogen addition to methanol.

Oxidation of methanol and hydrogen had been studied extensively—from the fundamentals to application-specific studies. However, the combustion behaviour of methanol-blended gasoline and hydrogen added methanol had been studied to a lesser extent. A limited number of studies explored the prospect of methanol-blended gasoline in engine combustion. Wei [10] found that methanol-blended gasoline significantly improved the brake thermal efficiency (BTE) of the engine. Shenghua [11] investigated the performance and emission characteristics of a gasoline engine with 10–30% methanol blended in gasoline. Shenghua's results supported Wei's findings that with methanol addition, the BTE improved with the increase of methanol percentage in the fuel blend, but the engine power and available torque somewhat dropped. Bilgin [12] varied the methanol percentage in the methanol–gasoline blend to find that a 5% methanol blend produced the highest brake mean effective pressure (BMEP). Altun [13] examined emissions characteristics of a gasoline engine running on a fuel blend with 10% methanol and observed 15% and 10% drop in UHC and CO emissions, respectively, with a slight increase in carbon dioxide (CO_2). Sapre [14] found that the benefit of increased thermal efficiency soon started to diminish with a higher amount of methanol, 70% and above, in the fuel blend. Bardaie [15] reported a 5% power loss from an engine operating using methanol instead of gasoline. Abu-Zaid [16] suggested that methanol–gasoline blends were more suitable for engines with higher compression ratios due to the enhanced octane rating of the fuel blend due to methanol addition. Overall, 5–30% of methanol addition had proved to be beneficial in improving the thermal efficiency and reducing UHC and CO emissions.

Similarly, extensive studies were carried out, focussing on the combustion characteristics of hydrogen-blended with a wide variety of fuels. Hu [17] found that the addition of hydrogen to low calorific value fuels accelerated the flame propagation and that in turn affected the flame stability. Several studies aimed to examine the combustion characteristics of hydrogen-blended natural gas (HCNG). Huang [18] observed an exponential rise in flame speed with an increased hydrogen percentage in HCNG. Ren [19] speculated that the radicals from hydrogen were responsible for a higher adiabatic flame temperature of HCNG. Hu [20] explained the role of hydrogen-generated radicals, H, OH, and O, in promoting the laminar flame speed of methane. Al-Hamamre [21] investigated the flame speed enhancement of hydrogen-blended acetylene and ethanol. The majority of the studies were fundamental in nature that primarily explored the improvement of laminar combustion behaviour of fuel blends due to hydrogen enrichment. However, the effect of hydrogen addition to methanol had hardly been studied. A handful of studies explored the engine performance and emissions behaviour with methanol/hydrogen blend [22–25]. In summary, engine performance was found to improve with hydrogen addition leading to an increased in-cylinder pressure and temperature, superior indicated thermal efficiency and faster burning while reducing the emission of formaldehyde, oxides of nitrogen (NOx), CO, and unburnt methanol.

The aforementioned studies evidently indicated the benefits of adding methanol to gasoline and hydrogen to methanol. However, an in-depth understanding of fuel blends behaviour and comprehensive engine testing are required before introducing them to production engines. Future automotive efficiency gains will require the current gasoline engines to move towards challenging lean/dilute regimes. A lean-burn engine suffers from cycle-to-cycle variation, reduced efficiency, and increased unburned hydrocarbon emissions due to difficulty in the ignition. A significant challenge faced by a highly diluted or high tumble combustion system is to sustain the flame from extinction. Hence, the ignition and extinction behaviour of fuel blends are of significant importance to the combustion and engine community.

The present study investigated the fundamental processes and chemical kinetics behind ignition and extinction behaviour of two fuel blends—20% methanol in gasoline, and 20% hydrogen in methanol, at 1 bar, 300 K. Fundamental laminar combustion properties, such as laminar flame speed, flame thickness, minimum ignition energy, thermal and chemical flame structure were discussed in detail. The effect of fuel blending on ignition and extinction strain rates was examined. To compare the performance of fuel blends to their base fuel, the laminar flame characteristics of methanol, gasoline, and hydrogen were also included in this study. Finally, a sensitivity analysis of fuel blends was performed to identify the critical reactions and intermediate species associated with the ignition and extinction phenomena.

2 Computational Methodology

Perfectly stirred reactor, premixed counterflow, and premixed freely propagating laminar flame [26] modules of ChemkinPro [27] were used to solve the conservation equations of mass, momentum, species concentrations, and energy for three base fuels, methanol, gasoline, and hydrogen, and two fuel blends, gasoline blended with 20% methanol, and methanol mixed with 20% hydrogen by volume. A surrogate fuel blend was defined for the research-grade gasoline RD587 with a research octane number (RON) of 92.1 and an octane sensitivity (OS) of 7.3 at the Sandia National Laboratories based on the actual fuel composition and was validated with the experimental findings [28]. The composition of RD587 resembles regular E10 gasoline [29, 30]. The mole fraction of each class of hydrocarbon present in the RD587 was included in the surrogate fuel based on a detailed hydrocarbon analysis. Different components of RD587 and the surrogate fuel are listed in Table 1.

Chemical simulations were carried out using a detailed reaction kinetic mechanism for gasoline surrogates with methanol chemistry [31] from Lawrence Livermore

Table 1 Components of the surrogate fuel formulation for RD587

RD587		Surrogate fuel	
Hydrocarbon class	Mole percentage (%)	Species	Mole percentage (%)
Paraffins	16.4	**n-Pentane** **n-Heptane**	9 9
Iso-paraffins	23.6	**Iso-octane**	29
Aromatics	21.1	**Toluene**	20
Oxygenated	19.9	**Ethanol**	20
Cyclo-paraffins	12.2	**Cyclo-pentane**	7
Olefins	5.8	**1-Hexene**	6

National Laboratory (LLNL). The mechanism consisted of 2027 species and 8619 reactions, including 215 reactions for NOx chemistry. The chemical mechanism used in this study is an updated edition of the open-source version available on the LLNL website [32], which had been validated widely against experimental data [33].

Minimum ignition poses a fundamental limit on the ignitibility of a fuel/air mixture. Minimum ignition energy, $E_{ign,min}$ was estimated by turning on a point heat source of power $\dot{Q}$ for a time-period of t, supplying energy, $E = \int_0^t \dot{Q}dt$ to the fuel/air mixture. The time-period was gradually increased until ignition occurred in the mixture. The minimum ignition energies were recorded for different fuels at various equivalence ratios. In practice, the ignition source has finite volume, e.g. spark or laser ignition. Hence, $E_{ign,min}$ can be depicted as the bare minimum energy supplied by the finite volume of hot inert gases for ignition to occur. To simplify further, assume the external heat source to be spherical. Moreover, it could safely be presumed that the characteristic diameter of the ignition source scaled with flame thickness [34]. Then minimum ignition energy can be written as

$$E_{ign,min} = \rho_{mix} l_3^f c_{p_{mix}} (T_{ad} - T_{mix}) \quad (1)$$

where ρ is density, l_f is flame thickness, c_p is constant pressure specific heat, T is temperature, subscript mix, and ad denotes the fuel/air mixture, and adiabatic property, respectively. Chemical equilibrium was used to calculate the adiabatic flame temperature, T_{ad} and the equilibrium product compositions. The premixed freely propagating flame configuration was used to compute the laminar flame speed, s_L. The laminar flame thickness was estimated using the following relation [35]:

$$l_f = \frac{\lambda_b}{c_p \rho_u s_L} \quad (2)$$

where λ is thermal conductivity, subscripts u and b denote unburned and burned, respectively.

Ignition and extinction strain rates were computed numerically using different chemical reactor models in ChemkinPro. The ignition states were estimated using two different methods. A perfectly stirred reactor model was used to achieve the ignition of a premixed fuel/air mixture by varying the initial temperature and residence time. Another method to attain ignition was via the premixed freely propagating model. The inlet velocity of the premixed fuel/air mixture was reduced slowly until the ignition was achieved. To ensure that a stable ignition state had arrived, the temperature of the burned products was equated with the adiabatic flame temperature from the chemical equilibrium calculations.

Extinction strain rates were determined using premixed counterflow configuration, as shown schematically in Fig. 1. A detailed description of the counterflow problem and governing equations can be found in [36]. The counterflow configuration was modified to include asymmetric boundary conditions and thermal radiation at the optically thin limit [37]. Both the opposing nozzles issued a premixed fuel/air mixture at 1 bar, 300 K, to establish twin flat flames sitting on either side of the stagnation plane. Based on the assumption that the radial velocity changed linearly in the radial direction, the fluid properties were assumed to vary only with the axial distance. Hence, the three-dimensional problem was reduced to a one-dimensional model solving for species, temperature, and velocity distributions within the flow

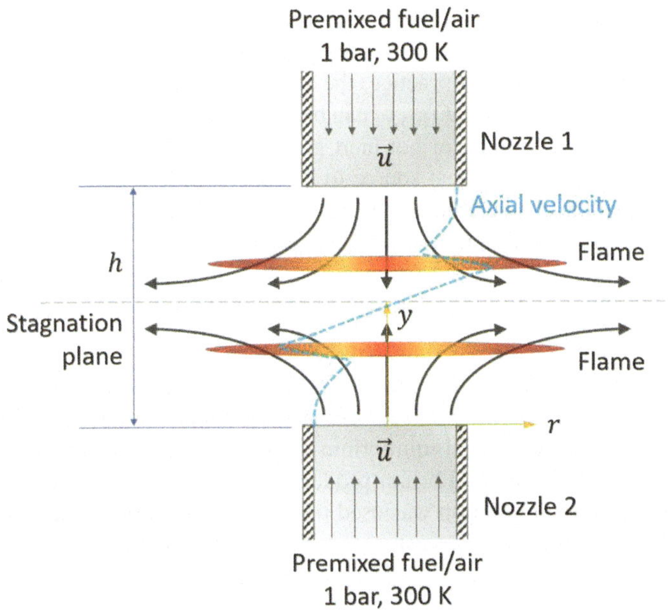

Fig. 1 Schematic of the premixed counterflow flame. The stagnation plane is shown by the grey dashed line, a typical axial velocity profile is indicated by the sky blue dashed line, and the twin flames are displayed by the red/orange shaded region

between the nozzles. The edge-effect or the effect of unsteady strain was not considered in this study. The extinction was achieved in two stages. First, a strongly burning flame was generated for a given equivalence ratio, pressure, temperature, and nozzle exit velocity in the premixed counterflow configuration with a mass flux ranging between 0.5 and 5 kg/m^{3}s. The nozzle exit velocities were slowly increased until the flame extinction was achieved, keeping all other parameters fixed. To accurately capture the extinction state, a two-point arc-length continuation method [38] was employed. The global strain rate, $\vec{u}/h$ where h is the distance between the opposed nozzle, increased with the increase in nozzle exit velocity.

A sensitivity analysis was performed to identify the most significant reactions that influenced ignition and extinction strain rates. The logarithmic sensitivity coefficients, S_i were estimated as

$$S_i = \frac{k_i}{\xi_i}\frac{\partial \xi_i}{\partial k_i} \tag{3}$$

where k is the reaction rate constant, ξ is the strain rate, and subscript i denotes the ith species. The detailed chemical mechanism consisting of 8619 reactions was too large for the sensitivity analysis. Hence, a skeleton reaction mechanism [32] with 538 species and 784 reactions derived from the original detailed mechanism was employed for the sensitivity analysis.

The errors and uncertainty associated with the computational results were investigated carefully. First, experimentally measured flame speed data available in the literature had been utilized to validate the computed flame speed of methanol, gasoline, and hydrogen. The computed results showed an excellent match with the existing experimental data. The variation of computed results due to the different chemical mechanisms (detailed versus skeleton) and solver settings (multicomponent versus species-averaged diffusion model) had been explored in detail. The variation in results was below 1.5% near stoichiometry, while slowly increased by nearly 5% at the sufficiently lean or rich side.

3 Results and Discussion

The results and discussion is divided into five sections. The first section discussed the laminar flame characteristics—flame speed and flame thickness of different base fuels and fuel blends. The second section examined how the addition of methanol and gasoline affects the minimum ignition energies of different fuel blends. The third section explored the equivalence ratio dependence of ignition and extinction strain rates. An in-depth discussion about the thermal and chemical flame structure, distribution of key products, and intermediate radical species from different fuel blends were presented in the fourth section. The concluding section explored the important reactions and the sensitivity of ignition and extinction strain rates towards these reactions. The idea was to find out the major reactions and intermediate species that play a leading role in altering the laminar flame properties of fuel blends.

3.1 Flame Speed and Flame Thickness

Figure 2 illustrates the variation of laminar flame speed and flame thickness with the equivalence ratio for methanol, hydrogen, gasoline, and methanol fuel blends at an initial condition of 1 bar, 300 K obtained using the detailed chemistry. Both species-transport models—mixture-average and multicomponent models—were employed in the freely propagating premixed flame computation. These results were virtually independent of the species-transport model utilized. Except for the hydrogen

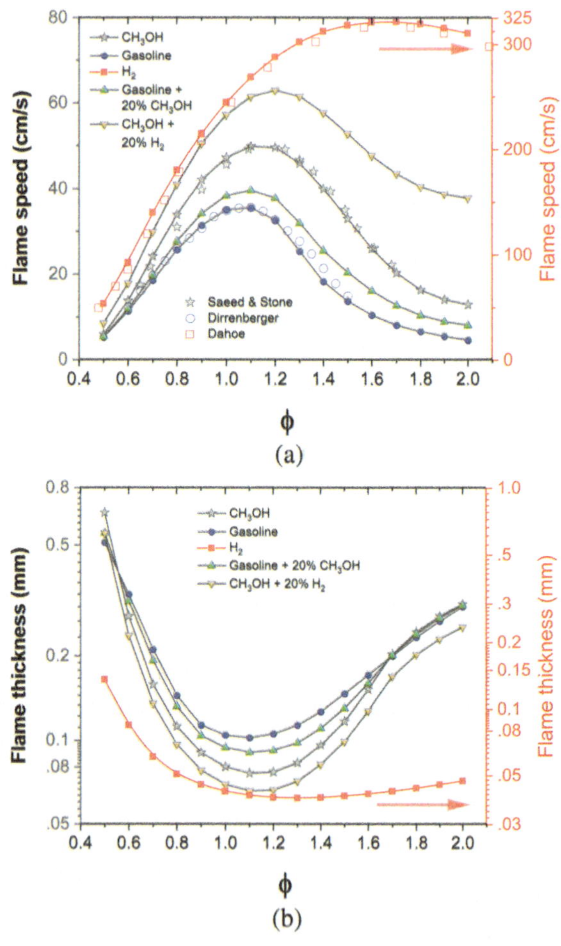

Fig. 2 Laminar (**a**) flame speed and (**b**) flame thickness of methanol, gasoline, hydrogen, and methanol fuel blends at 1 bar, 300 K. Computed flame speed data are compared with experimental measurements available in the literature, Dahoe [41], Saeed and Stone [42], Dirrenberger [43]

flame, the results were independent of constant-Lewis-number approximation as well. Moreover, with the multicomponent transport model, the flame speeds were nearly identical with and without the thermal-diffusion effect of different chemical species, indicating that the Soret effect was negligible for methanol, gasoline, and methanol fuel blends.

The numerically computed flame speeds of methanol, hydrogen, and gasoline matched within 4% of the most recent experimental measurements reported in the literature [39–41]. Except for hydrogen, the flame speed of methanol, gasoline, and methanol fuel blends reached its peak near $\phi = 1.1 \pm 0.04$ and dropped on the either side of the peak value. The maximum flame speed of methanol, 50 cm/s, was nearly 41% higher compared to gasoline, 35.5 cm/s. However, adding a small amount of methanol into gasoline improved the flame speed of the fuel blend. The peak flame speed of gasoline mixed with 20% methanol was 40 cm/s—12.7% higher relative to gasoline. The effect of methanol addition to gasoline on laminar flame speed was more prominent for a rich mixture. For example, the difference in flame speed of methanol-blended gasoline and neat gasoline was 2.4% at $\phi = 0.5$ compared to 43% at $\phi = 2$. The hydrogen exhibited a maximum flame speed of 322 cm/s at $\phi = 1.7$. Hence, it was of interest to examine the effect of hydrogen addition to methanol. Hydrogen addition enhanced methanol flame speed significantly. With a 20% hydrogen addition, the peak flame speed of methanol increased from 50 cm/s to 62.9 cm/s, nearly a 26% increase. The increase in the flame speed due to hydrogen addition was substantially higher in the rich regime. This increase in flame speed due to hydrogen addition can be attributed to the highly diffusive nature of hydrogen. The laminar flame speed can be related to Lewis number, Le via, $s_L \sim \sqrt{Le}$. The Lewis number for sufficiently lean and rich hydrogen/air mixtures are $Le \ll 1$ and $Le \gg 1$, respectively. Hence, the effect of Le reduces flame speed on the lean side while increases on the rich side. This shifted the peak flame speed of hydrogen and hydrogen-blended methanol towards the rich side.

Flame thickness is a key metric for laminar flames that affects ignition and extinction. Figure 2 (b) shows the flame thickness of different fuels and fuel blends. Like the flame speed, the flame thickness reduced to a minimum at a similar value of $\phi = 1.1$, and the thickness increased with the lean or rich mixtures. The flame thickness of methanol at stoichiometry was 0.08 mm, 20% thinner than the stoichiometric flame thickness of gasoline, 0.1 mm. Adding methanol to gasoline or hydrogen to methanol further reduced the flame thickness of the fuel blend. The effect of blending on flame thickness was particularly noticeable for $0.7 \leq \phi \leq 1.7$. Outside this range of equivalence ratio, the effect of blending started to diminish. Compared to other fuels and fuel blends, the hydrogen flame was extremely thin [41]. At 1 bar, 300 K, hydrogen flame thickness was 40–136 microns for an equivalence ratio of $0.5 \leq \phi \leq 2$. Moreover, after the hydrogen flame reached a minimum thickness of 40 microns at $\phi = 1.3$, the thickness increased only to 47 microns at $\phi = 2$.

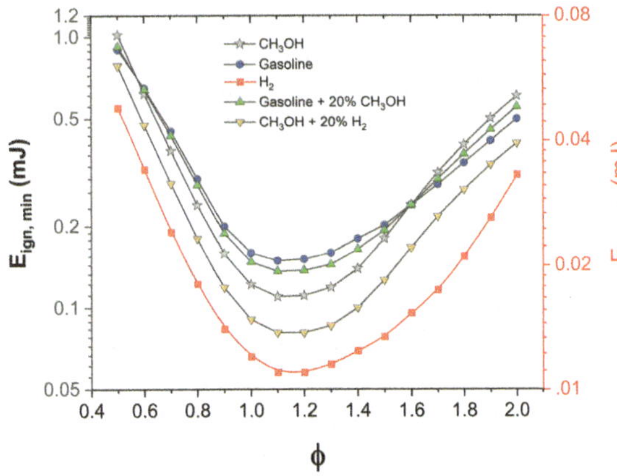

Fig. 3 Equivalence ratio dependence of minimum ignition energy of methanol, gasoline, hydrogen, and methanol fuel blends at 1 bar, 300 K

3.2 *Minimum Ignition Energy*

Minimum ignition energy is a critical parameter for a fuel/air mixture. $E_{ign,min}$ of different base fuels and fuel blends as a function of equivalence ratio is shown in Fig. 2. Note that Fig. 3 is plotted on a semi-logarithmic scale to exaggerate any minor differences of $E_{ign,min}$ values. The $E_{ign,min}$ for methanol varied between 0.11 and 1 mJ. The gasoline had a higher $E_{ign,min}$ than methanol, 0.18–1.1 mJ. However, with the addition of methanol into gasoline, the $E_{ign,min}$ of the fuel, blend dropped slightly. The largest change in $E_{ign,min}$ was found with the hydrogen addition to methanol. Hydrogen had extremely small $E_{ign,min}$ requirement, 0.01–0.048 mJ, nearly an order of magnitude lower than methanol. Adding hydrogen to methanol reduced the $E_{ign,min}$ of the fuel blend. At stoichiometry, $E_{ign,min}$ of hydrogen added methanol, 0.091 mJ, decreased 25% from methanol, 0.122 mJ. Hence, blended fuels demonstrated a higher reactivity with lower $E_{ign,min}$ compared to its neat fuel counterpart. These characteristics of blended fuels can be leveraged to alter the autoignition characteristics of gasoline.

3.3 *Ignition and Extinction Strain Rates*

Ignition and extinction of steady-state combustion can be correlated with the lower and upper turning points of S-curves. The S-curve concept is applied to study the response of a combustion system to an imposed parameter such as strain rate, external perturbation, or heat loss. Figure 4 exemplifies the S-curve for methanol/air at 1 bar, $\phi = 0.8$ with an inlet temperature of 1120 K. The lower and upper turning points of the S-curve denoted the ignition and extinction, respectively. The upper (shown in blue) branch indicated a vigorously burning flame, and the lower (shown in red) branch indicated a weakly reacting flame. The middle branch (shown by the dashed

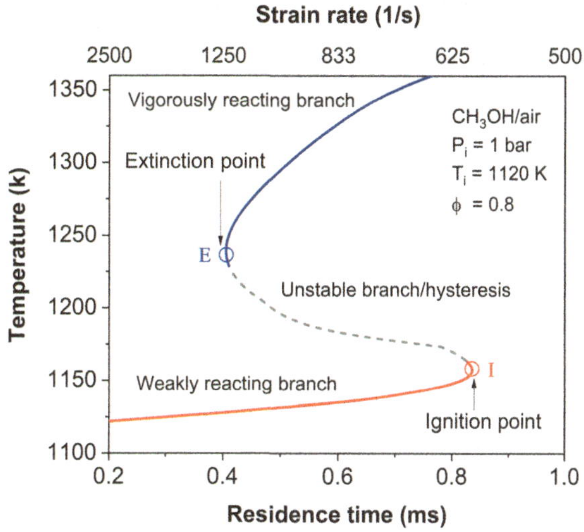

Fig. 4 Ignition and extinction states on an S-curve for $\phi = 0.8$ methanol/air mixture at initial pressure and temperature of 1 bar and 1120 K, respectively

grey line) showed the unstable branch or flame hysteresis. An additional axis for strain rate is also shown in Fig. 4. For every initial condition and mixture compositions, strain rates at ignition and extinction are computed.

Figure 5 illustrates the ignition and extinction strain rate as a function of equivalence ratio for methanol, and two fuel blends—methanol-blended gasoline and hydrogen added methanol. Ignition and extinction strain rates of gasoline are also added to this figure for comparison purposes. The peak ignition strain rate for methanol was 655 1/s at $\phi = 1.1$. The ignition strain rate increased with hydrogen addition. This implies that the addition of hydrogen enhanced the mixture reactivity that enabled the mixture to survive a higher strain rate, i.e. higher initial flow velocity or shorter residence time, prior to ignition. From an engine combustion standpoint, a fuel with a higher ignition strain rate could ignite in a challenging situation such as high tumble and excessive turbulent conditions. A similar improvement in ignition characteristics can be noticed for methanol-blended gasoline as well. The dashed grey line indicates the ignition strain rates of neat gasoline. However, with the addition of methanol, the ignition strain rates increased by nearly 15% at the peak strain rate condition of $\phi = 1.1$. Hence, the addition of methanol in gasoline and hydrogen to methanol facilitated a stronger flame kernel capable of sustaining a higher aerodynamic stretch during ignition.

Extinction strain rates for different base fuels and fuel blends show analogous behaviour similar to ignition strain rates. For methanol, the peak extinction strain rate of 1282 1/s was nearly two times higher than the ignition strain rate at $\phi = 1.1$. The ratio of extinction and ignition strain rates was ranged between 1.36 and 2.1, with an average of 1.8. Such characteristics of strain rates indicate that while the strongly burning flame branch can survive a higher inlet flow velocity (or lower residence time) prior to extinction, a weakly reacting state to jump to the strongly burning

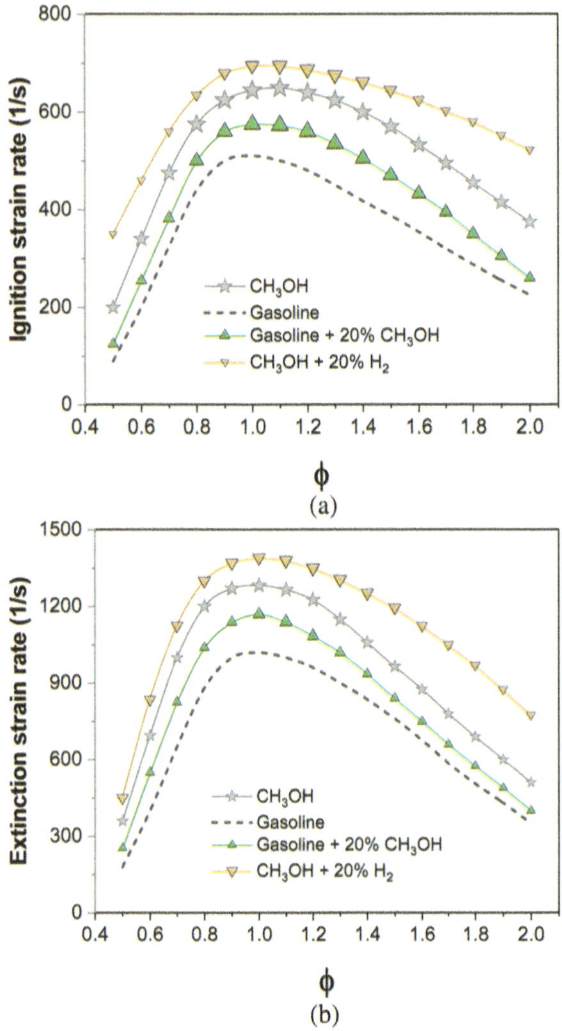

Fig. 5 **a** Ignition and **b** extinction strain rates of methanol and methanol fuel blends as a function of equivalence ratio at 1 bar, 300 K. The marker size is proportional to the maximum temperature of each test condition

state a flow velocity 1.8 times lower (or higher residence time) than previous was required. However, the addition of methanol and hydrogen showed a greater impact on the extinction strain rate. The extinction strain rate of methanol-blended gasoline was 38% higher compared to gasoline at $\phi = 1.1$.

For all different fuels and fuel blends, the ignition and extinction strain rates reached the maximum at the slightly rich side of stoichiometry, around $\phi = 1.1 \pm 0.03$. Compared to the rich mixture, the strain rates demonstrated a greater sensitivity to the equivalence ratio for lean mixtures. For hydrogen-blended methanol, the average slope in the lean portion of the extinction strain rate curve was four times higher than the average slope in the rich regime. This behaviour was pronounced for both the fuel blends during ignition as well as extinction.

The marker size in Fig. 5 is proportional to the temperature at ignition or extinction states. Two major observations can be drawn from this information. First, it is evident from the marker sizes that the temperature at extinction was lower compared to the temperature at the ignition for similar equivalence ratios. For different base fuels and fuel blends examined in this study, the extinction temperatures were 110–260 K lower than the ignition temperature (not shown here). This implies the hysteresis behaviour of laminar flames associated with the ignition and extinction states. Second, the relative changes in the marker sizes from the $\phi = 0.5$ to stoichiometry were more substantial for ignition compared to extinction cases. Hence, the ignition strain rates were extremely sensitive to the temperature compared to the extinction strain rates.

3.4 Thermal and Chemical Flame Structure

Previous sections have summarized that methanol addition to gasoline and hydrogen addition to methanol improves the laminar burning behaviour, ignition, and extinction characteristics. The present and the next section would take a closer look at the reaction kinetics to understand the reasons behind such improvements in fundamental behaviour. Figure 6 compares the thermal and chemical flame structure of methanol, gasoline, methanol-blended gasoline, and hydrogen added methanol with methanol flames at 1 bar, 300 K, $\phi = 0.8$. The total heat release (HR), and temperature distribution are shown in the top plot, and the middle and bottom plots show the stable reactants and products (methanol, CO_2, H_2O, H_2, CO, O_2) and intermediate radical species (OH, H, O, HO_2, HCO, CH), respectively. Even though the previous sections plotted both the lean and rich regimes, understanding the lean combustion bears more significance from an engine application standpoint. Hence, the $\phi = 0.8$ was chosen as the base condition to compare these two fuel blends.

At $\phi = 0.8$, the peak flame temperatures of methanol-blended gasoline (2097 K) and hydrogen added methanol (2061 K) were slightly higher than the methanol flame (2036 K). The maximum total HR followed a similar trend where the total HR by the blended fuels was slightly higher than the methanol. Note that the flame thickness of methanol at $\phi = 0.8$ was 0.12 mm. Flame thickness closely matched the full-width half maximum (FWHM) of the total HR. The peak flame temperature and flame thickness of gasoline were greater than methanol.

However, there were several key differences in the products and intermediate species among fuel blends. The middle row of Fig. 6 plots the stable species of reactants and products at $\phi = 0.8$. As expected, the fuel got completely consumed within a short expanse of ~0.5 mm. The length of the preheat zone of methanol was 12–16% longer compared to the fuel blends. The left-over oxygen in the product stream in all three cases indicated lean combustion. It was interesting to note the hydrogen concentrations in methanol, gasoline, and methanol-blended gasoline. Methanol produced 26 and 37% more hydrogen in the preheat zone than methanol-blended gasoline and

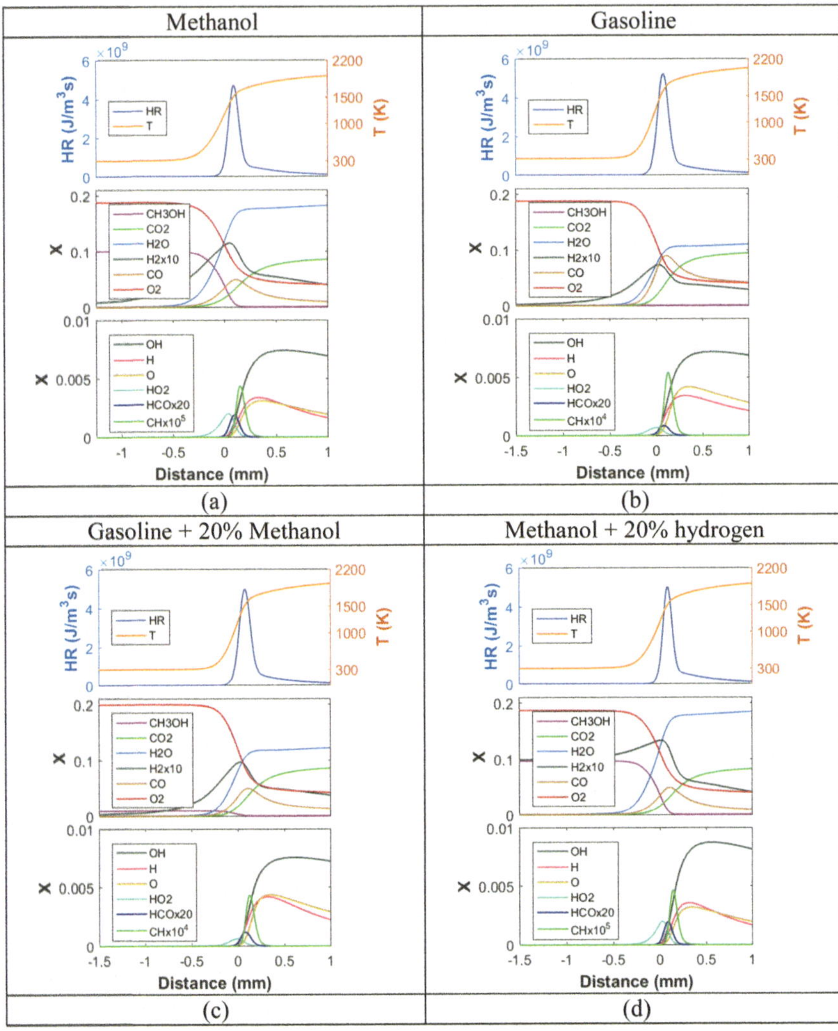

Fig. 6 Profiles of temperature, total heat release (top), stable reactants and products (middle), and intermediate radical species (bottom) of methanol, gasoline, and methanol fuel blends at 1 bar, 300 K, $\phi = 0.8$

neat gasoline, respectively. The reactions responsible for hydrogen production were

$$CH_3OH + H = CH_3O + H_2 \tag{4}$$

$$CH_3OH + H = CH_2OH + H_2 \tag{5}$$

Table 2 Chemical properties of the base fuels used in this study

Mass%	Methanol	Gasoline	Hydrogen
O [%]	49.9	~3.64	0
H [%]	12.6	~13.6	100
C [%]	37.5	~82.76	0
LHV [MJ/kg]	20.1	41.9	120
H/C	4	1.972	∞

Both these reactions (4) and (5) with nearly similar temperature sensitivity could occur within the preheat zone due to the presence of methanol. The presence of hydrogen before the reaction zone enhanced gasoline reactivity.

The distribution of intermediate radical species plotted in the bottom plot of Fig. 6 shows some interesting characteristics between the methanol, gasoline, and fuel blends. Note that the CH concentration was an order of magnitude higher for gasoline and methanol-blended gasoline compared to methanol. Adding 20% methanol to gasoline increased OH, O, and H concentrations compared to neat gasoline. Table 2 lists the important chemical properties of the base fuels. Fuel with a higher H/C ratio and lower heating value (LHV) is desired for greater heat release and faster flame propagation. Even though methanol possessed a lesser LHV, the H/C of methanol is twice compared to gasoline. Hence, adding 20% methanol improved the H/C ratio without hurting the LHV of the fuel blend.

Hydroperoxyl (HO_2) radicals played a crucial role in controlling the reactivity of the system. Methanol produced a significant amount of HO_2 compared to other fuel blends via the following reactions:

$$CH_3OH + O_2 = CH_3O + HO_2 \tag{6}$$

$$CH_3OH + O_2 = CH_2OH + HO_2 \tag{7}$$

For hydrogen added methanol, the most significant HO_2 production occurred via the three-body reaction

$$H_2 + O_2 + M = HO_2 + H + M \tag{8}$$

The concentration of aldehyde (HCO) radical produced during the combustion of methanol and methanol-blended fuels are different due to the difference in the molecular structure of the mixture composition. This observation was consistent with the study of Norton and Dryer [44]. HCO contributed to formaldehyde emissions. Hydrogen addition to methanol promoted the consumption of HCO via

$$HCO + H = CO + H_2 \tag{9}$$

Therefore, blending hydrogen to methanol had a positive consequence in reducing formaldehyde emissions.

The spatial distribution of three key radicals—hydroxyl (OH), atomic hydrogen (H), atomic oxygen (O) in methanol, methanol-blended gasoline, and hydrogen added methanol flames of $\phi = 0.8$ are shown in Fig. 7. Hydrogen added methanol produced 22% higher OH compared to methanol. However, the concentration of H and O were affected marginally with hydrogen addition. Peak O concentration of methanol-blended gasoline was 38% higher relative to methanol. Increased O concentration created a sustainable alternative pathway to generate OH. Another interesting observation for methanol-blended gasoline was that OH started forming slightly earlier compared to methanol. This observation can be correlated with the slightly thicker flame for methanol-blended gasoline ($l_f = 0.14$ mm) compared to methanol ($l_f = 0.12$ mm) at $\phi = 0.8$.

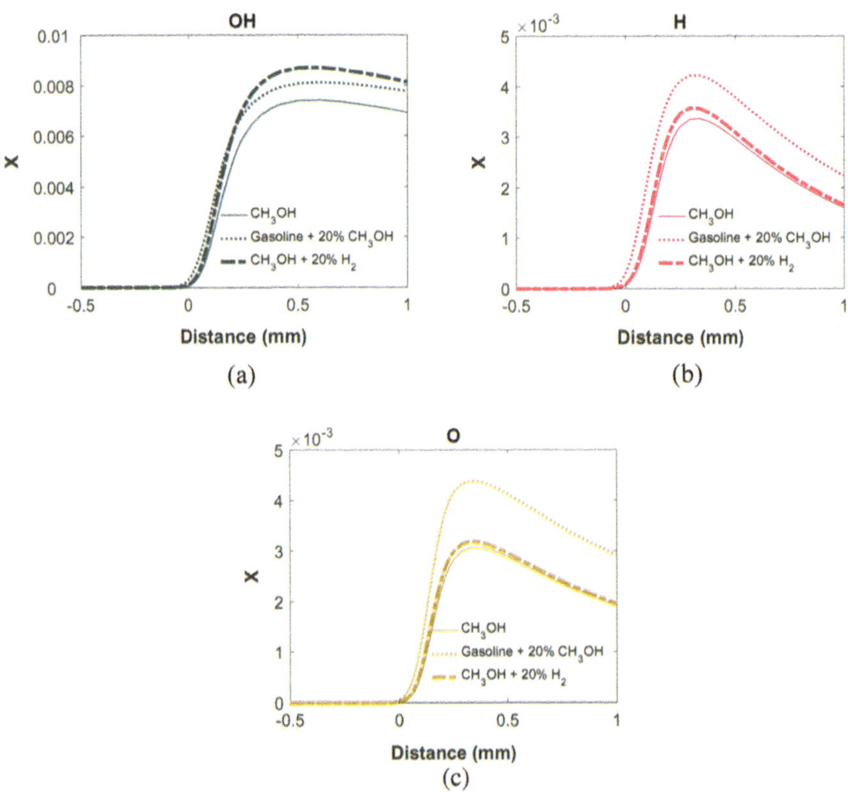

Fig. 7 Spatial distribution of **a** OH, **b** H, and **c** O of methanol and methanol fuel blends at 1 bar, 300 K, $\phi = 0.8$

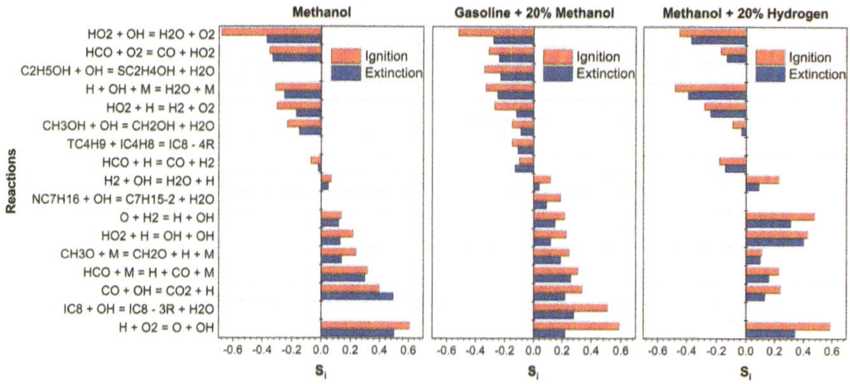

Fig. 8 Logarithmic sensitivity coefficients, S_i of ignition and extinction strain rates of methanol and methanol fuel blends at 1 bar, 300 K, $\phi = 0.8$

3.5 *Reaction Sensitivity Analysis*

Figure 8 illustrates the logarithmic sensitivity coefficients of ignition and extinction strain rates to different intermediate reactions for methanol and methanol fuel blends at 1 bar, 300 K, $\phi = 0.8$. While the most sensitive reactions associated with the ignition and extinction strain rates were similar, the following four reactions were only applicable for methanol-blended gasoline.

$$C_2H_5OH + OH = SC_2H_4OH + H_2O \quad (10)$$

$$TC_4H_9 + IC_4H_8 = IC_8 - 4R \quad (11)$$

$$NC_7H_{16} + OH = C_7H_{15-2} + H_2O \quad (12)$$

$$IC_8 + OH = IC_8 - 3R + H_2O \quad (13)$$

Hence, the sensitivity coefficients for methanol and hydrogen-blended methanol were zero for these reactions.

The sensitivity coefficients for ignition were greater compared to extinction indicated a bigger influence of reaction kinetics on the ignition process relative to extinction. Ignition strain rates were primarily sensitive to the kinetics of H_2, small hydrocarbons, and radicals such as OH, HO_2, HCO, etc. OH was a crucial radical that strongly influenced the ignition. The following intermediate reactions provided a steady source of OH:

$$H + O_2 = O + OH \quad (14)$$

$$HO_2 + H = OH + OH \tag{15}$$

$$O + H_2 = H + OH \tag{16}$$

With hydrogen addition to methanol, reaction (16) became a dominant pathway to produce OH.

As discussed in the previous section, a key intermediate species was HCO. HCO held a prominent role in both ignition and extinction. HCO, combined with O_2, provided CO production pathways.

$$HCO + O_2 = CO + HO_2 \tag{17}$$

HO_2 produced in reaction (17) got consumed reacting with OH

$$HO_2 + OH = H_2O + O_2 \tag{18}$$

Reaction (18) was a crucial reaction that substantially influenced both the ignition and extinction phenomena.

The sensitivity of methanol added gasoline was complex due to the presence of numerous large C4–C8 molecules such as iso-octane (IC_8), n-heptane (NC_7H_{16}), etc. Hence, there were several reaction pathways for radicals and intermediate species to interact with these large molecules. Moreover, several reactions exhibited strong temperature sensitivity. In an engine where the charge temperature varies considerably, reactions with strong temperature dependence would become critical. However, exploring the effect of engine-relevant temperature and pressure on blended methanol fuels is another exciting topic that would be addressed in the future.

4 Conclusions

The present study investigated the fundamental chemical kinetics processes behind the ignition and extinction characteristics of two fuel blends—20% methanol in gasoline, and 20% hydrogen in methanol, at 1 bar, 300 K. To compare the performance of fuel blends to their base fuel, methanol, gasoline, and hydrogen were also investigated in this study. Important laminar flame properties, such as flame speed, flame thickness, minimum ignition energy, the thermal and chemical flame structure of a lean flame, etc., were discussed thoroughly. The effect of fuel blending on ignition and extinction strain rates was assessed. A sensitivity analysis of strain rates was performed to identify the essential intermediate reactions and intermediate species associated with the ignition and extinction phenomena. Major findings are as follows:

- Fuel blends showed an improvement in flame speed compared to the base fuel. The flame speed of methanol-blended gasoline and hydrogen added methanol experienced an increase of 12.7 and 26% from neat gasoline and methanol, respectively. The effect of methanol and hydrogen addition on laminar flame speed was more prominent for a rich mixture.
- The minimum ignition energy requirement of fuel blends decreased from the base fuels. While the addition of methanol into gasoline showed a moderate 11% drop, hydrogen addition to methanol demonstrated a significant 25% reduction in the minimum ignition energy.
- The blended fuels could survive a higher ignition and extinction strain rates compared to the base fuel due to enhanced mixture reactivity. With added methanol, the ignition strain rates of gasoline increased by nearly 15% at $\phi = 1.1$. The strain rates demonstrated greater sensitivity to the equivalence ratio for lean mixtures. The extinction temperature was 110-260 K lower compared to the temperature at the ignition for similar equivalence ratios. Ignition strain rates were extremely sensitive to the temperature compared to the extinction strain rates.
- Hydrogen added methanol generated 22% higher OH compared to neat methanol. However, the concentration of H and O were affected slightly with hydrogen addition. The effect of hydrogen addition was manifested by the generation of OH instead of H. The large pool of H produced by the decomposition of hydrogen quickly got consumed through H consumption reactions pathways leading to a substantial increase in OH. OH, H, and O production in methanol-blended gasoline increased significantly. Peak O concentration of methanol-blended gasoline was 38% higher relative to methanol.
- The sensitivity coefficients for ignition were greater compared to extinction signifying a greater influence of chemical kinetics on the ignition process compared to extinction. With hydrogen addition to methanol, the intermediate reaction $O + H_2 = H + OH$ became a dominant pathway towards OH production. Hydrogen addition suppressed the formaldehyde emission by converting the aldehyde into carbon monoxide via reaction $HCO + H = CO + H$.

In this study, fuel blends demonstrated improved laminar combustion characteristics due to enhanced reaction kinetics. The individual components in the fuel blend not only affected the physical property of the fuel, such as heating value, H/C ratio, etc., but also altered the key intermediate reaction pathways that appear promising in designing tailored fuels. In the years to come, renewable fuels like methanol and hydrogen could create a potential pathway for developing low carbon tailored fuel blends with higher efficiency and lesser pollutant emission for automotive applications.

References

1. International Energy Agency (2019). https://www.iea.org/
2. Smil V (2017) Energy transitions: global and national perspectives. & BP Statistical Review of World Energy. Praege
3. Shurpali N, Agarwal AK, Srivastava VK (eds) (2019) Greenhouse gas emissions: challenges, technologies and solutions. energy, environment, and sustainability. Springer, Singapore
4. Beatrice C, Bertoli C, Giacomo ND (1998) New findings on combustion behavior of oxygenated synthetic diesel fuels. Combust Sci Technol 137:31–50
5. Agarwal AK, Gupta JG, Sharma N, Singh AP (eds) (2019) Advanced engine diagnostics. energy, environment, and sustainability. Springer, Singapore
6. Biswas S, Ekoto I (2019) Ozone added spark assisted compression ignition. advanced combustion techniques and engine technologies for the automotive sector, ed Singh AP, Sharma N, Agarwal RA, Agarwal AK. Springer, Singapore
7. Zhang C, Wu H (2016) Combustion characteristics and performance of a methanol fueled homogenous charge compression ignition (HCCI) engine. J Energy Inst 89(3):346–353
8. Rakopoulos CD, Kyritsis DC (2001) Comparative second-law analysis of internal combustion engine operation for methane, methanol, and dodecane fuels. Energy 26(7):705–722
9. Agarwal RA, Agarwal AK, Gupta T, Sharma N (2019) Pollutants from energy sources. energy, environment, and sustainability. Springer, Singapore
10. Wei J, Liu H, Li S, Yang R, Liu J, Wang Y (2008) Effects of methanol/gasoline blends on a spark ignition engine performance and emissions. Energy Fuels 22(2):1254–1269
11. Shenghua L, Clemente E, Tiegang H, Yanjv W (2007) Study of spark ignition engine fueled with methanol/gasoline fuel blends. Appl Therm Eng 27:1904–1910
12. Bilgin A, Sezer I (2008) Effects of methanol addition to gasoline on the performance and fuel cost of a spark ignition engine. Energy Fuels 22:2782–2788
13. Altun S, Oztop H, Oner C, Varol Y (2013) Exhaust emissions of methanol and ethanol-unleaded gasoline blends in a spark ignition engine. Thermal Sci 17(1):291–297
14. Sapre R (1989) Properties, performance and emissions of medium concentration methanol–gasoline blends in a single-cylinder, spark-ignition engine. SAE Trans 97(3):1105–1126
15. Bardaie Z, Janius R (1984) Conversion of spark-ignition engine for alcohol usage - comparative performance. Agric Mech Asia, Africa Latin Am 15(2):31–40
16. Abu-Zaid M, Badran O, Yamin J (2004) Effect of methanol addition on the performance of spark ignition engines. Energy Fuels 18(2):312–315
17. Hu Z, Zhang X (2019) Study on laminar combustion characteristic of low calorific value gas blended with hydrogen in a constant volume combustion bomb. Int J Hydrogen Energy 44(1):487–493
18. Huang Z, Zhanga Y, Zeng K, Liu B, Wang Q, Jiang D (2006) Measurements of laminar burning velocities for natural gas–hydrogen–air mixtures. Combust Flame 146(1–2):302–311
19. Ren F, Chu H, Xiang L, Han W, Gu M (2019) Effect of hydrogen addition on the laminar premixed combustion characteristics the main components of natural gas. J Energy Inst 92(4):1178–1190
20. Hu E, Huang Z, He J, Jin C, Zheng J (2009) Experimental and numerical study on laminar burning characteristics of premixed methane–hydrogen–air flames. Int J Hydrogen Energy 34(11):4876–4888
21. Al-Hamamre Z, Yamin J (2013) The effect of hydrogen addition on premixed laminar acetylene–hydrogen–air and ethanol–hydrogen–air flames. Int J Hydrogen Energy 38(18):7499–7509
22. Gong C, Li Z, Chen Y, Liu J, Liu F, Han Y (2019) Influence of ignition timing on combustion and emissions of a spark-ignition methanol engine with added hydrogen under lean-burn conditions. Fuel 235(1):227–238
23. Zhang B, Ji C, Wang S, Xiao Y (2014) Investigation on the cold start characteristics of a hydrogen-enriched methanol engine. Int J Hydrogen Energy 39(26):14466–14471

24. Zhen X, Li X, Wang Y, Liu D, Tian Z (2020) Comparative study on combustion and emission characteristics of methanol/hydrogen, ethanol/hydrogen and methane/hydrogen blends in high compression ratio SI engine. Fuel 267:
25. Ji C, Yang J, Liu X, Zhang B, Wang S, Gao B (2016) A quasi-dimensional model for combustion performance prediction of an SI hydrogen-enriched methanol engine. Int J Hydrogen Energy 41(39):17676–17686
26. Kee RJ, Rupley FM, Miller JA, Coltrin ME, Grcar JF, Meeks E, Moffat HK, Lutz AE, Dixon Lewis G, Smooke MD, Warnatz J, Evans GH, Larson RS, Mitchell RE, Petzold LR, Reynolds WC, Caracotsios M, Stewart WE, Glarborg P, Wang C, Adigun O (2000) CHEMKIN Collection, Release 3.6. In: Reaction Design, Inc., San Diego, CA
27. Reaction Design (2013) Reaction Workbench 15131 San Diego
28. Lopez Pintor D, Dec J, Gentz G (2019) Φ-sensitivity for LTGC engines: understanding the fundamentals and tailoring fuel blends to maximize this property. SAE Technical Paper 2019-01-0961, p 24
29. Biswas S, Ekoto I (2019) Detailed investigation into the effect of ozone addition on spark assisted compression ignition engine performance and emissions characteristics. SAE Technical Paper 2019-01-0966
30. Biswas S, Ekoto I (2019) Spark assisted compression ignition engine with stratified charge combustion and ozone addition. JSAE 20199089
31. Mehl M, Pitz W, Westbrook CK, Curran HJ (2011) Kinetic modeling of gasoline surrogate components and mixtures under engine conditions. Proc Combust Inst 33(1):193–200
32. Mehl M, Wagnon S, Tsang K, Kukkadapu G, Pitz WJ, Westbrook CK, Tsang Y, Curran HJ, Atef N, Rachidi MA, Sarathy MS, Ahmed A (2017) A comprehensive detailed kinetic mechanism for the simulation of transportation fuels. In: 10th US National combustion meeting, College Park, MD
33. Mehl M, Pitz WJ, Westbrook CK, Yasunaga K, Conroy C, Curran HJ (2011) Autoignition behavior of unsaturated hydrocarbons in the low and high temperature regions. Proc Combust Inst 33(1):201–208
34. Law CK (2006) Combustion physics, xviii edn. Cambridge University Press, Cambridge, New York, p 722
35. Peters N (2000) Turbulent combustion. Cambridge monographs on mechanics Cambridge, England, xvi edn. Cambridge University Press, New York, p 304
36. Kee RJ, Miller JA, Evans GH, Dixon-Lewis G (1989) A computational model of the structure and extinction of strained, opposed flow, premixed methane-air flames. In: Symposium (International) on combustion, vol 22(1), pp 1479–1494
37. Egolfopoulos FN (1994) Geometric and radiation effects on steady and unsteady strained laminar flames. Proc Combust Inst 25:1375–1381
38. Giovangigli V, Smooke MD (1989) Adaptive continuation algorithms with application to combustion problems. Appl Num Math 5(4):305–331
39. Vancoillie J, Verhelst S, Demuynck J (2011) Laminar burning velocity correlations for methanol-air and ethanol-air mixtures valid at SI engine conditions. SAE Technical Paper, 2011-01-0846
40. Hu E, Xu Z, Gao Z, Xu J, Huang Z (2019) Experimental and numerical study on laminar burning velocity of gasoline and gasoline surrogates. Fuel:256
41. Dahoe AE (2005) Laminar burning velocities of hydrogen–air mixtures from closed vessel gas explosions. J Loss Prev Process Ind 18(3):152–166
42. Saeed K, Stone CR (2004) Measurements of the laminar burning velocity for mixtures of methanol and air from a constant-volume vessel using a multizone model. Combust Flame 139:152–166
43. Dirrenberger P, Glaude PA, Bounaceur R, Le Gall H, Pires da Cruz A, Konnov AA, Battin-Leclerc F (2014) Laminar burning velocity of gasolines with addition of ethanol. Fuel 115:162–169
44. Norton TS, Dryer FL (1991) The flow reactor oxidation of C1 − C4 alcohols and MTBE. In: Symposium (International) on combustion, vol 23(1), pp 179–185

Measurement, Mechanism and Characteristics of Formaldehyde Emission from Methanol/Gasoline Blends Fueled Engine

Yanju Wei

Nomenclature

Ar	Argon
CO	Carbon monoxide
DNPH	Dinitrophenylhydrazine
DOAS	Differential optical absorption spectroscopy
EPA	Environmental Protection Agency
FEP	Fluoropolymer-ethylene-propylene
FID	Flame ionization detector
FTIR	Fourier transform infrared absorption
GC	Gas chromatography
HC	Hydrocarbon
HCHO	Formaldehyde
HPLC	High-performance liquid chromatography
IARC	The International Agency for Research on Cancer
M10	Fuel blend with $10\%_{vol.}$ methanol and $90\%_{vol.}$ gasoline
NOx	Nitrogen oxides
MS	Mass spectrometry
PDHID	Pulsed discharge helium ionization detector
PFR	Primary reference fuel
PM	Particulate matters
PN	Particle number
ROP	Rate of production
TDLAS	Tunable diode laser absorption spectroscopy
T_{tq}	Engine torque
T_{ex}	Temperature at engine exhaust port

Y. Wei (✉)
Xi'an Jiaotong University, Xi'an 710049, China
e-mail: weiyanju@xjtu.edu.cn

A. K. Agarwal et al. (eds.), *Methanol*, Energy, Environment, and Sustainability,
https://doi.org/10.1007/978-981-16-1224-4_10

T_{TWC}	Temperature at the inlet of TWC
TWC	Three-way catalytic converter
UV	Ultraviolet

1 Introduction

As one of the most important transportation tools to modern society, gasoline and diesel vehicles are not only the main body of oil consumption but also the important mobile sources of air pollution. With the rapid development of the automotive industry, China has an average CAGR (Compound Annual Growth Rate) of 6.4% in the past decade from 2010 to 2019, vehicle holdings in 2019 was 348 million, and there is no sign for this trend to decrease. The huge amount of vehicle holding results in two severe consequences. One is energy safety, the external dependence of crude oil has reached 70.8% in 2019, far exceeding the international alert line of 50%. And the other is air pollution. Automotive contributes about 20% of the airborne pollutants, especially for the PM (Particulate Matters), emission source analysis showed that the share of local motors in air pollution was around 30% in Beijing, Shanghai, Xi'an, and Guangzhou in 2017. So, it has been a big and immediate problem to control vehicle emissions. In the coming China VI vehicle emission standards [1, 2], which will have a legal effect on January 1, 2021, all the permitted amount of regular emission including hydrocarbons (HC), carbon monoxide (CO), nitric oxides (NOx), and particulate matters (PM) were stricter than China V standard by 1/3 (for CO and HC), 2/3 for PM and 77% for NOx, respectively. And moreover, the limits for the particle number (PN) and ammonia (NH3) are added. In addition, more stringent automatic monitoring requirements and joint control of emission fuel consumption are put forward.

The strict emission limit elevates the difficulty of the whole engine system including fuel properties, fuel supply, electric control, and emission reprocessing. Among the pollutants, the control of PM is the toughest. As we all know, PM is produced from the incomplete combustion of hydrocarbons in the circumstance of high temperature and fuel-rich conditions, the carbon atoms in C-C bonds are the source of PM emission. Not only the combustion of gasoline and diesel can, even methane (CH4) produces PM, and especially fine particles with a diameter around 10–100 nm [3, 4]. PM emission is inevitable in the combustion of hydrocarbons. Fortunately, the content of oxygen in the molecule could effectively reduce the production of PM.

Methanol, known as methyl alcohol, the simplest oxygenated hydrocarbon and thus PM free in combustion, has great potential to be used as the clean fuel for internal combustion engines for its environmentally friendly properties from production, combustion to emissions [5–7]. However, more formaldehyde (HCHO) will be emitted from methanol combustion [8–12]. As an important participant in photochemical pollution, formaldehyde is also classified as a probable human carcinogen by the U.S. Environmental Protection Agency (EPA). The International Agency for Research on Cancer (IARC) has determined that there is "sufficient evidence" that occupational exposure to formaldehyde causes nasopharyngeal cancer in humans [13, 14]. Methanol vehicle industry develops fast, especially in China, measures must be taken to control formaldehyde emissions in advance to avoid the potential threats to the public health.

2 Measurement of Formaldehyde

In terms of engine emission, formaldehyde is in the category of oxygenated hydrocarbon emission. However, as the simplest carbonyl organics, having only a carbonyl in its molecule, formaldehyde burns similarly with carbon monoxide, thus it cannot be detected by the ordinary hydrocarbon detector, i.e. flame ionization detector (FID). This puts the detection of formaldehyde into an embarrassing condition. Researchers have to develop various methods to make its either physical or chemical properties manifested and thus detected, via either direct or indirect approach. The detection methods could be summed up into five categories.

2.1 *Spectrophotography*

Spectrophotography is the simplest and most widely used method to detect aldehydes and acetones that contain carbonyl groups, due to its simple equipment, accurate measurement, and low cost. Spectrophotography is based on the absorption spectrum of materials to lights with certain wavelengths, each compound has its unique absorption spectrum, thus its concentration could be determined by evaluating the light absorbance [15, 16]. It needs to be noted that, this is an indirect method, formaldehyde should be converted to other chemical substances with chromogenic agents to produce colourful solutions, colours may be purple, yellow, red, blue, and so on, depending on the agents. Agents could be ① concentrated sulfuric acid (purple), ② acetylacetone (yellow), ③ phloroglucin (red), ④ phenylhydrazine hydrochloride (red), ⑤ fuchsin sulfite (blue), ⑥ potassium bromate—bromocresol (purple fading), ⑦ AHMT (aubergine), ⑧ phenol reagent (cyan), etc.

2.2 *Spectroscopy*

Differential optical absorption spectroscopy (DOAS). DOAS utilizes the different absorption characteristics of gaseous molecules toward the different bands of lights to detect and measure the concentration of airborne pollutants, it is one of the most common test methods in the research of air pollution.

Formaldehyde is identified by the three absorption peaks at 326.1 nm, 329.7 nm, and 339.0 nm, respectively. Although this method is still sensitive to the concentration even as low as 1×10^{-10}; the light source from a Xenon lamp should be introduced at 5 –10 km away, and collected by a telescope. So, the long optical length limits its application of formaldehyde detection in ordinary circumstances [18, 19].

Fourier transform infrared absorption (FTIR). The material absorbs radiation energy, like infrared light, and causes molecular vibrational energy level transitions, FTIR utilizes the absorption of the infrared light transmitting in the sample gas, where the intensity of the incident light is reduced from Ii to Io, then the absorbance is quantitated as $\mathrm{A} = \log(Ii/Io)$, where $\mathrm{A} = kbc$ according to Beer–Lambert Law, k is the absorption coefficient, b indicates the optical length and c infers the gas concentration. One may easily obtain the concentration of the target component with the given k and b.

Two wavelengths with their eigenvalues as 2779 and 2781.5 cm^{-1} are used to identify formaldehyde. The detection limit of this method is 4×10^{-9}. The detection limit is the minimum concentration of a chemical that the detector can respond to. The AVL FTIR SESAM spectrometer is widely used to measure the engine-out unregulated emissions, it is able to provide fast online measurement to tens of different components simultaneously by scanning the spectral information with a wide range of wavelength [8–10, 20, 21].

Tunable diode laser absorption spectroscopy (TDLAS). TDLAS is the most widely used method in atmospheric environmental monitoring. It detects the rotation-vibration signal of the target molecules exposed to the laser beam. Formaldehyde exhibits an absorption peak at the wavenumber 1740 cm^{-1}. The test limit of the TDLAS method is 2.5×10^{-10}, however, it can merely measure formaldehyde and the equipment are too expensive to be popularized for regular measurements, it is usually used to measure the atmospheric tropospheric formaldehyde in the airplane [22, 23].

2.3 Fluorescence Method

In this method, formaldehyde is converted with derivation agents to the third chemical that emit fluorescence. The fluorescence intensity is proportional to the concentration of formaldehyde, and thus it could be utilized to quantitate it. Formaldehyde-dihydropyridine derivatives generated from the Hantzsch reaction emit strong fluorescence while very weak from other aldehydes and acetone derivatives, so it can be used specifically to detect formaldehyde, especially extremely low-level formaldehyde, like the atmospheric formaldehyde at the North Pole [24, 25].

2.4 Electrochemical Method

Electrochemical method is based on the change of current (voltammetry), electric quantity (Coulomb method), or potential (potentiometric method) produced in a chemical reaction. There are two kinds of formaldehyde detection methods: oscillopolarography and selective electrode method. Oscillopolarography requires strict pretreatment of samples, and the usage of mercury-drop electrodes highlights the risk of contamination, it is usually used in formaldehyde detections in food and food packaging materials. The selective electrode method is under development and rarely applied in practice [26, 27].

2.5 Chemiluminescence Method

Gallic acid and H_2O_2 react with formaldehyde and emit chemiluminescence under alkaline conditions, the chemiluminescence intensity is proportional to the concentration of formaldehyde. The detection limit is 10×10^{-9}. This method is of rapid reaction, strong specificity, high sensitivity, and furthermore, hydrocarbons, SO_2, CO, CO_2, nitrogen oxides, acrolein, benzaldehyde, ozone have no interference on the determination. However, accompanying acetaldehyde and propionaldehyde cannot be distinguished and thus enlarge the result [28, 29].

2.6 Formaldehyde Sensors

The formaldehyde sensors usually include electrochemical sensor, optical sensor, and photo-biochemical sensor. The electrochemical sensor is relatively simple in structure and low in cost. The detection range and resolution can basically meet the requirements of indoor environmental detection. But the results are interfered with by many substances and the electrolyte is consumed at work, so its lifetime is short. The optical sensor is expensive and large in size, which is not suitable for online real-time analysis. Although photo-biochemical sensors improve their selectivity, the sensor is unstable and lacks practicability due to enzyme activity and other factors, and the price of formaldehyde gas sensors, in general, is too high to be popularized [30, 31].

2.7 Chromatography

Chromatography has powerful separation efficiency and is sensitive and accurate for complex samples, it is not easily disturbed by the sample substrate and the colour of the reagent. Formaldehyde is always firstly derivatized and then determined. The commonly used derivatives are 2,4-dinitrobenzhydrazine (2,4-DNPH), imidazole, ethiol, hydrazine sulfate, and so on.

Gas chromatography (GC). In this method, airborne formaldehyde is firstly adsorbed in a supporter coated with 2,4- dinitrophenylhydrazine (2,4-DNPH) 6201 to form a stable formaldehyde DNPH derivative, 2,4-dinitrophenylhydrazone. Then the derivative is eluted with carbon disulfide and sent to GC to separate by OV-1 chromatographic column and finally determined by flame ionization detector (FID) detector. The formaldehyde derivative is identified with retention time and quantitated with GC peak height. The lower detection limit of this method is 0.2 μg/mL (for 5 μL sample eluent). The derivative can also be detected by a mass spectrometry detector, namely gas chromatography-mass spectrometry (GC-MS). The "Method for Determination of Formaldehyde in Air in Public Places", National Standard of the China GB/T 18204.26-2000, uses GC-FID method [32, 33].

High-performance liquid chromatography (HPLC). In this method, formaldehyde in the air is collected by two large absorption tubes in series each with 5 ml of water as the absorption solution. Formaldehyde reacts with 2,4-DNPH and forms DNPH derivative in an acidic environment. Then DNPH derivative is extracted by dichloromethane, then concentrated and desiccated, at last, it is dissolved in methanol, and sent to HPLC to be separated by ODS2C18 column and detected by ultraviolet (UV) detector with the light wavelength of 254 nm. DNPH derivate is qualified by retain time and quantitated by peak area [35, 36].

The method is widely used for analyses in various fields for its high sensitivity, no interference, easy preservation of reagents, and good repeatability of test results, and it is thus adopted in the airborne formaldehyde detection method To-11A of the U.S. Environmental Protection Agency (EPA).

2.8 Problems for Engine-Out Formaldehyde

In the air, formaldehyde is the dominant carbonyl pollutants, and always much higher in concentration than other aldehydes and ketones. Thus, its detection in the air is much easier than that in the engine exhaust gas, which contains almost all types of low carbon organics, and the concentration of formaldehyde is relatively ignorable compared with other hydrocarbons, so the detection of engine-out formaldehyde is much complex than that in the air.

Although there are so many methods to detect the formaldehyde in air, all of them, except for FTIR, are indirect methods. Formaldehyde has to be dissolved in a solution, converted to other intermediate substances, and concentrated before detection. This pretreatment takes a very long time, which is around 100 min for the EPA To-11A method, for instance [34]. This is insufferable for the test of engine-out formaldehyde. FTIR is a direct and fast way and thus always applied in engine tests, however, it is easily interfered with by other aldehydes and acetones accompanying with formaldehyde and thus always results in a value higher than the fact 1. The complexity of the compositions in the exhaust gas of engines, containing always hundreds of different kinds of hydrocarbons, fuelled with ordinary fuels like gasoline and diesel are also the obstacle preventing the utilization of the ordinary testing methods.

How to separate formaldehyde in the exhaust gas is the first problem to be solved. GC or HPLC is the preferable way due to its good separation towards the compositions. Between which GC is simpler compared with HPLC with the comparative accuracy, since it does not need to convert gas into a solution and thus saves a step of pretreatment.

It is to be noted that, just because the FID detector, perfect in hydrocarbon detection, has no response on formaldehyde, researchers have to detour to develop so many indirect methods. Given a universal detector with a good response to formaldehyde for GC, the engine-out formaldehyde could be easily detected and quantitated with properly selected GC columns.

2.9 Gas Chromatographical Detection

The non-radioactive pulsed discharge helium ionization detector (PDHID) is a newly developed non-destructive high-sensitivity universal detector that can detect organic and inorganic substances. Except for the poor response to neon (ionization energy 21.6 eV), PDHID has a very highly sensitive response to all substances including carbonyl groups (e.g. formaldehyde), which is beyond the capacity of FID. This detector provides a good, easy, and direct way for formaldehyde detection.

Principle of PDHID. The ionization mode of PDHID is generally considered to consist of three parts. (1) Helium is excited as it flows through the discharge region, and the other substances are photoionized by the continuous spectrum of 17.7–13.5 eV emitted from the transition of excited atoms $He_2(1A^1\sum_u^+)$ back to free state; (2) the direct ionization of components by electrons accelerated by high voltage pulses generates signals, or direct ionization of carrier gas and impurities produces base flow signal; (3) metastable helium collides with components to produce ionization signals, or reacts with impurities to ionize to produce base flow signal. The ionization reactions that produce the signal are as follows [11, 37–39].

$$e + He \rightarrow He^{**} \rightarrow He^{*} + h\nu \tag{1}$$

$$h\nu + AB \rightarrow AB^{+} + e \tag{2}$$

$$e + AB \rightarrow AB^{+} + 2e \tag{3}$$

$$He^{*} + AB \rightarrow AB^{+} + He + e \tag{4}$$

This response mechanism is completely different from that of the FID detector. FID detector quantifies each carbon atom in different hydrocarbon molecules by converting them into the most basic, common response unit, CH4, and then generates a signal through the ionization of $CH + O \rightarrow CHO + e$ [11]. Since the carbonyl group contains $C = O$ double bonds, it can only be converted to CO, which has no response on FID, the detector does not respond to formaldehyde. While in the PDHID detector, formaldehyde molecules are directly ionized without destruction to form HCHO + ions, which are captured by the detector collecting poles to form the desired voltage signal.

Doping 2% argon in the discharged helium gas then the detector can operate in the selective Ar-PDHID mode. Mixing argon in helium reduces the energy of the radiation spectrum from 17.7–13.5 eV to 11.8–9.8 eV, which not only greatly enhances the detector's response sensitivity to formaldehyde (ionization energy 10.9 eV) and methanol (10.8 eV), but also avoids the interference of permanent gas, i.e. H_2O (12.6 eV), O_2 (12.1 eV), N_2 (15.6 eV), and CO_2 (13.8 eV), [12, 38]. This advantage makes the PDHID detector very suitable for the detection of formaldehyde in the engine exhaust gas.

Operating conditions of GC. Yanju Wei et al. [39] developed a gas chromatographic method to detect the formaldehyde in the exhaust gas of methanol/gasoline blends fuelled spark-ignition engine. In their method, the exhaust gas was collected by the fluoropolymer-ethylene-propylene (FEP) bags and kept at room temperature. The bags were then heated to 80 °C at detection in a thermostat to avoid the absorption of formaldehyde and methanol by condensed water. Then 100 μL sample was drawn into the GC by a vacuum pump and separated by a Gs-OxyPLOT capillary column. Finally, the components in the sample gas were detected by the PHHID working in Argon mode, where the discharge gas was helium doped with 2% Argon. The analysis duration of this method was within 5 min, and the detection limits of formaldehyde and methanol were 28.2×10^{-9} g/L and 89.3×10^{-9} g/L, respectively [39]. Figure 1 shows a typical chromatogram of the typical emissions before (black) and after (red) the three-way-catalytic converter.

3 Emission Mechanism of Engine-Out Formaldehyde

3.1 Formation Pathway

Formaldehyde (HCHO) is the most important oxygenated intermediate oxidation products of not only methanol, but also all hydrocarbons (HC). The detailed chemical kinetic mechanism of iso-octane/n-heptane (iC8H18/nC7H16) mixture, developed by Curran H. J. et al. [42], is always adopted as the primary reference fuel (PRF) to analyse the combustion mechanism of gasoline, and the effects of methanol on the oxidation route.

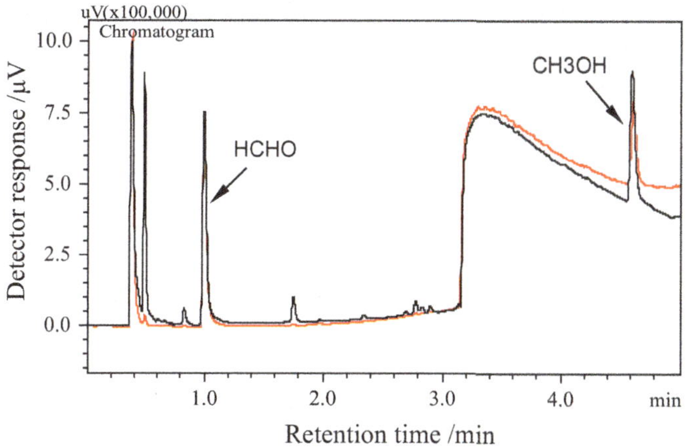

Fig. 1 Chromatograms of a typical exhaust gas of methanol/gasoline blend fuelled engine, black and red lines denote emissions before and after TWC, respectively

The formation of HCHO from high temperature (>1000 K) oxidation of iC8H18/nC7H16 mixture depends mainly on the reactions of CH_3, C_3H_5 and iC_4H_7 radicals:

$$\text{R3 } CH_4 + OH = CH_3 + H_2O \tag{5}$$

$$\text{R24 } CH_3 + CH_3 = C_2H_6 \tag{6}$$

$$\text{R207 } iC_4H_8 = iC_4H_7 + H \tag{7}$$

$$\text{R355 } C_5H_{10} = iC_4H_7 + CH_3 \tag{8}$$

$$\text{R1458 } CH_3 + O_2 = CH_2O + OH \tag{9}$$

$$\text{R1460 } C_3H_5 + O_2 = CH_2O + CH_3CO \tag{10}$$

while at the presence of methanol, HCHO is mainly formed from methanol reactions:

$$\text{R18 } CH_3OH + HO_2 = CH_2OH + H_2O_2 \tag{11}$$

$$\text{R78 } CH_3OH + OH = CH_2OH + H_2O \tag{12}$$

$$\text{R80 } CH_3OH + H = CH_2OH + H_2 \tag{13}$$

Rate of production (ROP) analysis shows that 62% of HCHO production in PRF oxidation is from CH_3O dehydrogenation and CH3 oxidation through the reactions R40 and R1458.

$$\text{R40 } CH_3O = CH_2O + H \tag{14}$$

$$\text{R1458 } CH_3 + O_2 = CH_2O + OH \tag{15}$$

While in M85 fuel blend (85%vol. PRF and 15%vol. CH_3OH), 84% of HCHO is produced from the oxidation of the radicals CH3O and CH2OH through reactions R83 and R40, which are the only two son-radical of methanol dehydrogenation.

$$\text{R83 } CH_2OH + O_2 = CH_2O + HO_2 \tag{16}$$

$$\text{R40 } CH_3O = CH_2O + H \tag{17}$$

Compared to the diversification of the production pathway, the consumption route of HCHO is unitary, it oxidizes all via the path of HCHO $\rightarrow$ HCO.

The pathway nets of the oxidations of PRF and methanol are visualized in Fig. 2, respectively. It can be clearly seen that HCHO from PRF oxidation has various production routes; it can be produced directly from C_1–C_4 paraffin radicals, among which, CH_3 radical plays a dominant role. The presence of methanol in PRF (as M85 fuel blend) provides a two-step shortcut to produce formaldehyde, i.e. $CH_3OH \rightarrow CH_2OH/CH_3O \rightarrow CH_2O$.

Figure 3 shows the detailed mechanism of methanol oxidation at 1000 K. The methanol oxidation undergoes two main paths, as shown in Fig. 3(a), one is $CH_3OH \rightarrow CH_2OH \rightarrow CH_2O \rightarrow HCO \rightarrow CO \rightarrow CO_2$, and the other is $CH_3OH \rightarrow CH_4 \rightarrow CH_3 \rightarrow CH_3O \rightarrow CH_2O \rightarrow HCO \rightarrow CO \rightarrow CO_2$. The former one is greatly influenced by the supply of O2 or O atom, and the latter one is affected by the supply of HO_2 radical, which is produced in route one. Results show that the former route has the priority over the second one. While the competitive routes exhibit the possible path for oxidation, the sensitive routes infer reactions that affect the completeness of oxidation. Results show that the decomposition of formaldehyde via the reactions $CH_2O = CO + H_2$ and $CH_2O = HCO + HO_2$. This infers that formaldehyde is the key intermediate product as well as the freight station that controls the oxidation of methanol.

3.2 *Formation of Engine-Out Formaldehyde*

As an intermediate product of HC oxidation, HCHO is produced and then oxidized, thus the HCHO emission is the survival part that escapes from further oxidation. In terms of neat production rate, HCHO could be regarded as firstly produced and

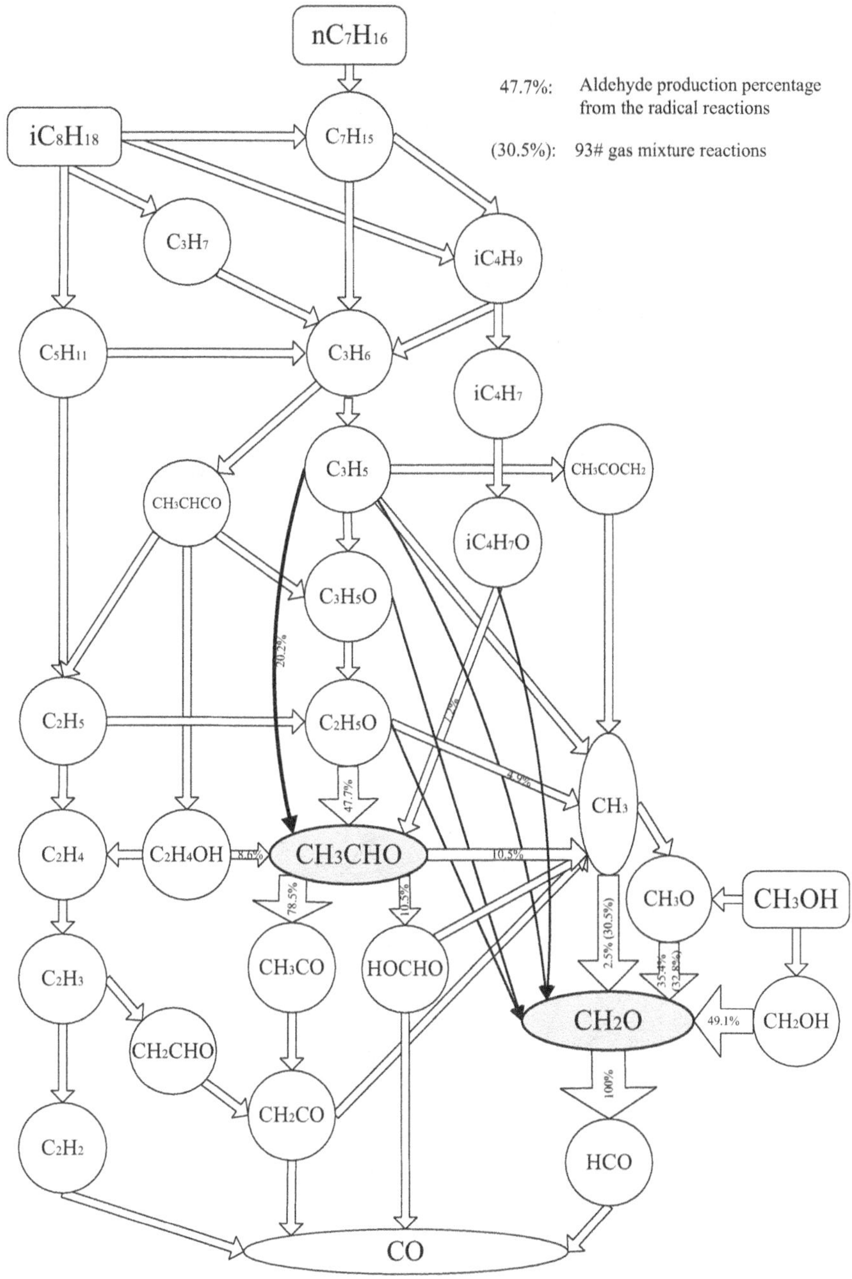

Fig. 2 HCHO production pathway from PRF and M85 oxidation [43]. [Reprinted with permission from (https://doi.org/10.1021/ef900721v). Copyright (2009) American Chemical Society.]

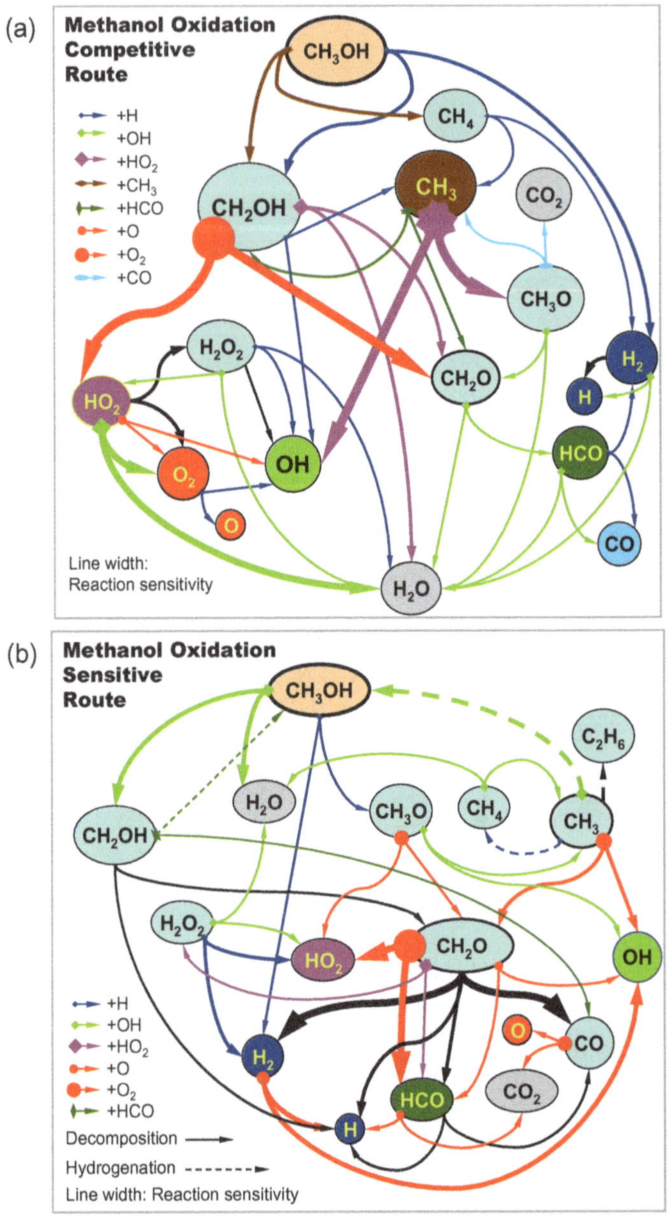

Fig. 3 **a** Competitive route and **b** sensitive route of methanol oxidation

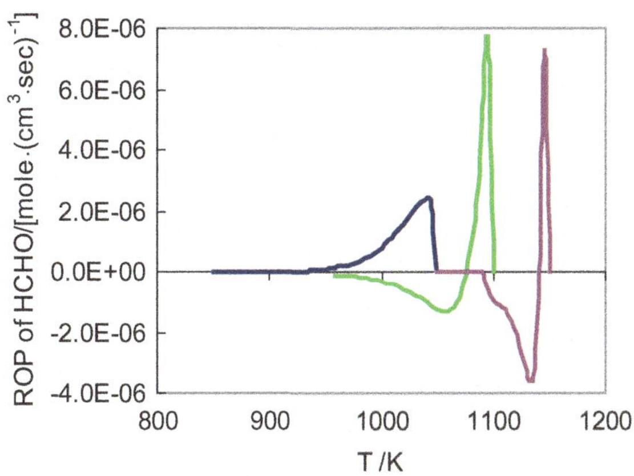

Fig. 4 Total ROP profiles of HCHO from M85 mixture oxidation in a flow reactor at different initial temperatures, flow velocity $v = 5$ m/s [43]. [Reprinted with permission from (https://doi.org/10.1021/ef900721v). Copyright (2009) American Chemical Society.]

then consumed, in other words, consumption comes after production. Knowing that high temperature corresponds to high reacting velocity, the consumption is as fast as production, thus no HCHO emits from ideal combustion of premixed air–fuel mixtures. Figure 4 shows the rate of production (ROP) of HCHO from M85 oxidation in a flow reactor with decrescent temperature (at the rate of 500 K/m, which is comparable to the fact of engine exhaust pipe), it can be clearly seen that the positive and negative ROPs infer the two stages of production and consumption, respectively. At high temperatures like initial temperature $T_0 > 1050$ K, the HCHO produced from methanol oxidation is soon consumed via the decomposition and dehydrogenation reactions, thus leaving no HCHO emission. However, when $T_0 < 1000$ K, the consumption is gradually frozen due to the excessively low temperature, in this case, all the produced HCHO will be released to the atmosphere and thus forms formaldehyde pollution.

Thus, in the engine exhaust pipe where the temperature decreases along with the distance to the exhaust valve, the HCHO emission is determined by the initial temperature and the gas flow velocity, that is to say, the residence time of the exhaust gas staying in hot environment. No HCHO will be released when the temperature exceeds a certain range, since HCHO will be consumed rapidly at high temperature and the generation reactions will be frozen at low temperature. A contribution zone exists for HCHO emission, as shown in Fig. 5, where the production of reactor-out (with the same temperature distribution as engine exhaust pipe) HCHO is gradually transmitted from the generation zone to the generation-consumption zone with the increasing initial temperature. Thus, the emission at the critical point, which is about 1050 K, has the maximum concentration. Acetaldehyde, also as the intermediate product of hydrocarbons, has the similar trend with that of formaldehyde, however, with a critical temperature at 1100 K. It is noted that all the intermediate products should have a contribution zone since they all have one in common; they undergo the generation and consumption process. The factors that improve the completeness of the oxidation process would move the emission towards the direction of higher temperature.

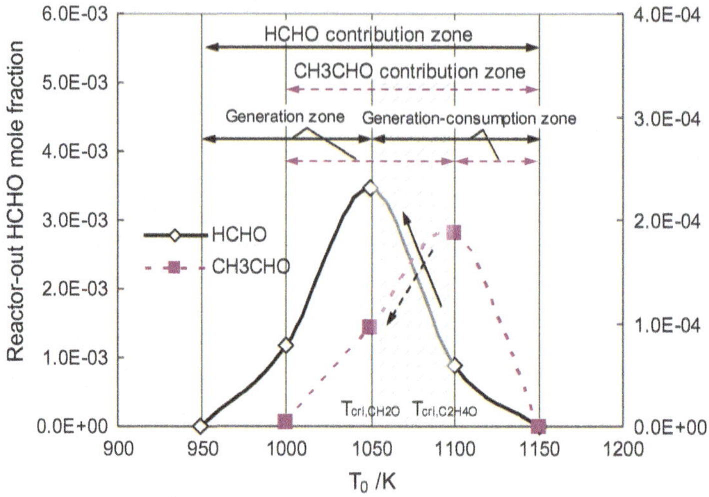

Fig. 5 Contribution zone of initial temperatures on aldehydes' emissions, flow velocity v = 5 m/s [43]. [Reprinted with permission from (https://doi.org/10.1021/ef900721v). Copyright (2009) American Chemical Society.]

4 Emission Characteristics of Methanol/Gasoline Blends Fuelled Engine

4.1 Effects of Methanol Fraction

Since formaldehyde is an intermediate product of carbon atom oxidation, so gasoline engine also emits formaldehyde, although this is seldomly mentioned to the public or reflected in the emission regulations previously, formaldehyde is to be taken into account in the latest emission regulation towards methanol vehicles.

For the gasoline engine, Fig. 6(a) shows that formaldehyde emission performs a U-shaped curve along with the increasing engine load. This is in accordance with the characteristics of HC emission, which is the result of the poor combustion due to the high content of residual gas in cylinder at low load, and the shortened residence time of exhaust gas in a hot environment at high load, respectively. Higher engine speed corresponds to higher exhaust temperature and shorter residence time of exhaust gas in a hot environment, and when the initial temperature lies in the range of generation-consumption zone in Fig. 5, HCHO emission would increase with engine speed. The emission concentration is no more than 80 μg/L under most conditions.

When methanol is blended into gasoline, the pathways of HCHO generation and oxidation are quite changed. Partially because the activation temperature of methanol oxidation is much lower than that of HCs [17], and partially because of the shorter reaction route from methanol to formaldehyde. Formaldehyde from M10 (fuel blend with 10%vol. methanol and 90%vol. gasoline) fuelled engine is mainly

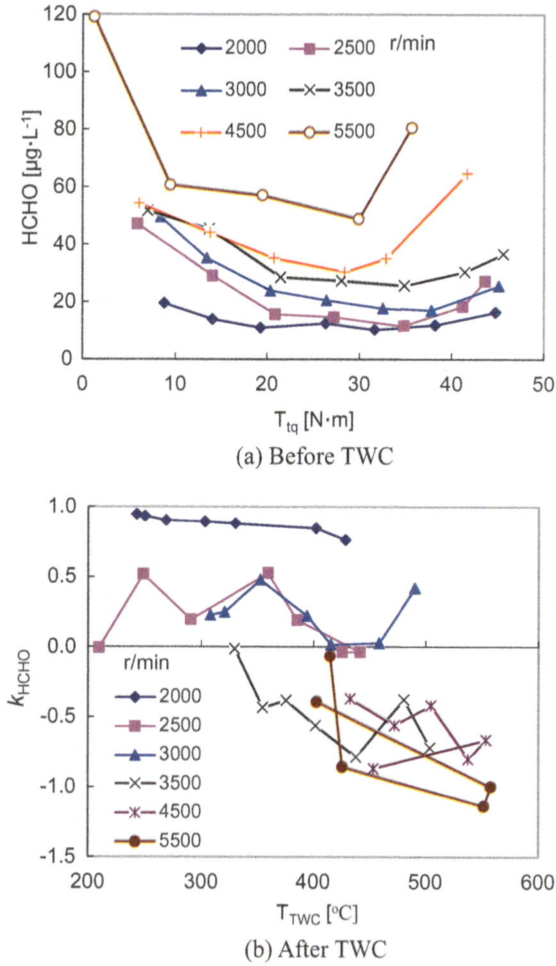

Fig. 6 Characteristics of formaldehyde emission from the gasoline engine [44]. [Reprinted with permission from (https://doi.org/10.1021/ef900175h). Copyright (2009) American Chemical Society.]

from methanol rather than HCs. Figure 7(a) shows that HCHO emission before three-way catalyst (TWC) rises almost parallell with engine speed and obtains the peak value 85 μg/L at rated speed. The increase rate accelerates obviously when the exhaust temperature exceeds 600oC. From Figs. 6(a) and 7(a) one knows that M10 engine-out formaldehyde is approximately 1 to 3 times of that from the gasoline engine.

The emission characteristics of M85 engine-out HCHO are similar to those of M10, however, with about 4 times higher HCHO emission, as shown in Fig. 8(a), the concentration at low load, 4500 r/min is around 250 μg/L. Since formaldehyde is the only and inevitable intermediate product in the oxidation of methanol, while it is merely one of the options in various routes in that of high-carbon hydrocarbons, thus the presence of methanol results in the increase of formaldehyde emission. as

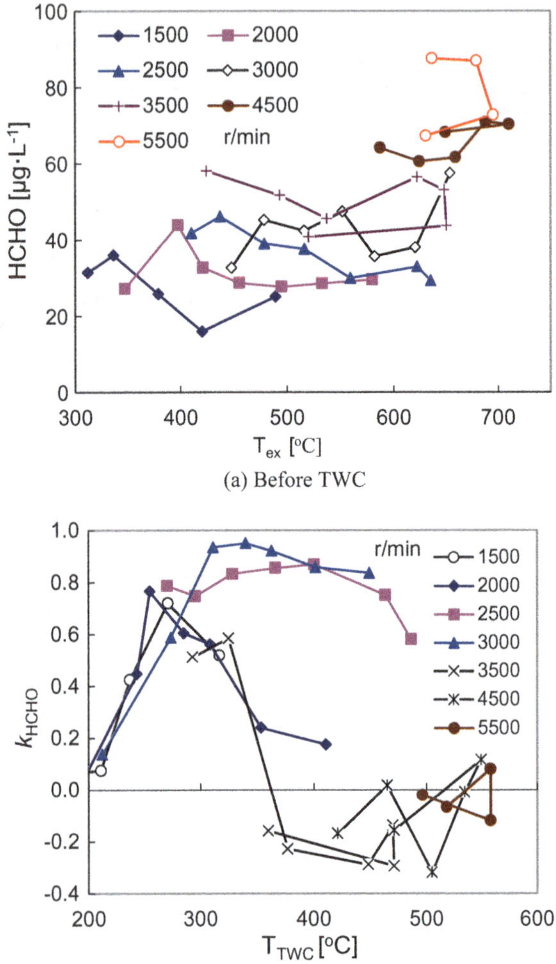

(a) Before TWC

(b) After TWC

Fig. 7 Characteristics of formaldehyde emission from M10 engine [44]. [Reprinted with permission from (https://doi.org/10.1021/ef900175h). Copyright (2009) American Chemical Society.]

shown in Fig. 9, HCHO emission is also approximately linear to the amount of cyclic supplied fuel methanol [9, 10, 44].

4.2 *Conversion of Formaldehyde in the Three-Way Catalyst*

TWC eliminates HC, CO, and NO_x emissions simultaneously by reducing the activation energy of thermodynamically feasible reactions and thus reduces the active temperature and promotes the reaction rates. It is able to accelerate the oxidation process, however, it can't change the balance state of the reactions. The essential

Fig. 8 Characteristics of formaldehyde emission from M85 engine [43]. [Reprinted with permission from (https://doi.org/10.1021/ef900721v). Copyright (2009) American Chemical Society.]

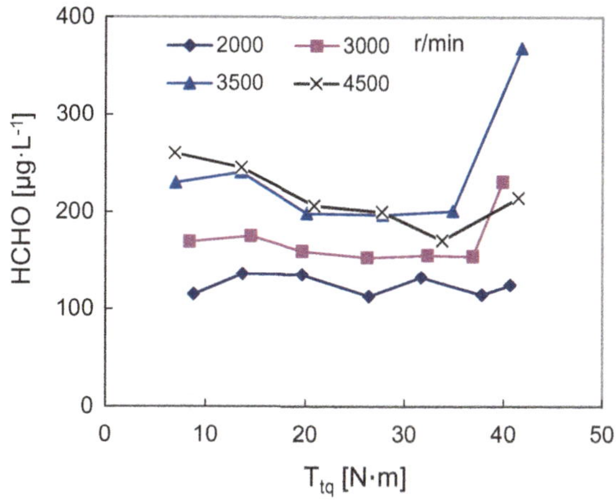

(a) Before TWC

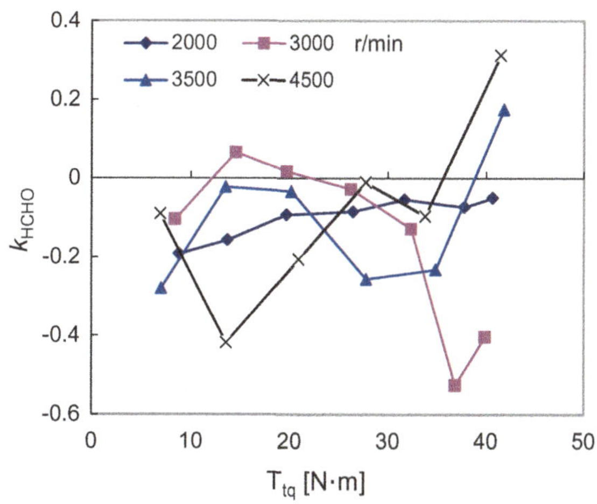

(b) After TWC

of the TWC conversion is to move the reactions from the reaction-frozen region at low temperature upward to the contribution regions at higher temperature along the curve in Fig. 5. With the abundant supply of methanol in the exhaust gas, the oxidation of methanol is reactivated in the TWC, however, with the increase of engine speed, which corresponds to the increase of exhaust temperature and decrease of the residence time of exhaust gas in TWC, the generation of the formaldehyde from methanol oxidation gradually exceeds its consumption, thus, the conversion efficiencies of HCHO (k_{HCHO}) decrease with the increasing engine speed. As can be seen in Figs. 6(b), 7(b) and 8(b), negative conversion efficiencies appear at high engine speeds, this is especially obvious for the M85 fuelled engine, TWC produces more

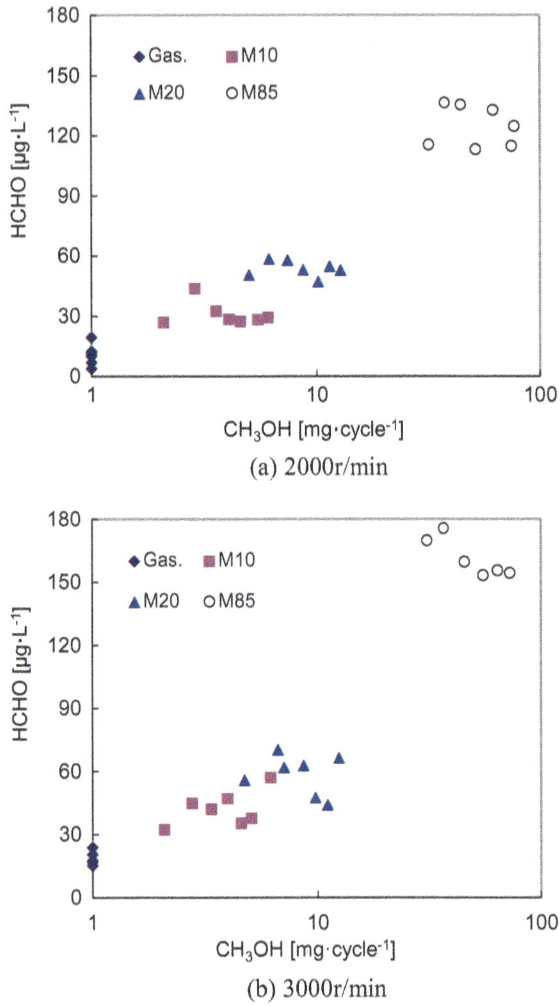

Fig. 9 Relationship between formaldehyde emission and cycle supplied methanol, Label Gas. In figure denotes gasoline [44]. [Reprinted with permission from (DOI: https://doi.org/10.1021/ef900175h). Copyright (2009) American Chemical Society.]

formaldehyde rather than eliminates it. This abnormal phenomenon needs particular concern in designing dedicated TWC for methanol vehicle. It is to be noted that, besides this, formaldehyde could also be converted to formic acid in TWC especially at cold start conditions when the TWC has not been initiated, the acid solves into the condensed water and forms an acidic solution, which corrodes the ceramic base of the TWC and thus washes the novel metal particles away and at last invalidates TWC.

5 Conclusions

Partially due to the low concentration and partially the poor response of formaldehyde to ordinary detectors such as flame ionization detector (FID), airborne formaldehyde has to be converted to various derivates with other agents, and the detect the derivates to indirectly quantitate formaldehyde. However, the composition of the engine exhaust gas is too complex for the engine-out formaldehyde to be simply measured as that in the air, it has to be firstly separated and then detected. The standard Measurement Methods for Non-Regulated Emissions From Methanol Fueled Vehicles of National Environmental Protection Standards of China (version requesting for comments) adopts DNPH-HPLC method, formaldehyde is firstly pretreated with 2,4-DNPH and detected by HPLC. Although complex in sample preparation, this method is the most mature and reliable by now. FTIR method and GC-PDHID method are the only two direct and fast methods that can detect and quantitate formaldehyde without complicated pretreatment for the sample gas. However, FTIR is relatively poor in accuracy and GC-PDHID is still under development.

Formaldehyde is on one of the paths of hydrocarbon oxidation, but on the hub location of methanol oxidation, thus, methanol vehicles release much more formaldehyde than that of the gasoline engine, its concentration is proportional to the cyclic injected fuel methanol. Formaldehyde is also highly affected by the exhaust temperature and residence time in a hot environment, thus reflecting on the engine, it increases with engine speed.

As an intermediate product, formaldehyde emission has a contribution zone of initial temperature in the range from about 950 K to 1050 K. TWC may transmit the formaldehyde production from the frozen zone to the generation zone, thus formaldehyde emission increases, rather than decreases as ordinary hydrocarbons, after the conversion of TWC at high engine speeds.

Acknowledgements This work in Xi'an Jiaotong University is financially supported by the National Nature Science Foundation of China (Grant Nos. 51576159 and 91741110), and Shaanxi Provincial Key R&D Program (Grant Nos. 2019ZDLGY15-10 and 2019ZDLGY15-07).

References

1. China Standard: limits and measurement methods for emissions from light-duty vehicles (China VI). GB18352.6-2016
2. China Standard: limits and measurement methods for emissions from diesel fuelled heavy-duty vehicles (China VI). GB17691-2018
3. Graves B, Olfert J, Patychuk B et al (2015) Characterization of particulate matter morphology and volatility from a compression-ignition natural-gas direct-injection engine. Aerosol Sci Technol 49(8):589–598
4. Stelmasiak Z, Larisch J, Pielecha J et al (2017) Particulate matter emission from dual fuel diesel engine fuelled with natural gas. Polish Maritime Res 24(2):96–104

5. Mishra PC, Gupta A, Kumar A, Bose A (2020) Methanol and petrol blended alternate fuel for future sustainable engine: a performance and emission analysis. Measurement 155:
6. Li J, Gong CM, Su Y, Dou HL, Liu XJ (2010) Effect of injection and ignition timings on performance and emissions from a spark-ignition engine fueled with methanol. Fuel 89(12):3919–3925
7. Wei Y, Sh L, Li H, Yang R, Liu J, Wang Y (2008) Effects of methanol/gasoline blends on a spark ignition engine performance and emissions. Energy Fuels 22(2):1254–1259
8. Yao DW, Ling XC, Wu F (2016) Experimental investigation on the emissions of a port fuel injection spark ignition engine fueled with methanol-gasoline blends. Energy Fuels 30(9):7428–7434
9. Zhang F, Zhang X, Shuai SJ, Xiao JH, Wang JX (2010) Unregulated emissions and combustion characteristics of low-content methanol – gasoline blended fuels. Energy Fuels 24(2):1283–1292
10. Geng P, Zhang H, Yang SC, Yao CD (2015) Comparative study on measurements of formaldehyde emission of methanol/gasoline fueled SI engine. Fuel 148:9–15
11. Valco Instruments Co. Inc. (2007) Pulsed Discharge Detector, Model D-4-I-SH17-R, Instruction Manual
12. Wu LJ (2005) Detection methods of gas chromatography, 2nd edn.. Beijing, Chemical Industry Press, pp 266–268, 326–327
13. National Cancer Institute. https://www.cancer.gov/about-cancer/causes-prevention/risk/substances/formaldehyde/formaldehyde-fact-sheet
14. International Agency for Research on Cancer (2012) Agents Classified by the IARC Monographs. 88:100F
15. Li QJ (1997) Experiment of determining methylalcohol and formaldehyde simultaneously by off-color acid spectrophotography. Environ Protect Oil Gas Fields 01:44–47
16. Wang LY, Zhao Z, Yuan R et al (2016) Detection and confirmation of free formaldehyde by acetyl acetone method. Chem Eng 30(09):22–24
17. Alzueta MU, Finestra M (2001) Methanol oxidation and its interaction with nitric oxide. Energy Fuels 15(3):724–729
18. Li YJ, Xie PH, Qin M et al (2009) Study of retrieving formaldehyde with differential optical absorption spectroscopy. Spectrosc Spectral Anal 29(1):196–200
19. Li X, Wang S, Zhou R et al (2014) Urban atmospheric formaldehyde concentrations measured by a differential optical absorption spectroscopy method. Environ Sci: Process Impacts 16(2):291–297
20. Kecskés T, Raskó J, Kiss J (2015) FTIR and mass spectrometric studies on the interaction of formaldehyde with TiO_2 supported Pt and Au catalysts. Appl Catal A: Gen 273(1–2):55–62
21. Zorin I, Kilgus J, Duswald K et al (2020) Sensitivity-enhanced Fourier transform mid-infrared spectroscopy using a supercontinuum laser source. arXiv:2003.00853
22. Harris GW, Mackay GI, Iguchi T et al (1989) Measurements of formaldehyde in the troposphere by tunable diode laser absorption spectroscopy. J Atmos Chem 8(2):119–137
23. Wert BP, Fried A, Henry BE et al (1996) Airborne measurements of tropospheric formaldehyde by tunable diode laser absorption spectroscopy. Proc SPIE Int Soc Opt Eng 2834:175–186
24. Hkfelt T, Ljungdahl K (1972) Modification of the Falck-Hillarp formaldehyde fluorescence method using the Vibratome: simple, rapid and sensitive localization of catecholamines in sections of unfixed or formalin fixed brain tissue. Histochemie 29(4):325–339
25. Li G, Han L (2013) Determination of formaldehyde in aquatic products by a sensitive catalytic fluorescence method. Anal Methods 6:426–432
26. Yang KH, Liu YC, Yu CC et al (2011) Fabrication of chitosan/silver nanocomposites based on electrochemical methods for removing formaldehyde in air. Mater Chem Phys 126(3):993–997
27. Kuang-Hsuan Y et al (2011) Fabrication of chitosan/silver nanocomposites based on electrochemical methods for removing formaldehyde in air. Mater Chem Phys 126(3):993–997
28. Motyka K, Onjia A, MikuKa P et al (2007) Flow-injection chemiluminescence determination of formaldehyde in water. Talanta 71(2):900–905

29. Zhou KW, Xu Z (2008) Determination of trace formaldehyde in air with static injection chemiluminescence's method. J Beijing Union Univ 01: 68-70 + 742008
30. Dirksen JA, Duval K, Ring TA (2001) NiO thin-film formaldehyde gas sensor. Sens Actuators B Chem 80(2):106–115
31. Lin S, Li D, Wu J et al (2011) selective room temperature formaldehyde gas sensor using TiO_2 nanotube arrays. Sens Actuators B Chem 156(2):505–509
32. China Standard: method for determination of formaldehyde in air in public places. GB/T 18204.26-2000
33. Kim S, Kim HJ (2005) Comparison of standard methods and gas chromatography method in determination of formaldehyde emission from MDF bonded with formaldehyde-based resins. Bioresour Technol 96(13):1457–1464
34. U.S. Environmental Protection Agency (1999) Determination of formaldehyde in ambient air using adsorbent cartridge followed by High Performance Liquid Chromatography (HPLC) [Active Sampling Methodology]. Compendium Method TO-11A
35. He RQ (2011) Assay of brain endogenous formaldehyde with 2,4-dinitrophenylhydrazine through UV-HPLC. Progr Biochem Biophys 38(12):1171–1177
36. Adrian-Romero M, Blunden G, Carpenter BG et al (1998) The use of HPLC for the detection and quantification of formaldehyde in the pteridophyta. Acta Biol Hung 49(2–4):303
37. Hunter MC, Bartle KD, Lewis AC et al (1998) The use of HID the helium ionization detector for gas chromatographic monitoring of trace atmospheric components. High Resol Chromatogr 21(2):75–80
38. Dojahn JG, Wentworth WE, Stearns SD (2001) Characterization of formaldehyde by gas chromatography using multipe pulsed-discharge photoionization detectors and a flame ionization detector. J Chromatogr Sci 39(2):54–58
39. Wei YJ, Liu SH, Liu FJ, Liu J, Zhu Z, Li GL (2010) Direct measurement of formaldehyde and methanol emissions from gasohol engine via pulsed discharge helium ionization detector. Fuel 89(9):2179–2184
40. Agilent Technology (2007) J&W GC Column Performance Summary: Gs-OxyPLOT
41. Hopkins JR, Still T, Al-Haider S, Fisher IR, Seaking PW (2003) A simplified apparatus for ambient formaldehyde detection via GC-pdHID. Atmos Environ 37:2557–2565
42. Curran HJ, Pitz WJ, Westbrook CK: Primary reference fuels (PRF): iso-octane/n-heptane mixtures. https://www-pls.llnl.gov/?url=science_and_technology-chemistry-combustion-prf
43. Wei YJ, Liu SH, Liu FJ, Liu J, Zhu Z, Li GL (2009) Aldehydes and methanol emission mechanisms and characteristics from a methanol/gasoline-fueled spark-ignition (SI) Engine. Energy Fuels 23(12):6222–6230
44. Wei YJ, Liu SH, Liu FJ, Liu J, Zhu Z, Li GL (2009) Formaldehyde and methanol emissions from a methanol/gasoline-fueled spark-ignition (SI) Engine. Energy Fuels 23(3):3313–3318

Application of Methanol as a Clean and Efficient Alternative Fuel for Passenger Cars

Harold Sun, Wesley Wang, and Kim-Pui Koo

1 Introduction

Methanol has been broadly utilized for racing vehicles because of its generally higher-octane number thus better knock resistance, power density, and brake thermal efficiency than gasoline. Methanol has long been studied as a clean and low-cost alternative fuel as it can be mass produced from coal or natural gas as low carbon synthetic fuel or even from bio-methane, algae, as an attractive renewable fuel at a competitive cost [27]. The methanol can even be produced from flare gas at an oil field, landfill, or biogas that would otherwise be burnt or vented to the environment [2, 4]. In 1981, Ford developed 40 dedicated methanol-fueled Escort vehicles for a fleet test in Los Angeles, California [25], followed by larger fleet tests in Michigan, USA as well as New Zealand, Sweden, Norway, United Kingdom, and Canada. Due to the cold start concern and knock resistance benefit, this methanol-fueled engines had compression ratio increased to 11.8:1, which helped torque, power, and thermal efficiency improvement. The methanol-fueled fleet tests at Ford between 1985 and 1992 further extended to larger engine platforms, such as 3L Taurus, 5L Crown Victoria, and 5L Econoline vans. These fleet tests demonstrated the practicality of methanol as a very competitive alternative fuel for passenger car applications. Since the late 1990, the rapid fall of oil prices has wiped out monetary motivations thus interrupted commercialization endeavors of methanol applications in US automotive industry [30, 32]. In recent years, methanol has been displaced by ethanol in US as an oxygenated alternative fuel or fuel additives, with heavy economic incentives and government policy guidelines. Since 2001, US government has promoted ethanol blending to

---

H. Sun (✉)
GAC R&D Center USA, 38555 Hills Tech Dr., Farmington Hills, MI 48323, USA
e-mail: harold.sun15@gmail.com

W. Wang · K.-P. Koo
Fuel Injection Technology, 17 Guantai Rd, Dongguan, China

A. K. Agarwal et al. (eds.), *Methanol*, Energy, Environment, and Sustainability,
https://doi.org/10.1007/978-981-16-1224-4_11

regular-grade gasoline, which is represented by Clean Air Act in 1990 [35] as well as the Energy Independence and Security Act of 2007 [36].

However, the exploration and limited vehicle fleet tests on methanol as a viable alternative fuel for light and heavy-duty automotive applications have never been halted in China. According to Zhao, Li and US Methanoal Institute [21, 37, 50], China government initiated a methanol-fueled vehicle pilot program in 2012 that included 1000 vehicles operating in various topographical regions in China. Over the past decades, the advantages and major technical barriers of methanol as a viable alternative fuel for passenger car application have been recognized and various potential solutions studied and demonstrated [14, 20, 22, 44, 47, 49, 50]. Obviously, the ever-stringent emission regulations may also pose new opportunities or technical challenges for methanol to be widely accepted as a clean and viable alternative fuel for automotive application. For example, the latest China VI emission regulation [6] claims "fuel neutral" and has the same NO_x emission limits for light-duty diesel and gasoline, which is extremely challenging for light-duty diesel engines (Table 1). While diesel engine still has advantages in fuel economy and low-end torque over the gasoline engine, the technical challenges and implementation cost associated with the sophisticated technical solutions due to the most stringent emission regulation for light-duty diesel in the world would likely prompt the OEMs to look for clean, efficient and low-cost alternative powertrain solutions such as methanol fueled engines. Besides the new opportunity for methanol-fueled engines coming with the China VIb emission regulations, the newly proposed, and surely unique, tailpipe methanol emission limit by China government [7, 8] would lead to revisiting of toxicity study of methanol emission, deterioration of methanol fuel system as results of corrosion, inadequate lubricity and material incompatibility by the industries and academia.

Table 1 China emission regulation for passenger cars (GVW < 2.5 Ton)

Stage	CO	HC + NOx	NOx	PM	CO	NOx	PM
		g/km Diesel				g/km Diesel	
China 3 (NEDC)	0.64	0.56	0.5	0.05	2.3	0.15	
China 4 (NEDC)	0.5	0.3	0.25	0.025	1	0.08	
China 5 (NEDC)	0.5	0.23	0.18	0.0045	1	0.06	0.0045
China 6a (WLTC)	0.7	0.068	0.06	0.0045	0.7	0.06	0.0045
China 6b (WLTC)	0.5	0.035	0.035	0.003	0.5	0.035	0.003

Source China VI Emission Regulation

2 Thermal Efficiency Benefit of Methanol Over Gasoline Fuel

Table 2 shows the comparison of properties between gasoline, ethanol and methanol fuels. As indicated by literature [28, 30, 32, 38], high octane number thus improved anti-knock capability of methanol will enable more advanced spark ignition timing thus better thermal efficiency for a methanol-fueled engine. The substantially high latent heat of evaporation of methanol as shown in Table 2 results in lower intake manifold temperature thus better volumetric efficiency [28], which, despite cold start difficulties, will benefit a methanol-fueled engine in terms of reduced pumping loss thus further improving the brake thermal efficiency when methanol is introduced at intake ports. The additional thermal efficiency advantage of methanol over gasoline is anticipated with a high level of cooled EGR thus more advanced spark timing is used, enabled by faster flame propagation speed (as shown in Table 2) of methanol, i.e. improved combustion stability when diluted by EGR, compared with gasoline.

Extensive experimental studies were conducted by Sun et al. [32] to compare the thermal efficiency of an SI engine when it was fueled by gasoline and neat methanol, respectively. Two natural aspirated production gasoline engines of the same vintage with multi-point port fuel injection from Geely were converted to methanol-fueled engines with the following modifications [32]:

- A separate port methanol injection system was added to the intake manifold, besides the original gasoline port injection system. The injection pressure was 3.5 bar for gasoline and 4 bar for methanol.
- A methanol fuel tank was added that was made of stainless steel to address methanol permeation and corrosion concerns; methanol fuel pump, fuel lines, and seals were all treated for alcohol application.

Table 2 Properties of different fuels

	Gasoline	Ethanol	Methanol
Octane (R + M)/2	88–91	115	119
Specific gravity (kg/L)	0.75	0.794	0.796
Stoich air fuel ratio	14.6	9	6.4
Lower heat value (MJ/kg)	43	27	20
Flash temperature (K)	230	285	280
Laminar flame speed (cm/s at 1 bar, 300 K)	33	41	50
Latent heat of evaporation (MJ/kg)	0.27	0.85	1.1
Peak flame temperature (K)	2303	2193	2143
CO_2 emissions (kg*CO_2/kg*fuel)	3.3	2.16	1.54
Min ignition energy (mJ)	0.8	0.65	0.14

Source [15, 34, 38, 51]

- The original gasoline pump, injectors, and a small gasoline tank were kept for cold start assistance, which was activated only for the initial warm-up period after the engine starts.
- A dedicated dual fuel injection control module was used to replace the original engine control module to manage both gasoline (during the cold start period) and methanol injections into the intake manifold, as well as spark ignition timing.
- The same spark plug for the gasoline engine was used for the methanol ignition.
- The intake and exhaust systems, including the three-way catalytic converter from the original gasoline engine, remained unchanged, even though the intake airflow system could have been downsized as the maximum airflow was reduced due to much lower air–fuel ratio for stoichiometric methanol combustion, compared with that for gasoline combustion (Table 2). This relatively oversized intake air system might have a small advantage due to lower inlet restriction for methanol combustion.

In most of these studies [31, 32, 50], both gasoline and methanol are multi-point injected into each intake port. Based on the engine cold start control strategy, either gasoline or methanol can be separately controlled by the engine control module to achieve their different air–fuel ratios for stoichiometric combustion. An electronic throttle valve controls the airflow into the engine while the fuel injection control system manages gasoline and methanol injection quantities accordingly to the intake ports at different engine speeds and loads to ensure stoichiometric combustion thus optimal conversion efficiency of the three-way catalyst, based on the signals from oxygen sensors before and after the downstream oxidation catalyst.

Due to the substantially different octane numbers between gasoline and methanol, the spark ignition timings were recalibrated for maximum brake torque (MBT) for both the gasoline and methanol-fueled engines to take advantage of the better anti-knock capability of the methanol that leads to better thermal efficiency.

The steady-state engine dynamometer test conducted by Sun et al. [32] reveals (Fig. 1) the difference of the part-load MBT spark ignition timings between methanol and gasoline at 1500 rpm and 2000 rpm, respectively. As expected, the methanol MBT timing is more advanced than that for gasoline at a higher load, due to the higher octane number. However, the MBT timing for methanol at light to medium load is a little retarded, relative to gasoline, possibly due to the fast flame speed of methanol fuel.

Since the methanol and gasoline have dramatically different heating values, typical fuel economy comparison will be replaced by a comparison between the brake thermal efficiencies, which is defined as

$$\eta = \frac{1}{\left(\frac{BSFC}{3600} * LHV\right)} \tag{1}$$

where BSFC stands for brake-specific fuel consumption (g/kWh); LHV stands for Low Heat Value (MJ/kg) of the fuels.

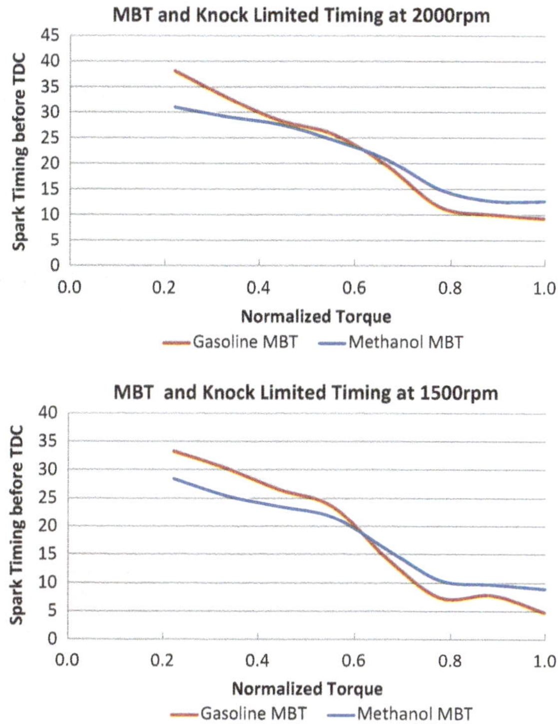

Fig. 1 Max brake torque spark ignition timing for methanol and gasoline *Source* Sun 2018

Combined with better brake thermal efficiency and the fact that the methanol itself is a low carbon fuel [30, 38], the methanol-fueled engine has low CO_2 emission over wide torque (Y-axis) and speed (X-axis) ranges, as shown in Fig. 3 [32], which is beneficial as future vehicles will be regulated based on CO_2 or greenhouse gas emissions. The CO_2 comparison in Fig. 3 does not include greenhouse gas emissions during the production process of the fuels. This CO_2 benefit or the brake thermal efficiency benefit as shown in Fig. 3 and Fig. 2 does not include additional benefits from potential engine downsizing since methanol fuel can operate without a knock at higher torque or power [32]

(a) Gasoline
(b) Methanol

Steady-state engine dynamometer test conducted by Sun et al. [32] indicated that methanol-fueled engine (Fig. 2b) has better brake thermal efficiency over wider operation area than that of the original gasoline engine (Fig. 2a), which can be attributed to higher octane number of methanol, thus advanced timing at high load, and faster flame speed. The better volumetric efficiency or lower pumping loss due to high latent heat of evaporation [28] and much lower flow losses as a result of lower air–fuel ratio for the methanol as an oxygenated fuel [30, 38] also contributed to the brake thermal efficiency improvement over a gasoline engine.

Fig. 2 Engine brake thermal efficiency distribution operating with gasoline and methanol. *Source* Sun [32]

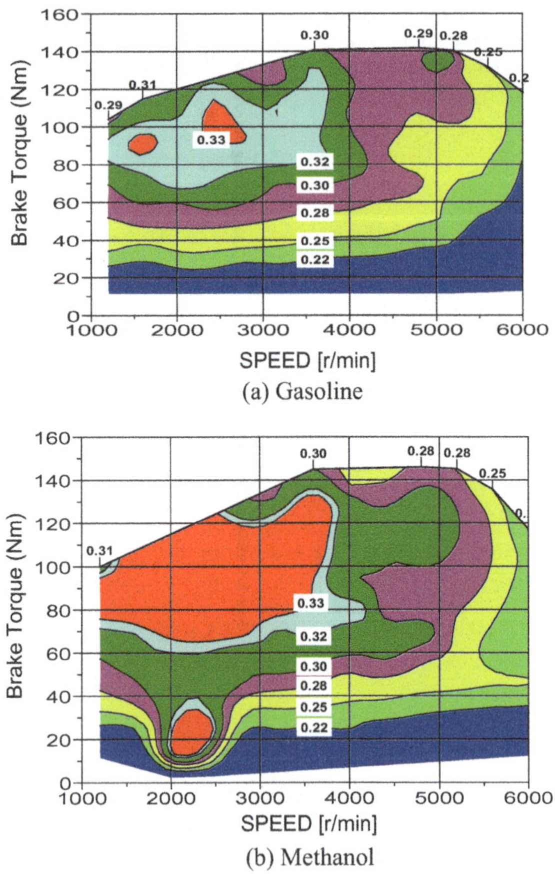

In the experimental study by Sun et al. [32], there was no exhaust gas recirculation (EGR) in the engine, even though it was anticipated [30, 38, 41] that more aggressive cooled EGR may further enhance the thermal efficiency advantage of methanol over gasoline, which is enabled by faster flame propagation speed of methanol due to improved combustion stability at highly EGR diluted conditions.

3 Technical Challenges to Meet Emission Regulations

Besides the advantages that can be utilized for potentially lower tailpipe emission and better brake thermal efficiency with methanol-fueled engine, there are also many technical challenges that need to be addressed to develop a production viable methanol-fueled engine [18, 30, 32, 38].

Fig. 3 Engine CO_2 emissions (g/kWh) when operating with gasoline and methanol *Source* Sun [32]

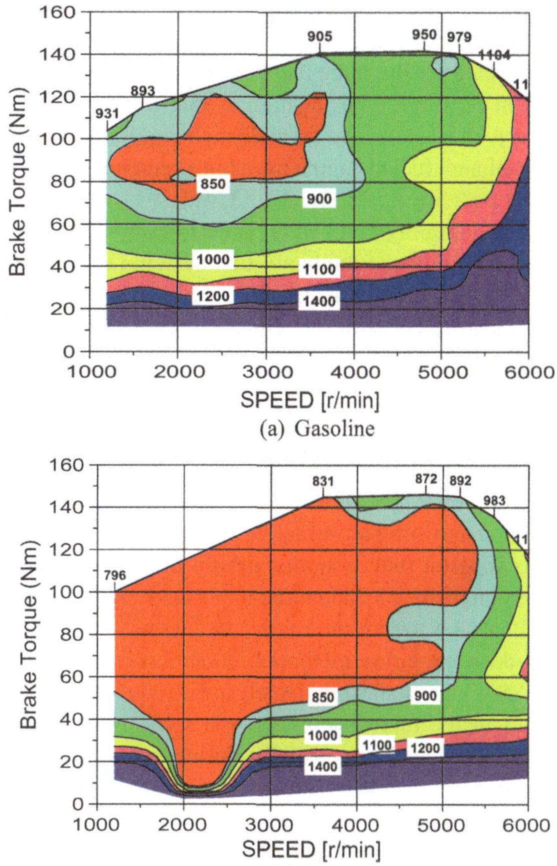

(a) Gasoline

(b) Methanol

- Peak flame temperature for methanol is lower than that for gasoline, i.e. methanol-fueled engine may potentially have lower engine NO_x emissions [5, 30, 38].
- Methanol has a much higher octane number than gasoline therefore the engine may operate at a higher compression ratio or advanced spark timing for better thermal efficiency without knock concerns [5, 30, 38].
- The high-octane number, combined with high flame speed for the methanol may extend bore diameters of an SI engine that is currently limited by knock risk when gasoline fuel is used [32].
- Methanol is an oxygenated fuel [30, 31, 38]; thus, smoke out of combustion, esp. for heterogeneous combustion like in-cylinder direct injection, spark, or compression-ignition engines, will be significantly reduced. The substantially lower stoichiometric air–fuel ratio for a methanol fueled engine means that a compressor on a turbocharged engine can and should be sized smaller to avoid surge when the methanol is introduced downstream of the compressor.

- The latent heat of evaporation of methanol is substantially higher than that of gasoline. Considering the difference in heating value, with a larger fuel quantity injected for the same thermal energy, the latent heat of evaporation for methanol will be seven times as high as that for gasoline [38]. As a result, the cold start as well as excessive formaldehyde emissions created during cold start period of methanol fueled engine have long been recognized as a major technical challenge for passenger cars [31, 32, 43, 47].
- The durability concerns related to corrosion, inadequate lubricity, and material compatibility of methanol fuel system [12, 18, 30–32, 38] need to be carefully addressed for consistent and reliable operation over a respective useful lifetime as required by different emission regulations, esp. the newly proposed methanol tailpipe emission limit by China [7, 8].

The combustion of a methanol-fueled engine can either be initiated by a spark plug or compression ignition. Like other oxygenated fuels, the methanol-fueled engine has better mixing with oxygen, thus more complete combustion that leads to low HC, CO, and particulate emissions [28, 32, 38]. Either neat methanol or methanol blended with gasoline may offer an attractive alternative solution for passenger car application due to its superior fuel economy, mostly low gaseous emission and basically smokeless emission [9, 32, 33], and more importantly, low greenhouse gas emission as methanol itself is a low carbon alternative fuel. However, despite repeated efforts over the past few decades, either methanol, ethanol, or other alternative fuels failed to replace gasoline as a mainstream fuel, mostly due to low petroleum oil prices, durability issues of the fuel system and valve seat wear as well as excessive formaldehyde emission concerns, especially during cold-start period. The increase in aldehyde emissions in methanol-fueled engines remains one of the major implementation concerns, according to most of the studies [11, 14, 20, 29, 40, 42, 44, 47, 49]. The impact of methanol on NO_x emission may be a little bit more completed. While some research reported NO_x emission increase on ECE test cycle [33, 49] or steady-state conditions [16] with methanol blending, most research reported a decrease in NO_x emission with methanol blending [29, 30]. Apparently different mechanisms on NO_x generation and oxidation such as ignition delay and flame temperature with methanol blending at different engine operating conditions will have different impacts on NO_x generation and oxidation [44].

Many studies [16, 38, 49] reported a consistent decrease in THC and CO emissions from methanol-fueled engines with various rates of methanol and gasoline blending, as shown in Fig. 4, which were attributed to lower carbon content in the methanol fuel that is oxygenated, promoting complete combustion. The experimental study by Zhao [49] was conducted on the city driving ECE cycle that is mostly light load where the Three-Way Catalyst was less efficient, especially for methanol-fueled or blended engines due to lower exhaust temperature as compared to a gasoline engine. As a result, the tailpipe NO_x emission of methanol blended (except the case with neat methanol) engine was slightly higher than that of a pure gasoline engine, which is also consistent with research by Svensson [33].

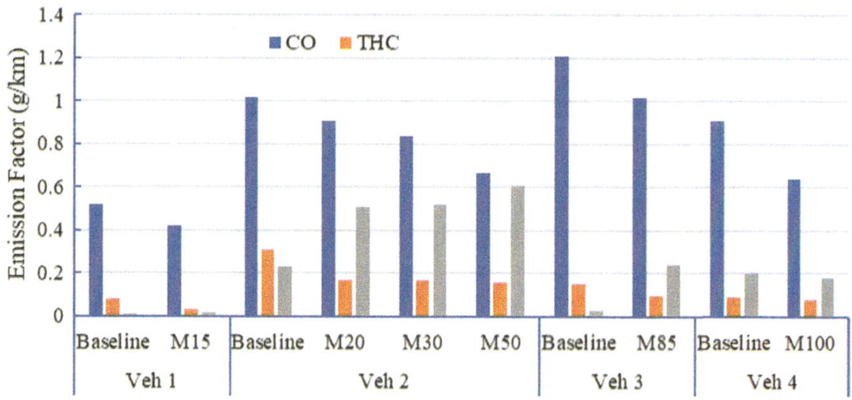

Fig. 4 Gaseous emissions of gasoline (baseline) and methanol blend on different vehicles on ECE test cycle. Reproduced from: Zhao [50]

While the HC, CO, and NOx emissions can be taken care of by a conventional three-way catalyst, aldehyde emission remains a major concern for methanol-fueled engine applications. Aldehyde is highly carcinogenic; therefore, emission regulations in the US, South Korea, Brazil, and lately China set a separate limit for formaldehyde tailpipe emission. It was pointed out by Nichols [24] that formaldehyde is created during the first 120 s after the cold start. Other studies [32, 33, 43, 45], also concluded that formaldehyde is generated during cold and incomplete combustion of methanol, and formaldehyde emission is a serious concern for methanol-fueled engines. Study shows that when neat methanol was used in spark-ignition combustion mode, the formaldehyde could be 4 times as high as the gasoline-fueled engine [40, 49].

As summarized in literatures [30, 32, 38], corrosion-related tailpipe emission deterioration has long been considered a barrier for the commercialization of methanol-fueled engine technology. For instance, thirty years ago, Karl Hellman reported [11] continued increase of HC, CO and aldehyde emissions over 6,000 miles on their methanol-fueled Toyota Carina, Volkswagen Rabbit vehicles even though the engine was equipped with an electrically heated catalyst for fast warm-up assistance, due to deteriorations of fuel system components and oxidation catalytic converters. Windawi also claimed that aldehyde emission is the primary concern for emission control of vehicles fueled with methanol and ethanol [46]. The aldehyde generation, oxidation, and effective catalytic conversion during the prolonged warm-up time for engine and after-treatment have been a major concern and a focus of research when the exhaust catalyst converter is integrated into an alcohol-fueled engine system.

4 Production Viable Solutions to Address Cold Start Related Technical Challenge

Aldehyde is an intermediate product of partial HC oxidation of methanol. It is widely reported that there was a dramatic increase in aldehyde emissions from methanol-fueled engines [10, 48]. Aldehyde is highly toxic and is newly regulated in China emission regulations as already regulated in the US, Brazil, and South Korea. Aldehyde has long been a concern for alcohol-fueled engines that needs to be addressed.

According to Windawi [32, 46], the maximum tailpipe aldehyde emission occurs as a result of partial oxidation of methanol prior to the three-way catalyst light-off, or when the fuel–air mixture is locally rich or lean. A typical engine cold start strategy may include fuel enrichment to ensure combustion stability before the engine and catalyst warm-up. Since aldehyde is generated due to partial oxidation of methanol at either low temperature or insufficient oxygen, a cold start strategy to avoid prolonged fuel enrichment due to high latent heat of evaporation of methanol will be beneficial to reduce tailpipe aldehyde and other emissions, as suggested by Windawi [32, 46].

Conventional gasoline three-way catalyst (TWC) will help to oxidize the formaldehyde to a certain extend. A modified TWC may improve the conversion efficiency of formaldehyde [49, 50]. As reported by Zhao [49], a specially designed TWC managed to marginally reduce the formaldehyde emission of a neat methanol-fueled engine from 5.96 to 4.97 mg/km, a 20% reduction, on the NEDC cycle.

According to the literature [43], the conversion efficiency of a three-way catalyst converter may deteriorate at high exhaust flow conditions due to short residence time in the catalyst. The tailpipe aldehyde emission is likely to increase at high exhaust volume flow conditions, especially under fuel enrichment conditions near peak power, and the formaldehyde could also be created from TWC at high-temperature operation condition with short residence time [38, 43], which could be mitigated via proper sizing of the turbocharger, after-treatment, and "right" sizing of engine displacement as well as advanced engine control to minimize fuel enrichment near peak power area.

The practical methods to address the cold start difficulty and reduce aldehyde and other regulated emissions during the warm-up period were reported to use gasoline or natural gas assistance before the engine and after-treatment were warmed up [3, 32]. In the gasoline-assisted cold start method [32, 38, 50] for methanol-fueled engines, the original gasoline multi-point port fuel injection system was maintained and responsible for cold start assistance during the warm-up period. The gasoline cold start assistance period was calibratable as a function of engine coolant temperature and altitude, which was managed through an engine control module. As illustrated in Fig. 5, the gasoline was used to start and warm the engine before it was switched to methanol when the engine coolant temperature was above a certain threshold when the tailpipe emissions such as aldehyde can be mitigated via a conventional TWC.

Similar to the gasoline cold start assistance method reported by Sun et al. [32], China Automobile Technology and Research Center (CATARC) tested three

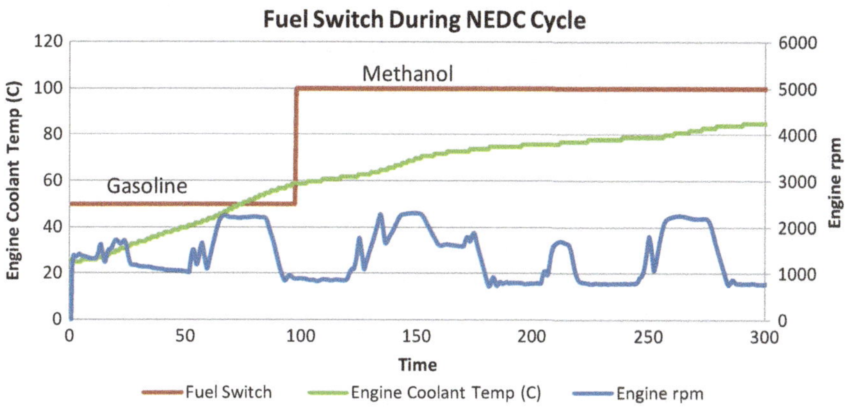

Fig. 5 Fuel switches from gasoline to methanol after the engine warms up *Source* Sun [32]

randomly selected methanol-fueled passenger cars with a mileage between 6,400 and 7,900 km [50]. These reasonably aged and methanol-fueled vehicles used conventional three-way catalysts and gasoline assistance during the warm-up period. Among the tailpipe emissions that are compliant with China V emission regulation as shown in Table 3, the formaldehyde emissions [50] were well below the newly proposed limit of 2.5 mg/km [7], which was consistent with an early study by Sun et al. [32] and recently published research [33].

These studies suggest that once the tailpipe aldehyde is mitigated with brief gasoline combustion assistance during the warm-up period of a transient cycle at low combustion temperature and fuel enrichment conditions, the aldehyde emission would not be a concern for newly proposed regulations even for aged vehicles [8, 32, 33, 50]. After the engine and three-way oxidation catalyst warmed-up, tailpipe aldehyde emissions out of methanol combustion can be mitigated effectively with a conventional three-way oxidation catalyst [19].

Before a highly efficient TWC to mitigate tailpipe formaldehyde emission for alcohol fuels like methanol or ethanol is available, a gasoline cold start assistance might be a viable production-ready approach [32, 33, 50]. The effectiveness of this method has been repeatedly demonstrated in transient emission tests on the NEDC cycle [32, 33, 50]. It turns out to be a practical solution to deal with the huge latent heat of evaporation of methanol, utilizing existing components and infrastructure, unlike those measures that had to rely on high energy ignition systems, engine block heater, heated fuel injector, electrically heated catalyst converter or higher compression ratio for ignition of methanol fueled engines [1, 11, 13, 17]. This gasoline-assisted cold start might be even more attractive for extremely low ambient conditions, given the HC emission limit at −7 °C in China VI and US Tier 3 emission regulations, to avoid an extended period of warm-up time thus excessive HC, CO, and aldehyde emissions.

Besides the challenge of tailpipe formaldehyde emission mitigation, corrosion and material compatibility and degradation issues using neat methanol in the fuel system

Table 3 Emission result of passenger cars. Reproduced from: Zhao [50]

Tailpipe emission result of aged passenger cars relative to China V limit

Vehicle no	Mileage (km)	HC (g/km)		CO (g/km)		NOx (g/km)		HCHO (g/km)	
		Limit	Result	Limit	Result	Limit	Result	Limit	Result
1	65030	0.10	0.04	1.00	0.39	0.06	0.05	2.5	0.737
2	79210	0.10	0.07	1.00	0.69	0.06	0.05	2.5	0.605
3	64552	0.10	0.05	1.00	0.60	0.06	0.04	2.5	0.733

have been major concerns as well [28, 30]. The early study cited by Kumar et al. [18] pointed out that components made of brass, copper, aluminum, and magnesium are prone to heavy corrosion, compared with steel and other ferrous metals when methanol fuel is used. The latest study by SGS Inspire [30] does not find major copper corrosion in methanol gasoline blends except "light tarnish". Even though methanol has a tendency to corrode both metals as well as elastomers used in the fuel supply system and engine seals, anode protection and careful material selection of the material of the component, either elastomers or metal alloys as summarized by Methanol Institute [23] may be very helpful to address the corrosion concerns.

Despite the lack of corrosion resistance surface processing study for methanol, there are lots of studies on corrosion resistance coating and seal for ethanol fuel [26, 30, 39]. The ethanol-rated metal and rubber fuel system components that were readily available in the market could be used on methanol-fueled engine as well [32]. To demonstrate the durability of these ethanol-rated components, Sun et al. repeatedly conducted a tailpipe emission test on the NEDC cycle, every 10,000 km, over a useful lifetime of 160,000 km, per China V emission regulation, on two methanol-fueled Geely passenger vehicles [32]. The study shows that tailpipe emissions consistently stay below China V emission limits for HC, NO_x, CO, and NMHC, demonstrating performance consistency of the ethanol-rated fuel system components during the journey.

Most recently, Su et al. [33] studied emission deteriorations of six taxi passenger vehicles operated neat methanol with mileages ranging from 166,000 to 225,000 km, well beyond what was required by China V emission regulation (160,000 km). These vehicles were also started with gasoline port injection during the warm-up period. Despite noticeable deterioration as widely expected in averaged CO, THC, NOx, methanol, and HCHO of the six vehicles, mostly due to the aging of catalyst, possibly wearing and corrosion of fuel system components, Figure 6 indicates that the aged methanol-fueled vehicles were still China V emission compliant even after the useful life of 160,000 km. It may be worth noting that Su et al. [33] also reported that among these six road aged methanol-fueled vehicles, one of them had excessive unburned

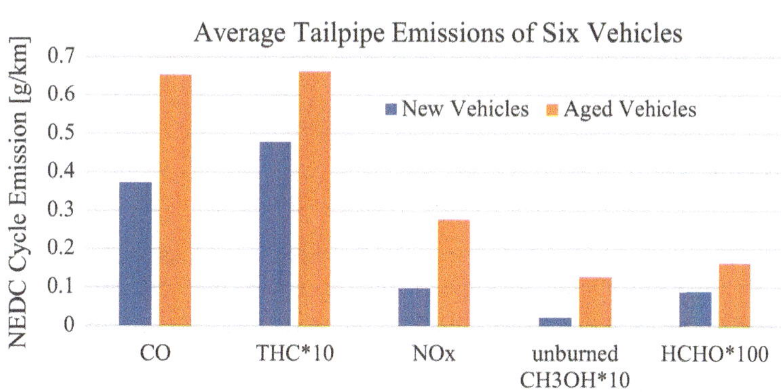

Fig. 6 Vehicle tailpipe emissions measured before/after aging. Reproduced from: Su [33]

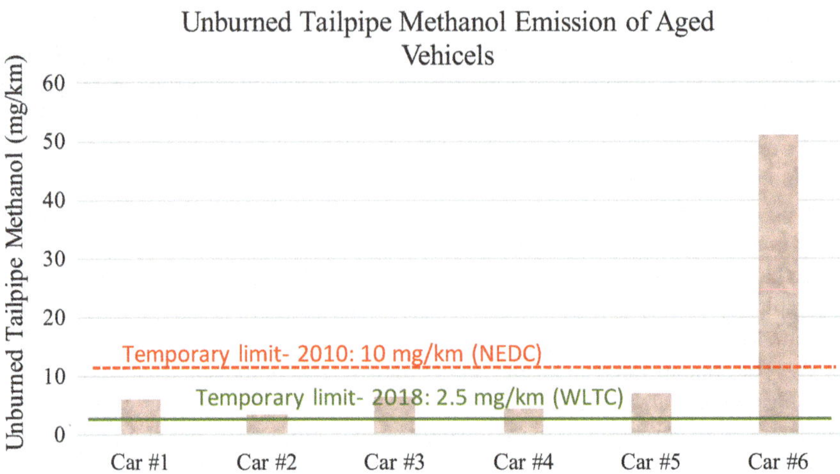

Fig. 7 Vehicle tailpipe methanol emissions measured after aging. Reproduced from: Su [33]

tailpipe methanol emission as revealed in Fig. 7, after operating for 198,000 km, possibly due to fuel components failure. The fuel system component deterioration could be the results of inadequate lubricity or corrosion issues in the fuel system.

Wang et al. [40] conducted a toxicity study of vehicles with different powertrains and concluded that the toxicity of methanol fueled engines is slightly below gasoline engine. However, methanol is still widely considered toxic. China government [7, 8] has proposed further regulations to limit tailpipe methanol emission of methanol-fueled vehicles, besides formaldehyde tailpipe emission limit that has been regulated by the USA, South Korea, and Brazil, even though the methanol has already been included in currently regulated tailpipe HC emissions. According to Su et al. [32] the rationale to single out methanol from the tailpipe total hydrocarbon emission standard [8] is because the oxygenated hydrocarbons could dominate THC emissions, which may potentially miscalculate the THC mass flow when the engine runs into misfire, according to latest study [32], as the oxygen occupies 50% of molecular weight of methanol in the exhaust.

5 Summaries

Many analytical and experimental studies have been conducted over the past few decades on methanol-fueled engines for light duty passenger car applications that lead to the following conclusions:

- Methanol is a viable low carbon alternative fuel for passenger car applications. Even with the combustion system and the three-way oxidation catalyst that were optimized for a gasoline engine, the methanol-fueled engines demonstrated superior brake thermal efficiencies and much lower CO_2 emission over a wide operation range over conventional port fuel injection gasoline engines, mostly due to better volumetric efficiency, the higher octane number of methanol and faster flame speed thus advanced and faster combustion.

- Conventional gasoline three-way catalyst is capable of mitigating HC, CO, and NO_x emissions of methanol-fueled passenger cars in warm-up conditions. Retaining a gasoline port fuel injection system for cold start assistance has consistently demonstrated its potential to be a practical low-cost solution to address the cold start emission concerns for methanol-fueled engines, esp. formaldehyde emission.
- Ethanol-rated fuel system components can be used for methanol fuel as well and studies show that methanol-fueled vehicles demonstrated emissions of HC, CO, and NO_x consistently below China V limits for which these vehicles were certified, with mileages well beyond what was required for the emission regulation.
- One latest study does show that one of the six methanol-fueled vehicles had dramatically increased methanol emission after operating for 198,000 km, indicating potential deterioration of fuel system components because of inadequate lubricity or corrosion.
- The newly proposed tailpipe methanol emission limit by China government seems to set the direction for future study on commercialization endeavors of methanol-fueled vehicles, i.e. corrosion, lubricity and material compatibility of materials of the methanol fuel system to ensure reliable and consistent performances of the fuel system components throughout the useful lifetime.

References

1. Baranescu R et al (1989) Prototype development of a methanol engine for heavy-duty application-performance and emissions. SAE Paper 891653. https://doi.org/10.4271/891653
2. Bazaluk O, Havrysh V, Nitsenko V, Baležentis T, Streimikiene D, Tarkhanova E (2020) Assessment of green methanol production potential and related economic and environmental benefits: the case of China. Energies 13: https://doi.org/10.3390/en13123113
3. Breidenstein W, Sun H, Wang Y, Shan (2019) US Patent US20190353111A1. Tri-fuel electric power generator
4. Breidenstein W (2020) https://gastechno.com/gastechno-solutions-section.html
5. Brusstar M, Stuhldreher M, Swain D, Pidgeon W (2002) High efficiency and low emissions from a port-injected engine with neat alcohol fuels. SAE 2002-01-2743, https://doi.org/10.4271/2002-01-2743
6. China VI Emission Regulation (2016) http://transportpolicy.net/index.php?title=China:_Light-duty:_Emissions
7. China MIIT (2019) Methanol and formaldehyde emission regulation (in Chinese). http://www.miit.gov.cn/n1146295/n1652858/n1652930/n3757016/c6684042/content.html
8. China MEE (2019) Irregular emission measurement of methanol fueled vehicles (in Chinese). http://www.mee.gov.cn/xxgk2018/xxgk/xxgk06/202006/W020200619562210135811.pdf

9. Cowart JS, Boruta WE, Dalton JD, Dona FF, Rivard FL, Furby RS, Piontkowski JA, Seiter RE, Takai RM (1996) Powertrain development of the 1996 Ford flexible fuel Taurus. SAE 952751. doi.org/10.4271/952751
10. Ge Y, He L, Li L (2013) Investigation of automotive emission characteristics with and application status of alternative fuels in China. Report to William and Flora Hewlett Foundation (in Chinese)
11. Hellman KH (1990) Durability testing of an M100-fuelled Toyota LCS-M Carina and a resistively heated catalyst converter," EPA/AA/CTAB/90-04, Tech Report, US-EPA, 24 Sept 1990
12. Heywood J (1988) Internal combustion engines fundamentals
13. Hikino K, Suzuki T (1989) Development of methanol engine with autoignition for low NOx emission and better fuel economy. SAE Paper 891842. https://doi.org/10.4271/891842
14. Hu TG, Wei YJ, Liu SH, Zhou LB (2007) Improvement of spark-ignition (SI) engine combustion and emission during cold start, fueled with methanol/gasoline blends. Energy Fuels 21(1):71–75. https://doi.org/10.1021/ef0603479
15. Hunwartzen I (1982) Modification of CFR test engine unit to determine octane numbers of pure alcohols and gasoline-alcohol blends. SAE 820002. https://doi.org/10.4271/820002
16. Iliev S (2018) Comparison of ethanol and methanol blending with gasoline using engine simulation. Open access peer-reviewed chapter. https://doi.org/10.5772/intechopen.81776
17. Krenus R et al (2014) Ethanol flex fuel system with Delphi heated injector application. SAE 2014-01-1369. https://doi.org/10.4271/2014-01-1369
18. Kumar V, Agarwal A (2019) Material compatibility aspects and development of methanol-fueled engines. In: Advanced combustion techniques and engine technologies for the automotive sector. Springer. https://doi.org/10.1007/978-981-15-0368-9_3
19. Lemel M, Hultqvist A, Vressner A, Nordgren H et al (2005) Quantification of the formaldehyde emissions from different HCCI engines running on a range of fuels. SAE 2005-01-3724. https://doi.org/10.4271/2005-01-3724
20. Li J, Gong CM, Wang EY, Yu XM, Wang Z, Liu XJ (2010) Emissions of formaldehyde and unburned methanol from a spark-ignition methanol engine during cold start. Energy Fuels 24(2):863–870. https://doi.org/10.1021/ef9009982
21. Li C, Negnevitsky M, Wan X (2019) Review of methanol vehicle policies in China: current status and future implications. Energy Proc 160:324–331. https://doi.org/10.1016/j.egypro.2019.02.164
22. Liu SH, Clemente ER, Hu TG, Wei YJ (2007) Study of spark ignition engine fueled with methanol/gasoline fuel blends. Appl Therm Eng 27(11–12):1904–1910. https://doi.org/10.1016/j.applthermaleng.2006.12.024
23. Methanol Institute (2020) Materials compatibility. https://www.methanol.org/materials-compatibility/
24. Nichols RJ, Clinton EL, King ET, Smith CS, Wineland RJ (1988) A view of flexible fuel vehicle aldehyde emissions. J Fuels Lubric 97(Section 3):422–429. https://www.jstor.org/stable/i40187948
25. Nichols RJ (2003) The methanol story: a sustainable fuel for the future. J Sci Ind Res 62:97–105
26. Nie XW et al (2007) Corrosion behavior of metallic materials in ethanol-gasoline alternative fuels. Mater Sci Forum 546–549:1093–1100
27. Olah GA (2005) Beyond oil and gas: the methanol economy. Angew Chem Int Ed 44:2636–2639. https://doi.org/10.1002/anie.200462121
28. Prayogi Y, Syaiful, Sinaga N (2019) Performance and exhaust gas emission of gasoline engine fueled by gasoline, acetone and wet methanol blends. Mater Sci Eng 535. https://doi.org/10.1088/1757-899x/535/1/012013
29. Qi DH, Liu SQ, Liu JC, Zhang CH, Bian YZh (2005) Properties, performance, and emissions of methanol–gasoline blends in a spark ignition engine. In: Proceedings of the IMechE, vol 219 Part D: J. Automobile Engineering. https://doi.org/10.1243/095440705x6659
30. SGS Inspire (2020) Methanol: Properties and Uses. Methanol Institute Report, 2020, https://www.methanol.org/wp-content/uploads/2020/03/SGS-INSPIRE-Methanol-Properties-and-Uses.pdf

31. Su S, Ge Y, Wang X, Zhang M, Hao M, Tan J, Shi F, Guo D, Yang Z (2020) Evaluating the in-service emissions of high-mileage dedicated methanol-fueled passenger cars: regulated and unregulated emissions. Energies 13:2680. https://doi.org/10.3390/en13112680
32. Sun H, Wang W, Koo KP (2018) The practical implementation of methanol as a clean and efficient alternative fuel for automotive vehicles. Int J Engine Res 20:350–358. https://doi.org/10.1177/1468087417752951
33. Svensson E, Li C, Shamun S, Johansson B, Tuner M, Perlman C, Lehtiniemi H, Mauss F. Potential levels of soot, NOx, HC and CO for methanol combustion. SAE 2016-01-0887. https://doi.org/10.4271/2016-01-0887
34. US EPA (2014). https://www.epa.gov/sites/production/files/2015-07/documents/emission-factors_2014.pdf
35. US EPA (1990) The clean air act. https://www.epa.gov/clean-air-act-overview/clean-air-act-highlights-1990-amendments
36. US EPA (2007) The Energy Independence and Security Act, https://www.epa.gov/laws-regulations/summary-energy-independence-and-security-act
37. US Methanol Institute (2019) A brief review of China's methanol vehicle pilot and policy. https://www.methanol.org/wp-content/uploads/2019/03/A-Brief-Review-of-Chinas-Methanol-Vehicle-Pilot-and-Policy-20-March-2019.pdf
38. Verhelst S, Turner J, Sileghem L, Vancoillie J (2019) Methanol as a fuel for internal combustion engines. Prog Energy Combust Sci 70:43–88. https://doi.org/10.1016/j.pecs.2018.10.001
39. Vu NSH et al (2019) Improved corrosion resistance of steel in ethanol fuel blend by Titania nanoparticles and Aganonerion polymorphum leaf extract. ACS Omega 4(1):146–158. https://doi.org/10.1021/acsomega.8b02084
40. Wang J, Zhang P, Zhang C, Zheng J (2020) Performance and emissions of a premixed combustion engine fueled by methanol-gasoline blends. Res Square. https://doi.org/10.21203/rs.3.rs-32148/v1
41. Wang Q, Wei L, Pan W, Yao C (2015) Investigation of operating range in a methanol fumigated diesel engine. Fuel 140:164–170. https://doi.org/10.1016/j.fuel.2014.09.067
42. Wei L, Yao C, Wang Q, Pan W, Han G (2015) Combustion and emission characteristics of a turbocharged diesel engine using high premixed ratio of methanol and diesel fuel. Fuel 140:156–163. https://doi.org/10.1016/j.fuel.2014.09.070
43. Wei Y, Liu S, Liu F, Liu J, Zhu Z, Li G (2009) Aldehydes and methanol emission mechanisms and characteristics from a methanol/gasoline-fuelled spark-ignition (SI) engine. Energy Fuels 23:6222–6230. https://doi.org/10.1021/ef900721v
44. Wei YJ, Liu SH, Li HS, Yang R, Liu J, Wang Y (2008) Effects of methanol/gasoline blends on a spark ignition engine performance and emissions. Energy Fuels 22(2):1254–1259. https://doi.org/10.1021/ef7003706
45. Wei YJ, Liu SH, Liu FJ (2009) Formaldehyde and methanol emissions from a methanol/gasoline-fueled spark-ignition (SI) engine. Energy Fuels 23:3313–3318. https://doi.org/10.1021/ef900175h
46. Windawi H (1992) Controlling the exhaust emissions from alternative fuel vehicles-new opportunity for platinum metals catalysts. Platin Met Rev 36(4):185–195
47. Yao C, Cheung CS, Cheng C, Wang Y, Chan TL, Lee SC (2008) Effect of Diesel/methanol compound combustion on Diesel engine combustion and emissions. Energy Convers Manag 49(6):1696–1704. https://doi.org/10.1016/j.enconman.2007.11.007
48. Zhao H, Ge Y, Hao C, Han X, Fu M, Yu L, Shah AN (2010) Carbonyl compound emissions from passenger cars fuelled with methanol/gasoline blends. Sci Total Environ 408:3607–3613. https://doi.org/10.1016/j.scitotenv.2010.03.046
49. Zhao H, Ge Y, Tan J, Yin H, Guo J, Zhao W, Dai P (2011) Effects of different mixing ratios on emissions from passenger cars fueled with methanol/gasoline blends. J Environ Sci 23(11):1831–1838. https://doi.org/10.1016/S1001-0742(10)60626-2

50. Zhao K (2019) A brief review of China's methanol vehicle pilot and policy. Methanol Institute Report. https://www.methanol.org/wp-content/uploads/2019/03/A-Brief-Review-of-Chinas-Methanol-Vehicle-Pilot-and-Policy-20-March-2019.pdf
51. Zipfel L, Fournel P (2000) HFC-365mfc blends: status and development as blowing agent for high performance insulation foams. In: Polyurethanes Conference 2000. https://doi.org/10.1177/0021955x02038001189

The manufacturer's authorised representative in the EU is Springer Nature Customer Service Centre GmbH, Europaplatz 3, 69115 Heidelberg, Germany. If you have any concerns regarding our products, please contact ProductSafety@springernature.com

Printed and bound by CPI Group (UK) Ltd, Croydon, CR0 4YY

30/07/2026

02182003-0001